Symmetry and Separation
of Variables

ENCYCLOPEDIA OF MATHEMATICS
and Its Applications

GIAN-CARLO ROTA, Editor
Department of Mathematics
Massachusetts Institute of Technology
Cambridge, Massachusetts

Editorial Board

GIAN-CARLO ROTA, *Editor*

ENCYCLOPEDIA OF MATHEMATICS AND ITS APPLICATIONS

Volume 4

Section: Special Functions
Richard Askey, *Section Editor*

Symmetry and Separation of Variables

Willard Miller, Jr.

School of Mathematics
University of Minnesota
Minneapolis, Minnesota

With a Foreword by
Richard Askey

University of Wisconsin

1977

Addison-Wesley Publishing Company
Advanced Book Program
Reading, Massachusetts

London·Amsterdam·Don Mills, Ontario·Sydney·Tokyo

Library of Congress Cataloging in Publication Data

Miller, Willard.
 Symmetry and separation of variables.

 (Encyclopedia of mathematics and its applications;
v. 4: Section, special functions)
 Bibliography: p.
 Includes index.
 1. Symmetry (Physics) 2. Functions, Special.
3. Differential equations, Partial—Numerical solutions.
I. Title. II. Series: Encyclopedia of mathematics and
QC174.17.S9M54   530.1'5'55   77-12572
ISBN 0-201-13503-5

American Mathematical Society (MOS) Subject Classification Scheme (1970):
22E70, 33A75, 35C10, 81A78

Printed in the United States of America

ABCDEFGHIJ-HA-7987

Contents

Editor's Statement

A large body of mathematics consists of facts that can be presented and described much like any other natural phenomenon. These facts, at times explicitly brought out as theorems, at other times concealed within a proof, make up most of the applications of mathematics, and are the most likely to survive changes of style and of interest.

This ENCYCLOPEDIA will attempt to present the factual body of all mathematics. Clarity of exposition, accessibility to the non-specialist, and a thorough bibliography are required of each author. Volumes will appear in no particular order, but will be organized into sections, each one comprising a recognizable branch of present-day mathematics. Numbers of volumes and sections will be reconsidered as times and needs change.

It is hoped that this enterprise will make mathematics more widely used where it is needed, and more accessible in fields in which it can be applied but where it has not yet penetrated because of insufficient information.

Anyone who has ever had to solve a differential equation is familiar with separation of variables. Mostly, this method is remembered as a bag of tricks at the borderline of mathematics.

Professor Miller has given the first systematic treatment of this method. He shows how separation of variables relates to one of the central fields of today's mathematics and mathematical physics; namely, the theory of Lie algebras.

This volume is the first in the Section dealing with the theory of those special functions which occur in the practice of mathematics.

GIAN-CARLO ROTA

Foreword

This is the first in a series of books that will try to show how and why special functions arise in many applications of mathematics. The elementary transcendental functions, such as the exponential, its inverse, the logarithm, and the trigonometric functions, form part of the working tools of all mathematicians and most users of mathematics. There was a time when knowledge of some of the higher transcendental functions was almost as widespread. For example, there were a surprisingly large number of books written on elliptic functions in the last half of the nineteenth century, and esoteric facts about Bessel functions and Legendre functions were regularly set as tripos problems. However, knowledge of these functions and the few other very useful special functions is no longer as widespread, and it has even been possible for important special functions to arise in applications and be studied for twenty-five years or more without any of the people studying them being aware that some of the results they rediscovered were found about a hundred years earlier. This has occurred in the last forty years with what are called $3-j$ symbols. These functions occur when studying the decomposition of the direct product of two irreducible representations of $SU(2)$. Since a knowledge of hypergeometric series was not as widespread as it should be, it has only recently been realized that one of the orthogonality relations for $3-j$ symbols is the same as the orthogonality for a set of polynomials that was found by Tchebychef in 1875 and that Tchebychef had some useful formulas for these polynomials that still have not been published in the physics literature, where most of the development of $3-j$ symbols has occurred. Similarly, a symmetry relation for the $3-j$ symbols that was found by Regge in 1958 had been given by Whipple in 1923 and even earlier by Thomae in 1879. The first of the symmetries for these functions was stated by Kummer in 1836. If these results were easy to derive and

ENCYCLOPEDIA OF MATHEMATICS and Its Applications, Gian-Carlo Rota (ed.).
Vol. 4: Willard Miller, Jr., Symmetry and Separation of Variables

could be found by anyone who needs them, then there would be no reason to worry about old results getting lost. However, this is often not the case, and was definitely not true with respect to Regge's symmetry. The $3-j$ symbols had been studied by many people in the period from 1930 to 1958, and every one of them had missed this symmetry.

The lack of information transfer between mathematicians and users of mathematics goes both ways; this can also be illustrated by a similar example. In 1942 Racah published an important orthogonality relation for functions that are now called $6-j$ symbols or Racah coefficients. He also found an important representation for these functions as a single sum. They arise naturally as a fourfold sum. When Racah's single-sum expression is used in his orthogonality relation, and a transformation formula of Whipple is used on the $6-j$ symbol (Racah had rediscovered this transformation), a new set of orthogonal polynomials arises which had been completely missed in the mathematics literature. In fact the situation was worse than that; not only had this set of orthogonal polynomials not been discovered, there were a number of theorems that seemed to say that the existing set of orthogonal polynomials in one variable were all the orthogonal polynomials in one variable that would have useful explicit formulas. This was not true, as the polynomials buried in Racah's work show.

The most important lesson to be learned from this example is that people with different backgrounds need to talk to each other, since mathematics does not come in isolated parts that are unrelated. This series of books is one attempt to try to show how one part of mathematics relates to other parts, and how it can be used to solve problems of interest to scientists with many different backgrounds. The rest of the Foreword will be a short outline of our current view of special functions. Since there are a number of important special functions, this summary will be given in approximately the order in which the functions were discovered. One fact which will surprise many people is that our current view of some subjects has changed only slightly since the first deep results. We write in a more modern language, but most of the ideas we use were present at a very early time.

When applications are considered, the most important special functions are the hypergeometric functions. A generalized hypergeometric series is a series

$$\sum_{n=0}^{\infty} a_n$$

with a_{n+1}/a_n a rational function of n. This rational function is usually factored as

$$\frac{a_{n+1}}{a_n} = \frac{(n+a_1)(n+a_2)\cdots(n+a_p)}{(n+b_1)(n+b_2)\cdots(n+b_q)} \frac{x}{(n+1)},$$

ISBN-0-201-13503-5

so that

$$a_n = \frac{(a_1)_n \cdots (a_p)_n}{(b_1)_n \cdots (b_q)_n} \frac{x^n}{n!}.$$

The shifted factorial $(a)_n$ is defined by

$$(a)_n = a(a+1) \cdots (a+n-1), \qquad n=1,2,\ldots,$$
$$(a)_0 = 1.$$

The series $\sum_{n=0}^{\infty} a_n$ is written as

$$_pF_q\left(\begin{matrix} a_1,\ldots,a_p \\ b_1,\ldots,b_q \end{matrix}; x\right) = \sum_{n=0}^{\infty} \frac{(a_1)_n \cdots (a_p)_n}{(b_1)_n \cdots (b_q)_n} \frac{x^n}{n!}.$$

This series converges for all complex x when $p \leqslant q$ and $|x| < 1$ when $p = q+1$. Among the special cases are

$$\exp(x) = {}_0F_0(x) = \sum_{n=0}^{\infty} \frac{x^n}{n!};$$

$$(1-x)^{-a} = {}_1F_0\left(\begin{matrix} a \\ - \end{matrix}; x\right) = \sum_{n=0}^{\infty} \frac{(a)_n}{n!} x^n \qquad (|x|<1);$$

$$\sin x = x\,{}_0F_1\left(\begin{matrix} - \\ \tfrac{3}{2} \end{matrix}; \frac{-x^2}{4}\right);$$

$$\cos x = {}_0F_1\left(\begin{matrix} - \\ \tfrac{1}{2} \end{matrix}; \frac{-x^2}{4}\right);$$

$$\log(1+x) = x\,{}_2F_1\left(\begin{matrix} 1,1 \\ 2 \end{matrix}; -x\right) \qquad (|x|<1);$$

$$\arctan x = x\,{}_2F_1\left[\begin{matrix} \tfrac{1}{2},1 \\ \tfrac{3}{2} \end{matrix}; -x^2\right] \qquad (|x|<1);$$

$$\arcsin x = x\,{}_2F_1\left[\begin{matrix} \tfrac{1}{2},\tfrac{1}{2} \\ \tfrac{3}{2} \end{matrix}; x^2\right] \qquad (|x|<1);$$

$$\cos \pi x = {}_2F_1\left(\begin{matrix} x, -x \\ \tfrac{1}{2} \end{matrix}; 1\right).$$

The last example is particularly important, for it suggests that the parameters that occur in hypergeometric series can play a more important role in the study of hypergeometric series than that of just enabling us to distinguish between different series. Gauss was probably the first person to realize this; we will return to his results after describing some earlier work of Wallis and Euler that is necessary before we can see why this last formula holds.

The factorial function $n! = 1 \cdot 2 \cdot \cdots \cdot n$ occurs as soon as one considers the binomial theorem. The easiest extension of $n!$ is the shifted factorial $(a)_n$ defined earlier. Clearly, $n! = (1)_n$. However, it does not solve an interesting question, what is $(\frac{1}{2})!$? This question was solved by Euler when he introduced $\Gamma(x)$. His original expression was an infinite product, but he also gave an integral representation which is equivalent to

$$\Gamma(x) = \int_0^\infty t^{x-1} e^{-t} \, dt.$$

This integral is our usual starting point when developing properties of $\Gamma(x)$, but there is something to be said for Euler's product or other definitions which define the gamma function directly for all x and not just for $\mathrm{Re}\, x > 0$, as in the integral. One such definition is

$$\frac{1}{\Gamma(x)} = xe^{\gamma x} \prod_{n=1}^{\infty} \left(1 + \frac{x}{n}\right) e^{-x/n}$$

where

$$\gamma = \lim_{n \to \infty} \left(1 + \frac{1}{2} + \cdots + \frac{1}{n} - \log n\right).$$

Another, which was found by Euler but is usually attributed to Gauss, is

$$\frac{1}{\Gamma(x)} = \lim_{n \to \infty} \frac{(x)_n}{(1)_n} n^{1-x}.$$

Euler used the gamma function to evaluate the beta function integral

$$B(x,y) = \int_0^1 t^{x-1} (1-t)^{y-1} \, dt$$

as

$$B(x,y) = \frac{\Gamma(x)\Gamma(y)}{\Gamma(x+y)}.$$

It is very easy to see that this implies $\Gamma(\frac{1}{2}) = \sqrt{\pi}$. In fact, Euler's original

ISBN-0-201-13503-5

definition of the gamma function reduces (after using some simple algebra) to Wallis's infinite product for π when $x=\frac{1}{2}$.

In the nineteenth century many different integral representations were given for $\Gamma(x)$, and Hankel proved that it did not satisfy a differential equation with algebraic coefficients. It does satisfy the difference equation $\Gamma(x+1)=x\Gamma(x)$, but this is not a sufficiently strong condition to determine $\Gamma(x)$. A natural condition that forces the solution to be unique was found by Bohr and Mollerup: $\log\Gamma(x)$ is convex for $x>0$. The current generation of mathematicians is quite interested in structural conditions, and this theorem is a beautiful example of the type of result that is held in high regard by contemporary mathematicians. It is a very pretty theorem, and it is useful; but we must not forget that the real reason the gamma function is studied is because it is so useful. It occurs so often that we are forced to consider it. This is just one of many examples of the way mathematical aesthetics and utility force us in the same path. Why this happens is still a mystery.

Study of the factorial and the gamma function has led to the development of a number of general mathematical ideas that have been useful elsewhere. One of the most fruitful is the notion of an asymptotic expansion. Stirling found a method of computing $n!$ for large n. The series he obtained does not converge, but it can still be used to obtain very accurate values of $n!$. Euler's formula

$$\Gamma(x)\Gamma(1-x)=\frac{\pi}{\sin\pi x}$$

can be used to give an analytic continuation of the gamma function from $\mathrm{Re}\,x>0$ to $\mathrm{Re}\,x<1$, $x\neq0$, $-1,\ldots$. When used with one of the infinite products for $\Gamma(x)$ it gives Euler's product

$$\frac{\sin\pi x}{\pi x}=\prod_{n=1}^{\infty}\left(1-\frac{x^2}{n^2}\right).$$

This product, the one given earlier for $1/\Gamma(x)$, and some further products for elliptic functions and theta functions which will be mentioned later led Weierstrass to his theorem on canonical products for entire functions, and the logarithmic derivatives of the product formulas led Mittag-Leffler to his expansion theorem for meromorphic functions.

To return to hypergeometric series, Gauss evaluated

$$\sum_{n=0}^{\infty}\frac{(a)_n(b)_n}{(c)_n n!}={}_2F_1\left(\begin{matrix}a,b\\c\end{matrix};1\right)=\frac{\Gamma(c)\Gamma(c-a-b)}{\Gamma(c-a)\Gamma(c-b)},\qquad \mathrm{Re}(c-a-b)>0.$$

ISBN-0-201-13503-5

When $c=\frac{1}{2}$, $a=x$, $b=-x$, this is

$$_2F_1\left(\begin{matrix}x,-x\\\frac{1}{2}\end{matrix};1\right)=\frac{\left[\Gamma(\frac{1}{2})\right]^2}{\Gamma(\frac{1}{2}-x)\Gamma(\frac{1}{2}+x)}=\sin\pi\left(\tfrac{1}{2}+x\right)=\cos\pi x.$$

Euler was the first to study $_2F_1\left(\begin{matrix}a,b\\c\end{matrix};x\right)$ in the general case. He found a second-order differential equation it satisfies, and gave the transformation formula

$$_2F_1\left(\begin{matrix}a,b\\c\end{matrix};x\right)=(1-x)^{c-a-b}\,_2F_1\left(\begin{matrix}c-a,c-b\\c\end{matrix};x\right)$$

and the integral representation

$$_2F_1\left(\begin{matrix}a,b\\c\end{matrix};x\right)=\frac{\Gamma(c)}{\Gamma(b)\Gamma(c-b)}\int_0^1(1-xt)^{-a}t^{b-1}(1-t)^{c-b-1}dt.$$

Pfaff, in the course of editing some of Euler's posthumous papers, found two more transformation formulas. He stated both only in the case in which the series terminates, but one extends immediately to the non-terminating case. They are

$$_2F_1\left(\begin{matrix}a,b\\c\end{matrix};x\right)=(1-x)^{-a}\,_2F_1\left(\begin{matrix}a,c-b\\c\end{matrix};\frac{x}{x-1}\right)$$

and

$$_2F_1\left(\begin{matrix}-n,b\\c\end{matrix};x\right)=\frac{(c-b)_n}{(c)_n}\,_2F_1\left(\begin{matrix}-n,b\\b-n+1-c\end{matrix};1-x\right),\qquad n=0,1,\ldots.$$

The first contains a number of examples Euler gave of transformations of series that speed convergence. For example, when $x=-1$, the series $_2F_1\left(\begin{matrix}a,b\\c\end{matrix};-1\right)$ converges slowly, whereas $_2F_1\left(\begin{matrix}a,c-b\\c\end{matrix};\frac{1}{2}\right)$ converges much more rapidly. In an age when computation is easy and relatively inexpensive, we find it hard to realize how many mathematical developments were stimulated by the desire to compute something. These transformation formulas, along with Euler's transformation, were the first of a limited number of transformation formulas between generalized hypergeometric series which have been discovered in the last two hundred years. Another one is Regge's symmetry of the $3-j$ symbols mentioned earlier. Gauss found the correct extension of Pfaff's second transformation in the case when the series does not terminate. The factor

$$\frac{(c-b)_n}{(c)_n}=\frac{\Gamma(n+c-b)\Gamma(c)}{\Gamma(n+c)\Gamma(c-b)}$$

ISBN-0-201-13503-5

becomes

$$\frac{\Gamma(c-a-b)\Gamma(c)}{\Gamma(c-a)\Gamma(c-b)}$$

when $-n$ is replaced by a, as one might suspect, but that is not the only change. There is another term that must be added.

Gauss also considered a different type of result. He defined two hyper-geometric series to be contiguous if all their parameters are the same with one exception, and if they differ by one in this parameter. He showed that a general ${}_2F_1\left({a,b \atop c};x\right)$ and any two ${}_2F_1$ series that are contiguous to it are linearly dependent. There are nine such relations after the symmetry of ${}_2F_1\left({a,b \atop c};x\right)$ in a and b is used. These contiguous relations can be iterated, so any three functions ${}_2F_1\left({a+j,b+k \atop c+l};x\right)$ with j,k,l integers are linearly dependent. Since

$$\frac{d}{dx}{}_2F_1\left({a,b \atop c};x\right) = \frac{ab}{c}{}_2F_1\left({a+1,b+1 \atop c+1};x\right),$$

it is easy to see that Euler's differential equation for ${}_2F_1\left({a,b \atop c};x\right)$ can be written as one of these iterated contiguous relations. At the end of his one published paper on hypergeometric series, Gauss stated this difference equation. His second paper, which remained unpublished during his life, treated this equation as a differential equation and he found most of the explicit formulas that can be derived directly from this equation. This includes the quadratic transformations, which play a fundamental role in a number of problems. To get a better perspective on these transformations, two other important eighteenth-century discoveries need to be considered.

One was the study of elliptic integrals by Fagnano, Euler, Landen, and Legendre, and of the arithmetic-geometric mean by Lagrange and Gauss. The other was the introduction of spherical harmonics and Legendre polynomials by Legendre and Laplace. The first of these developments led to elliptic functions, a subject that was studied extensively in the last three quarters of the nineteenth century by Abel, Jacobi, Eisenstein, Weierstrass, Hermite, and many others. The second is directly tied up with some of the algebraic approaches to special functions which have been developed in the last fifty years. A good historical summary of the early work on elliptic integrals was given by Mittag-Leffler (see [6]); Landen's transformation in the form given by Lagrange is described there (the reference to Enneper on page 291 should be to page 357, not page 307). Gauss's quadratic transformation of the elliptic integral of the first type is also given by Mittag-Leffler. Lagrange was motivated by a desire to compute the value of an

ISBN-0-201-13503-5

important integral. Gauss first considered the sequences $a_{n+1}=(a_n+b_n)/2$, $b_{n+1}=(a_nb_n)^{1/2}$; observed that they converge, and was able to recognize the value they converge to when $a_0=2^{1/2}$, $b_0=1$; and finally evaluated the limit in general. Gauss then used this result to lead him to two further results, the introduction of the lemniscate functions, which are special elliptic functions, and of the two general quadratic transformations of the ordinary hypergeometric function $_2F_1(a,b;c;x)$ with different restrictions on one of the parameters. These functions form a very important subclass of the general $_2F_1$, for after they have been multiplied by an appropriate algebraic function, they are exactly the class of hypergeometric series that we call Legendre functions.

Legendre polynomials were studied extensively by Legendre and Laplace in the 1780s. They can be introduced in the following way. The function $(c^2-2cr\cos\theta+r^2)^{-1/2}$ represents the potential in an inverse square field at a point P of a source at C, where r and c are the distances from P and C to a fixed point O, and θ is the angle between the segments PO and OC. This function can be expanded in a power series in r to give

$$(c^2-2cr\cos\theta+r^2)^{-1/2}=\sum_{n=0}^{\infty}P_n(\cos\theta)r^nc^{-n-1}.$$

$P_n(x)$ is a polynomial of degree n in x which we call the Legendre polynomial. Legendre and Laplace discovered the following facts about these polynomials.

$$\int_{-1}^{1}P_n(x)P_m(x)\,dx=0,\quad m\neq n;\qquad \int_{-1}^{1}(P_n(x))^2\,dx=\frac{2}{2n+1}. \quad (L.1)$$

$$\int_{0}^{\pi}P_n(\cos\theta)P_m(\cos\theta)\sin\theta\,d\theta=0,\qquad m\neq n;$$

$$\int_{0}^{\pi}[P_n(\cos\theta)]^2\sin\theta\,d\theta=\frac{2}{2n+1}. \qquad (L.1a)$$

$$P_n(\cos\theta)=(1/\pi)\int_{0}^{\pi}[\cos\theta+i\sin\theta\cos\varphi]^n\,d\varphi. \qquad (L.2)$$

$$P_n(\cos\theta)P_n(\cos\varphi)=(1/\pi)\int_{0}^{\pi}P_n(\cos\theta\cos\varphi+\sin\theta\sin\varphi\cos\Psi)\,d\Psi. \quad (L.3)$$

$$(1-x^2)y''-2xy'+n(n+1)y=0,\qquad y=P_n(x). \qquad (L.4)$$

$$P_n(\cos\theta\cos\varphi+\sin\theta\sin\varphi\cos\Psi)=P_n(\cos\theta)P_n(\cos\varphi)$$

$$+2\sum_{k=1}^{n}\frac{(n-k)!}{(n+k)!}P_n^k(\cos\theta)P_n^k(\cos\varphi)\cos k\Psi. \qquad (L.5)$$

ISBN-0-201-13503-5

The associated Legendre functions are defined by

$$P_n^k(x) = (-1)^k (1-x^2)^{k/2} \frac{d^k}{dx^k} P_n(x), \quad -1 < x < 1, \ k=1,\dots,n. \quad \text{(L.6)}$$

Earlier Lagrange had come across these same polynomials as solutions to a difference equation

$$(2n+1)xP_n(x) = (n+1)P_{n+1}(x) + nP_{n-1}(x). \quad \text{(L.7)}$$

Each of the foregoing results is only one of an extensive class of formulas for more general special functions. To see what they are, the corresponding results for trigonometric functions will be given next and then a description of the natural setting for these formulas will be given. Since $\cos n\theta$ is a polynomial of degree n in $\cos\theta$, we will consider $T_n(x)$, which is defined by $T_n(\cos\theta) = \cos n\theta$.

$$\int_{-1}^{1} T_n(x) T_m(x)(1-x^2)^{-1/2}\,dx = 0, \quad m \neq n,$$

$$\int_{-1}^{1} [T_n(x)]^2 (1-x^2)^{-1/2}\,dx = \begin{cases} \pi, & n=0, \\ \dfrac{\pi}{2}, & n=1,2,\dots. \end{cases} \quad \text{(T.1)}$$

$$\int_{0}^{\pi} \cos n\theta \cos m\theta\,d\theta = 0, \quad m \neq n,$$

$$\int_{0}^{\pi} \cos^2 n\theta\,d\theta = \begin{cases} \pi, & n=0, \\ \dfrac{\pi}{2}, & n=1,2,\dots. \end{cases} \quad \text{(T.1a)}$$

$$\cos n\theta = \frac{e^{in\theta} + e^{-in\theta}}{2}. \quad \text{(T.2)}$$

$$\cos n\theta \cos n\varphi = \frac{1}{2}\left[\cos n(\theta+\varphi) + \cos n(\theta-\varphi)\right]. \quad \text{(T.3)}$$

$$(1-x^2)y'' - xy' + n^2 y = 0, \quad y = T_n(x). \quad \text{(T.4)}$$

$$u''(\theta) + n^2 u(\theta) = 0, \quad u = \cos n\theta. \quad \text{(T.4a)}$$

$$\cos n(\theta+\varphi) = \cos n\theta \cos n\varphi - \sin n\theta \sin n\varphi. \quad \text{(T.5)}$$

$$\frac{d\cos n\theta}{d\theta} = -n\sin n\theta,$$

$$\frac{d\cos n\theta}{d\cos\theta} = \frac{dT_n(x)}{dx} = \frac{n\sin n\theta}{\sin\theta} = nU_{n-1}(x), \quad x = \cos\theta. \quad \text{(T.6)}$$

$$2\cos\theta\cos n\theta = \cos(n+1)\theta + \cos(n-1)\theta; \quad \text{(T.7)}$$

$$xT_n(x) = \tfrac{1}{2}T_{n+1}(x) + \tfrac{1}{2}T_{n-1}(x), \quad n=1,2,\dots,$$

$$xT_0(x) = T_1(x). \quad \text{(T.7a)}$$

ISBN-0-201-13503-5

The orthogonality relations (L.1) and (T.1) are fundamental. Since both $P_n(x)$ and $T_n(x)$ are polynomials, they are orthogonal polynomials. Any set of polynomials in one variable that is orthogonal with respect to a positive measure satisfies a three-term recurrence relation

$$xp_n(x) = A_n p_{n+1}(x) + B_n p_n(X) + C_n p_{n-1}(x)$$

with $A_{n-1} C_n > 0$, B_n real. Conversely, any set of polynomials that satisfy this recurrence relation are orthogonal with respect to a positive measure when $A_{n-1} C_n > 0$ and B_n is real. If $A_{n-1} C_n > 0$, $(n = 1, 2, \ldots, N)$, $A_N C_{N+1} = 0$, then the polynomials are orthogonal with respect to a positive measure that has only finitely many points of support. This recurrence relation reminds one of Gauss's contiguous relations for $_2F_1$'s, and there are a number of cases when the recurrence relation can be shown to be an instance of one of Gauss's formulas or an iterate of these formulas. In other cases there are other hypergeometric series, either $_3F_2\left(\begin{matrix} a,b,c \\ d,e \end{matrix}; 1\right)$ or $_4F_3\left(\begin{matrix} -n, n+a, b, c \\ d, e, f \end{matrix}; 1\right)$ where $a + b + c + 1 = d + e + f$, which satisfy more general contiguous relations that lead to orthogonal polynomials. Now the polynomial variable is in one or more of the parameter spots, rather than a power series variable. As a result of this and our too exclusive interest in power series, these polynomials were not studied and applied as early and as often as they should have been.

One of the reasons $\cos\theta$ and $\sin\theta$ are so useful is their connection with the circle. The addition formula (T.5) is most easily proved by using a rotation of the circle. Cauchy gave this proof. Similarly, the addition formula for $P_n(x)$, which is (L.5), arose by considering the rotation group acting on the sphere in R^3.

To understand the setting, consider first the circle. A function $f(\theta)$, $0 \leqslant \theta \leqslant 2\pi$, $f(0) = f(2\pi)$, can be expanded in a Fourier series

$$f(\theta) \sim \frac{a_0}{2} + \sum_{n=1}^{\infty} (a_n \cos n\theta + b_n \sin n\theta)$$

where

$$a_n = \frac{1}{\pi} \int_{-\pi}^{\pi} f(\theta) \cos n\theta \, d\theta, \qquad b_n = \frac{1}{\pi} \int_{-\pi}^{\pi} f(\theta) \sin n\theta \, d\theta.$$

One application of this expansion is to the construction of a harmonic function $u(x,y)$ for $x^2 + y^2 < 1$, which assumes a given boundary value. For let

$$u(x,y) = \frac{a_0}{2} + \sum_{n=1}^{\infty} r^n [a_n \cos n\theta + b_n \sin n\theta],$$

ISBN-0-201-13503-5

$x = r\cos\theta$, $y = r\sin\theta$. Then $u(x,y)$ is harmonic; that is,

$$\frac{\partial^2 u}{\partial x^2} + \frac{\partial^2 u}{\partial y^2} = 0$$

and

$$\lim_{r \to 1^-} u(r\cos\theta, r\sin\theta) = f(\theta)$$

when $f(\theta)$ is continuous for $0 \leqslant \theta \leqslant 2\pi$.

A similar problem exists for three variables, and it is solved in a similar way. First, one must find a set of functions that satisfies Laplace's equation

$$\frac{\partial^2 u}{\partial x^2} + \frac{\partial^2 u}{\partial y^2} + \frac{\partial^2 u}{\partial z^2} = 0$$

for $x^2 + y^2 + z^2 < 1$. This is done by introducing spherical coordinates $x = r\cos\varphi\sin\theta$, $y = r\sin\varphi\sin\theta$, $z = r\cos\theta$, $0 \leqslant \varphi \leqslant 2\pi$, $0 \leqslant \theta \leqslant \pi$, and finding solutions of Laplace's equation of the form $a(r)b(\theta)c(\varphi)$. One can take $a(r) = r^n$, $c(\varphi) = \cos k\varphi$ or $\sin k\varphi$, and then $b(\theta)$ can be taken to be $P_n^k(\cos\theta)$. The functions $r^n\cos n\theta = \mathrm{Re}(x+iy)^n$ and $r^n\sin n\theta = \mathrm{Im}(x+iy)^n$ are homogeneous polynomials of degree n in x and y. Similarly, $r^n P_n^k(\cos\theta)\cos k\varphi$ and $r^n P_n^k(\cos\theta)\sin k\varphi$, $k = 0, 1, \ldots, n$, are homogeneous harmonic polynomials of degree n in x, y, and z which are linearly independent. There are $2n+1$ of these, and this is the same number that appears the denominator in (L.1). Similarly, the functions $r^n\cos n\theta$ and $r^n\sin n\theta$ are linearly independent when $n = 1, 2, \ldots$, and there are two of them, while for $n = 0$ there is only one. This explains the denominators in (T.1). These homogeneous harmonic polynomials are then used to construct a harmonic function for $x^2 + y^2 + z^2 < 1$ with given boundary values in exactly the same way as in the case of the circle, since the functions $P_n^k(\cos\theta)\cos k\varphi$ and $P_n^k(\cos\theta)\sin k\varphi$, $k = 0, 1, \ldots, n$, $n = 0, 1, \ldots$, form a complete orthogonal set of functions.

Formula (L.3) is the basic functional equation satisfied by the zonal spherical harmonic of degree n on S^2. Zonal means independence from the angle φ. We call the zonal spherical harmonics spherical functions. The general setting for such spherical functions is a space with a distance function and a group G operating on this space. The space is homogeneous in that any point can be mapped to any other point by the group. Also, the space should have the property that if $d(x_1, y_1) = d(x_2, y_2)$, then there is $g \in G$ with $g(x_1) = x_2$, $g(y_1) = y_2$. Such spaces are said to be two-point homogeneous. In addition to the sphere in R^3, spheres of any dimension, projective spaces over the reals, complexes, and quaternions, and a two-dimensional projective space over the Cayley numbers are compact two-point homogeneous spaces which are Riemannian manifolds. In each of

ISBN-0-201-13503-5

these cases the spherical functions are orthogonal polynomials in a variable depending on the distance. Each of these orthogonal polynomials is also a hypergeometric function of the type $_2F_1\left(\begin{matrix} -n, n+a \\ b \end{matrix}; t\right)$ for some a and b. The measure $\sin\theta\,d\theta$ in the case (L.1a) comes from looking at the size of an orbit of a small arc $d\theta$ under a rotation that leaves the north pole fixed.

There are other very important two-point homogeneous spaces which are compact. The easiest to visualize is the set of vertices of the unit cube in R^N. Again the spherical functions are orthogonal polynomials, and they are orthogonal with respect to the symmetric binomial distribution $\binom{N}{x}2^{-N}$, $x=0, 1,\ldots,N$, since this is the size of the orbit of any point with x zeros and $N-x$ ones under the octahedral group acting on this space and leaving $(0,0,\ldots,0)$ fixed. These orthogonal polynomials are also hypergeometric functions, $_2F_1\left(\begin{matrix} -n, -x \\ -N \end{matrix}; 2\right)$, x, $n=0, 1,\ldots,N$, and their three-term recurrence relation is one of the Gauss contiguous relations. They are called Krawtchouk polynomials, though they were introduced almost one hundred years ago by Gram, and they play an important role in coding theory, a subject covered in Volume 3 (*The Theory of Information and Coding*) in this Encyclopedia.

The differential equations (L.4) and (T.4), (T.4a) arise when Laplace's equation is solved by separation of variables. The addition formulas (L.5) and (T.5) are among the most important facts known about these functions. In most of the cases of two-point homogeneous spaces where explicit formulas have been found for the spherical functions there is an addition formula that contains the functional equation as the constant term in an orthogonal expansion. For example, if (L.5) is integrated on $[0,\pi]$ with respect to $d\Psi$ and (T.1a) is used, the result is (L.3). The most natural way to derive addition formulas of this type is to use a group acting on the space, and this is essentially the method used by Laplace and Legendre almost two hundred years ago.

Another important class of functions that were introduced in the eighteenth century are Bessel functions. The Bessel function of the first kind $J_\alpha(x)$ can be defined by

$$J_\alpha(x) = \sum_{n=0}^{\infty} \frac{(-1)^n(x/2)^{2n+\alpha}}{\Gamma(n+\alpha+1)n!} = \frac{(x/2)^\alpha}{\Gamma(x+1)}\,_0F_1\left(\begin{matrix} - \\ \alpha+1 \end{matrix}; \frac{-x^2}{4}\right).$$

After the elementary transcendental functions, these are the class that have been studied most extensively and they have been used in many fields where mathematics has been applied. They have a strong connection with Legendre functions which has been studied by many people. A simple example is Mehler's formula

$$\lim_{n\to\infty} P_n\left(\cos\frac{z}{n}\right) = J_0(z).$$

ISBN-0-201-13503-5

This can be interpreted by considering Legendre polynomials as spherical functions on a sphere of large radius and seeing what happens in a neighborhood of the north pole. The sphere becomes flat and this suggests that $J_0(z)$ should play a role in R^2 similar to that of $P_n(\cos\theta)$ on S^2. The analogues of zonal functions are called radial functions, those functions which depend only on the distance from the origin. One important fact is due to Poisson. If $f(x_1,x_2)=g((x_1^2+x_2^2)^{1/2})$ and

$$F(y_1,y_2)=\int_{-\infty}^{\infty}\int_{-\infty}^{\infty}f(x_1,x_2)\exp\left[i(x_1y_1+x_2y_2)\right]dx_1\,dx_2,$$

then

$$F(y_1,y_2)=G\left((y_1^2+y_2^2)^{1/2}\right)$$

and

$$G(t)=2\pi\int_0^{\infty}g(r)rJ_0(rt)\,dr.$$

The next big development in special functions was the introduction of elliptic functions and theta functions by Abel and Jacobi. Again Mittag-Leffler's paper gives a good historical summary. There have been a few other developments since then that allow us to view the subject with a slightly different perspective. One important development was when Heine introduced what are now called basic hypergeometric series. Recall that a hypergeometric series is a series Σa_n with a_{n+1}/a_n a rational function of n. A basic hypergeometric series is a series Σa_n with a_{n+1}/a_n a rational function of q^n for some fixed q. The role played by the shifted factorial $(a)_n$ in hypergeometric series is now played by $(a;q)_n=(1-a)(1-aq)\cdots(1-aq^{n-1})$. If $|q|<1$, then $(a;q)_\infty=\prod_{n=0}^{\infty}(1-aq^n)$ and $(a;q)_n=(a;q)_\infty/(aq^n;q)_\infty$ is defined for noninteger values of n as long as $aq^{n+k}\neq 1$, $k=0,1,\ldots$. Euler had evaluated two basic hypergeometric series:

$$\sum_{n=0}^{\infty}\frac{x^n}{(q;q)_n}=\frac{1}{(x;q)_\infty};\qquad \sum_{n=0}^{\infty}(-1)^n\frac{q^{\binom{n}{2}}x^n}{(q;q)_n}=(x;q)_\infty.$$

These are special cases of the q-binomial theorem

$$\sum_{n=0}^{\infty}\frac{(a;q)_n}{(q;q)_n}x^n=\frac{(ax;q)_\infty}{(x;q)_\infty},$$

a result which is attributed to many different people. Heine discovered it when he introduced the basic analogue of $_2F_1\left(\begin{smallmatrix}a,b\\c\end{smallmatrix};x\right)$ in 1847. Cauchy

ISBN-0-201-13503-5

had published a proof a few years before this, and Jacobi gave a reference to an 1820 book of Schweins. This formula is given by Schweins, but he refers to an earlier work of Rothe. Unfortunately I have not seen Rothe's book and so cannot confirm the 1811 date given by Schweins. However, it seems very likely that he is correct. Gauss published related formulas in 1811.

One of the most important basic hypergeometric series is the theta function

$$\sum_{-\infty}^{\infty} q^{n^2}x^n = (q^2;q^2)_\infty(-qx;q^2)_\infty(-qx^{-1};q^2)_\infty.$$

This was not the first instance of a bilateral series (one which is infinite in both directions), for

$$\pi \cot \pi z = \lim_{n\to\infty} \sum_{m=-n}^{n} \frac{1}{z-m}$$

$$= \sum_{m=-\infty}^{\infty} \left(\frac{1}{z-m} - \frac{1}{\frac{1}{2}-m}\right) = \sum_{-\infty}^{\infty} \frac{(\frac{1}{2}-z)}{(m-z)(m-\frac{1}{2})}$$

and

$$\frac{\pi^2}{(\sin \pi z)^2} = \sum_{n=-\infty}^{\infty} \frac{1}{(z-n)^2}.$$

However, it was a very fruitful discovery. Initially Jacobi, in his treatment of elliptic functions in *Fundamenta Nova Theoriae Functionum Ellipticarum*, 1829, derived results about theta functions as consequences of results on elliptic functions. Later he reversed this procedure and used theta functions to derive facts about elliptic functions. The function $\sum_{-\infty}^{\infty} q^{n^2}x^n$ had occurred in Fourier's work on the heat equation, and Poisson derived a very important transformation of this function, but the realization that this function was fundamental is due to Jacobi. Recently a new setting for this function has been discovered which ties it up with a group-theoretic approach similar to that outlined above. The relevant group is the three-dimensional Heisenberg group, the group of matrices

$$\begin{bmatrix} 1 & z & y \\ 0 & 1 & x \\ 0 & 0 & 1 \end{bmatrix}.$$

(See Cartier [3] and Auslander and Tolimieri [1].) Other basic hypergeometric series are orthogonal polynomials that arise as spherical functions

ISBN-0-201-13503-5

on discrete two-point homogeneous spaces with certain Chevalley groups acting on these spaces. It is too early in the development of these ideas to say how important they will be, but I am reasonably sure that some important results will be obtained from them. In the nineteenth century elliptic functions were studied exhaustively and they seemed to have a secure place in the mathematics curriculum. Many ideas arose out of scholars' efforts to understand them. However, they themselves have not been as useful as one would have hoped, and as a result their place in the standard curriculum was taken by other subjects that were thought to be more useful, and for decades the knowledge of elliptic functions was largely restricted to number theorists and some applied mathematicians and engineers. Now many people who study and use combinatorial arguments are going to have to learn something about basic hypergeometric series. This will include statisticians interested in block designs and many people studying and using computer algorithms. They play a central role in the study of partitions, the subject of Volume 2 (*The Theory of Partitions*) in this Encyclopedia.

Another development in the study of special functions in the last century was the introduction of differential equations with more than three regular singular points. Riemann observed that Euler's differential equation

$$x(1-x)y'' + \left[c - (a+b+1)x \right] y' - aby = 0, \qquad y = {}_2F_1\left({a,b \atop c}; x \right),$$

has regular singular points at $x = 0$, 1, ∞, and by a linear fractional transformation these singular points can be put at three arbitrary points. The resulting differential equation is determined by the location of these singularities and certain parameters which determine the nature of the solutions in neighborhoods of the singular points. He showed how to obtain the results of Gauss, Kummer, and some of Jacobi's work on hypergeometric series in a simple way and found a cubic transformation which is still not really understood. However, the real importance of this work was the realization that the singularities of a differential equation determine much more about the solutions than one would have thought. Other differential equations were introduced; Heun, Mathieu, Lamé, and those for spheroidal wave functions are examples. These often arose when the wave equation or Laplace's equation was reduced to ordinary differential equations by separating variables. The solutions to these equations give interesting special functions which are much more complicated than hypergeometric functions. It is still not clear how best to study these functions, and one hopes that the algebraic methods used by Miller in this book will allow us to really understand these important functions.

Appell introduced hypergeometric functions of two variables and found analogues of some of the useful facts about the ordinary hypergeometric function for them. However, a real understanding of hypergeometric

ISBN-0-201-13503-5

functions in two variables still lies in the future, though we now have a few ways of treating parts of this subject that have been fruitful.

Pincherle, and later Mellin and Barnes, introduced a new way of treating hypergeometric series and functions. They integrated quotients of gamma functions and were able to perform analytic continuations with ease. The type of integrals they considered has arisen in many different places, from Mehler's earlier work on electrical problems with conical symmetry to Bargmann's work on representations of the Lorentz group.

Poincaré discovered an important extension of elliptic functions when he introduced automorphic functions. They have been extended to several variables in a number of ways. One of the most fruitful is due to Siegel and involves functions of matrix argument. Matrix argument gamma functions were introduced somewhat earlier by Ingham, who was led to them by work of statisticians. In terms of the special functions that are useful in applied mathematics, the main usefulness of automorphic functions will probably be the methods that were used to develop a theory in several variables; for although the theory of hypergeometric and basic hypergeometric functions in several variables is largely undeveloped, enough results have been obtained to indicate that many deep results can be found. A recent example is the work of Macdonald on identities similar to the triple product for the theta function, which he derived from affine root systems of the classical Lie algebras. Feynman integrals can be considered as multivariate hypergeometric functions [4], as are the $3n-j$ symbols which are used to decompose tensor products of representations of $SU(2)$ [2]. Both of these are very useful, yet far from understood. So this area of mathematics is no different than most other parts of mathematics; the pressing problems are to understand what happens in several variables.

One term which has not been defined so far is "special function." My definition is simple, but not time invariant. A function is a special function if it occurs often enough so that it gets a name. There are a number of very important special functions which do not fit into the framework outlined earlier—for example, the Riemann zeta function, which plays a central role in the study of primes and many other number-theoretic problems. Other examples are the Bernoulli polynomials and Bernoulli numbers. Bernoulli numbers were introduced to aid in the computation of series, and they now appear in many unexpected places.

Harry Bateman had a list of over a thousand special functions. Although many of these were special hypergeometric series with no reason for having a separate name, since everything that was known about them was a special case of facts known about a more general hypergeometric series, there are clearly too many functions to make it worthwhile to write books on each of them. However, some of them have so many interesting properties and occur so often that it is essential that each generation of mathematicians consider them anew and record their results for others to

ISBN-0-201-13503-5

use. It is too early to say exactly what books on special functions will appear in this series, but at present there is no adequate treatment of hypergeometric and basic hypergeometric series. There are a few books that treat special functions from an algebraic point of view [5, 7, 8], but none of these contain the very interesting work on the unitary group which led to addition formulas for Jacobi and Laguerre polynomials and for the disk polynomials, an important class of orthogonal polynomials in two variables. The discrete orthogonal polynomials have also not been treated in an adequate way. Books on all these topics will be written.

There have also been some very interesting applications of special functions to combinatorial problems that have only partially been treated in previously mentioned Volumes 2 and 3 in this Encyclopedia. Beyond this we will have to wait and see what develops. If experience is a guide, we will be surprised by the next development. Such developments are predictable in retrospect, but not before the fact.

RICHARD ASKEY
General Editor, Section on
Special Functions

References

1. L. Auslander and R. Tolimieri, *Abelian Harmonic Analysis, Theta Functions and Function Algebras on a Nilmanifold* (Lecture Notes in Math. No. 436), Springer, Berlin, 1975.
2. L. C. Biedenharn and H. VanDam, *Quantum Theory of Angular Momentum*. Academic Press, New York, 1965.
3. P. Cartier, "Quantum mechanical commutative relations and theta functions," *Proc. Symp. Pure Math.* IX, Amer. Math. Soc., Providence, R.I. (1965), 363–387.
4. V. A. Golubeva, "Some problems in the analytic theory of Feynman integrals," *Russian Math. Surveys* 31 (1976), 139–207; *Uspehi Mat. Nauk* 31 (1976), 135–202.
5. W. Miller, Jr., *Lie Theory and Special Functions*. Academic Press, New York, 1968.
6. G. Mittag-Leffler, "An introduction to the theory of elliptic functions," *Ann. of Math. Ser.* 2, 24 (1923), 271–351. (Translation from a paper originally published in 1876.)
7. J. D. Talman, *Special Functions, A Group Theoretic Approach*. W. A. Benjamin, New York, 1968.
8. N. Ya. Vilenkin, *Special Functions and the Theory of Group Representations*. Amer. Math. Soc., Providence, R.I., 1968.

ISBN-0-201-13503-5

Preface

This book is concerned with the relationship between symmetries of a linear second-order partial differential equation of mathematical physics, the coordinate systems in which the equation admits solutions via separation of variables, and the properties of the special functions that arise in this manner. It is an introduction intended for anyone with experience in partial differential equations, special functions, or Lie group theory, such as group theorists, applied mathematicians, theoretical physicists and chemists, and electrical engineers. We will exhibit some modern group-theoretic twists in the ancient method of separation of variables that can be used to provide a foundation for much of special function theory. In particular, we will show explicitly that all special functions that arise via separation of variables in the equations of mathematical physics can be studied using group theory. These include the functions of Lamé, Ince, Mathieu, and others, as well as those of hypergeometric type.

This is a very critical time in the history of group-theoretic methods in special function theory. The basic relations between Lie groups, special functions, and the method of separation of variables have recently been clarified. One can now construct a group-theoretic machine that, when applied to a given differential equation of mathematical physics, describes in a rational manner the possible coordinate systems in which the equation admits solutions via separation of variables and the various expansion theorems relating the separable (special function) solutions in distinct coordinate systems. Indeed for the most important linear equations, the separated solutions are characterized as common eigenfunctions of sets of second-order commuting elements in the universal enveloping algebra of the Lie symmetry algebra corresponding to the equation. The problem of expanding one set of separable solutions in terms of another reduces to a problem in the representation theory of the Lie symmetry algebra.

Although this method is simple, elegant, and very useful, it has as yet been applied to relatively few differential equations. (At the time of this writing, the wave equation $(\partial_{tt} - \Delta_3)\Psi = 0$ is still under intensive study.) Moreover, few theorems have yet been proved that delineate the full scope of the method. It is the author's hope that the present work, which is aimed at a general audience rather than at specialists, will convince the reader that group-theoretic methods are singularly appropriate for the study of separation of variables and special functions. It is also hoped that this work will encourage others to enter the field and solve the many interesting problems that remain.

The ideas relating Lie groups, special functions, and separation of variables spring from a number of rather diverse historical sources. The first deep work on the relationship of group representation theory and special functions is commonly attributed to E. Cartan (27). However, the first detailed use of the relationship for computational purposes is probably found in the papers of Wigner. Wigner's work on this subject began in the 1930s and is given an elementary exposition in his 1955 Princeton lecture notes. These notes were later expanded and updated in a book by Talman (124).

A second major contributor to the computational theory is Vilenkin, who wrote a series of papers commencing in 1956 and culminating in his book (128). This encyclopedic treatise was strongly influenced by the explicit constructions of irreducible representations of the classical groups due to Gel'fand and Naimark (e.g., (41)). Vilenkin (and Wigner) obtain special functions as matrix elements of operators defining irreducible group representations.

Another precursor of our theory is the factorization method. The method was discovered by Schrödinger and applied to solve the time-independent Schrödinger equation for a number of systems of physical interest (e.g., (117)). This useful tool for computing eigenvalues and recurrence relations for solutions of second-order ordinary differential equations was developed by several authors, including Infeld and Hull (52), who summarized the state of the theory as of 1951. An independent and somewhat different development was given by Inoui (53).

The author contributed to this theory by showing, in 1964 (80), that the factorization method was equivalent to the representation theory of four Lie algebras.

Another approach to the subject matter of this book is contained in three remarkable papers by Weisner (133–135), the first appearing in 1955. Weisner showed the group-theoretic significance of families of generating functions for hypergeometric, Hermite, and Bessel functions. In these papers are also found examples of separable coordinate systems characterized in terms of Lie algebra symmetry operators. Weisner's theory is extended and related to the factorization method in the author's monograph (82). This monograph is primarily devoted to the representation theory of local Lie groups rather than the theory of global Lie groups, which is treated in the works of Talman and Vilenkin.

We should also mention Truesdell's monograph on the F equation (126), which demonstrated how generating functions and integral representations for special functions can be derived directly from a knowledge of the differential recurrence relations obeyed by the special functions. By 1968 it was recognized that Truesdell's technique fits comfortably into the group-theoretic approach to special functions (82).

A major theme in the present work is that separable coordinate systems for second-order linear partial differential equations can be characterized in terms of sets of second-order symmetry operators for the equations. This idea is very natural from a quantum-mechanical point of view. Moreover, since the work of Lie, it has been known to be correct for certain simple coordinates, such as spherical, cylindrical, and Cartesian (i.e., subgroup) coordinates. For a few important Schrödinger equations, such as the equation for the hydrogen atom, operator characterizations of a few nonsubgroup coordinates were well known (9, 30). However, the explicit statement of the relationship between symmetry and separation of variables appeared for the first time in the 1965 paper (138) by Winternitz and Fris. These authors gave group-theoretic characterizations of the separable coordinate systems corresponding to the eigenvalue equations for the Laplace-Beltrami operators on two-dimensional spaces with constant curvature. This work was extended by Winternitz and collaborators in (74, 106, 139, 140). Finally, the author in collaboration with C. P. Boyer and E. G. Kalnins has classified group theoretically the separable coordinate systems for a number of important partial differential equations and investigated the relationship between the classification and special function theory. One interesting feature of this work, primarily due to Kalnins, has been the discovery of many new separable systems that are not contained in such standard references as (97). A second feature has been the development of a group-theoretic method that makes it possible to derive identities for nonhypergeometric special functions, such as Mathieu, Lamé, spheroidal, Ince, and anharmonic oscillator functions, as well as for the more familiar hypergeometric functions.

Prerequisites for understanding this book include some acquaintance with Lie groups and algebras (i.e., homomorphism and isomorphism of groups and algebras) such as can be found in (43) and (85). However, the examples treated here are very explicit and can be understood with only a minimal knowledge of Lie theory. Secondly, it is assumed that the reader has some experience in the solution of partial differential equations by separation of variables, in, say, rectangular, polar, and spherical coordinates.

Due to limitations of space, time, and the author's competence, it has been found necessary to omit certain topics. The most important among these is the theory of spherical functions on groups. This topic, a generalization of the theory of spherical harmonics, has an extensive literature (e.g., [47, 130]). Moreover, spherical functions were recently used to derive an addition theorem for Jacobi polynomials (68, 119). However, spherical functions are always associated with subgroup coordinates, and even for the most elementary equations considered in this book, they fail to encompass all of the special functions that arise via separation of variables.

Boundary value problems have also been omitted, even though symmetry methods are important for their solution (see [16]). This last reference, as well as (105) and (38) contain discussions of symmetry techniques for finding solutions of nonlinear partial differential equations, a subject that has been omitted here because its ultimate forum is not yet clear.

I should like to thank Paul Winternitz for helpful discussions leading to the basic concepts relating symmetry and separation of variables. Finally, I wish to thank Charles Boyer and Ernie Kalnins, without whose research collaboration this book could not have been written.

WILLARD MILLER, JR.

CHAPTER 1 _____

The Helmholtz Equation

1.0 Introduction

The main ideas relating the symmetry group of a linear partial differential equation and the coordinate systems in which the equation admits
separable solutions are most easily understood through examples. Perhaps
the simplest nontrivial example that exhibits the features we wish to
illustrate is the Helmholtz, or reduced wave, equation

$$(\Delta_2 + \omega^2)\Psi(x,y) = 0 \qquad (0.1)$$

where ω is a positive real constant and

$$\Delta_2 \Psi = \partial_{xx}\Psi + \partial_{yy}\Psi.$$

(Here $\partial_{xx}\Psi$ is the second partial derivative of Ψ with respect to x.)

In this chapter we will study the symmetry group and separated solutions of (0.1) and related equations in great detail, thereby laying the
groundwork for similar treatments of much more complicated problems in
the chapters to follow.

For the present we consider only those solutions Ψ of (0.1) which are
defined and analytic in the real variables x,y for some common open
connected set $\mathcal{D}$ in the plane R^2. (For example, $\mathcal{D}$ can be chosen as the
plane itself.) The set of all such solutions Ψ forms a (complex) vector space
$\mathcal{F}_0$; that is, if $\Psi \in \mathcal{F}_0$ and $a \in \mathbb{C}$, then $(a\Psi)(x,y) \equiv a\Psi(x,y) \in \mathcal{F}_0$, and
$(\Psi_1 + \Psi_2)(x,y) \equiv \Psi_1(x,y) + \Psi_2(x,y) \in \mathcal{F}_0$ whenever $\Psi_1, \Psi_2 \in \mathcal{F}_0$. Considering
$\mathcal{D}$ as fixed throughout our discussion, we call $\mathcal{F}_0$ the *solution space* of (0.1).

Let $\mathcal{F}$ be the vector space of all complex-valued functions defined and

ENCYCLOPEDIA OF MATHEMATICS and Its Applications, Gian-Carlo Rota (ed.).
Vol. 4: Willard Miller, Jr., Symmetry and Separation of Variables

ISBN-0-201-13503-5

real analytic on $\mathcal{D}$ and let Q be the partial differential operator

$$Q = \Delta_2 + \omega^2 \tag{0.2}$$

defined on $\mathcal{D}$. Clearly, $Q\Phi \in \mathcal{F}$ for $\Phi \in \mathcal{F}$, and $\mathcal{F}_0$ is that subspace of $\mathcal{F}$ which is the kernel or null space of the linear operator Q.

1.1 The Symmetry Group of the Helmholtz Equation

It is a well-known fact that if $\Psi(\mathbf{x})$, $\mathbf{x} = (x,y)$, is a solution of (0.1), then $\Psi^\wedge(\mathbf{x}) = \Psi(\mathbf{x} + \mathbf{a})$ where $\mathbf{a} = (a_1, a_2)$ is a real two-vector and $\Psi^{\wedge\wedge}(\mathbf{x}) = \Psi(\mathbf{x}O)$ where

$$O(\theta) = \begin{pmatrix} \cos\theta & -\sin\theta \\ \sin\theta & \cos\theta \end{pmatrix}, \qquad 0 \leqslant \theta \leqslant 2\pi,$$

are also solutions. (However, $\mathbf{x}$ must be chosen so that $\mathbf{x} + \mathbf{a}$ and $\mathbf{x}O$ lie in $\mathcal{D}$ in order for $\Psi^\wedge$ and $\Psi^{\wedge\wedge}$ to make sense when evaluated at $\mathbf{x}$.) Thus translations in the plane and rotations about the origin map solutions of (0.1) into solutions. These translations and rotations generate the group $E(2)$, the *Euclidean group*, whose elements are just the rigid motions in the plane. As we shall show, exploitation of this Euclidean symmetry of (0.1) yields simple proofs of many facts concerning the solutions of the Helmholtz equation. In the following paragraphs we rederive the existence of Euclidean symmetry for (0.1) and show that in a certain sense $E(2)$ is the maximal symmetry group of this equation.

We say that the linear differential operator

$$L = X(\mathbf{x})\partial_x + Y(\mathbf{x})\partial_y + Z(\mathbf{x}), \qquad X, Y, Z \in \mathcal{F} \tag{1.1}$$

is a *symmetry operator* for the Helmholtz equation provided

$$[L, Q] = R(\mathbf{x})Q, \qquad R \in \mathcal{F}, \tag{1.2}$$

where $[L, Q] = LQ - QL$ is the commutator of L and Q, and the analytic function $R = R_L$ may vary with L. Recall that Q is the operator (0.2). (We interpret the relation (1.2) to mean that the operators on the left- and right-hand sides yield the same result when applied to any $\Phi \in \mathcal{F}$.)

Let $\mathcal{G}$ be the set of all symmetry operators for the Helmholtz equation.

THEOREM 1.1. *A symmetry operator L maps solutions of (0.1) into solutions; that is, if $\Psi \in \mathcal{F}_0$, then $L\Psi \in \mathcal{F}_0$.*

Proof. If $\Psi \in \mathcal{F}_0$, we have $\Psi \in \mathcal{F}$ and $Q\Psi = 0$. Then from (1.2), $QL\Psi = LQ\Psi - RQ\Psi = 0$, so $L\Psi \in \mathcal{F}_0$. ∎

ISBN-0-201-13503-5

Furthermore, it is not difficult to show that if an operator L of the form (1.1) maps solutions Ψ of $Q\Psi=0$ into solutions, then L satisfies the commutation relation (1.2) for some $R \in \mathcal{F}$. (However, it is not known whether this statement is true for an arbitrary linear differential equation of second order.)

THEOREM 1.2. *The set $\mathcal{G}$ of symmetry operators is a complex Lie algebra; that is, if $L_1, L_2 \in \mathcal{G}$, then*

(1) $a_1 L_1 + a_2 L_2 \in \mathcal{G}$ *for all* $a_1, a_2 \in \mathcal{C}$,

(2) $[L_1, L_2] \in \mathcal{G}$.

Proof. Since $L_1, L_2 \in \mathcal{G}$, these operators satisfy the equations $[L_j, Q] = R_j(\mathbf{x})Q$ where $R_j \in \mathcal{F}$, $j = 1, 2$. A simple computation shows that the first-order operator $L = a_1 L_1 + a_2 L_2$ satisfies (1.2) with $R = a_1 R_1 + a_2 R_2$. Similarly, $L = [L_1, L_2]$ is a first-order operator that satisfies (1.2) with $R = \tilde{L}_1 R_2 - \tilde{L}_2 R_1$ where $\tilde{L} = \tilde{L} + Z(\mathbf{x})$, (1.1). ∎

Note: It is not excluded that $\mathcal{G}$ is an infinite-dimensional Lie algebra, although for the example considered here $\dim \mathcal{G} = 4$.

We now explicitly compute the symmetry algebra of (0.1). Substituting (0.2) and (1.1) into (1.2) and evaluating the commutator, we find

$$2X_x \partial_{xx} + 2(X_y + Y_x)\partial_{xy} + 2Y_y \partial_{yy} + (X_{xx} + X_{yy} + 2Z_x)\partial_x$$
$$+ (Y_{xx} + Y_{yy} + 2Z_y)\partial_y + (Z_{xx} + Z_{yy}) = -R(\partial_{xx} + \partial_{yy} + \omega^2). \tag{1.3}$$

For this operator equation to be valid when applied to an arbitrary $\Phi \in \mathcal{F}$, it is necessary and sufficient that the coefficients of ∂_{xx}, ∂_{yy}, and so on be the same on both sides of the equation:

(a) $2X_x = -R = 2Y_y$, $X_y + Y_x = 0$,

(b) $X_{xx} + X_{yy} + 2Z_x = 0$, $Y_{xx} + Y_{yy} + 2Z_y = 0$,   (1.4)

(c) $Z_{xx} + Z_{yy} = -R\omega^2$.

From equations (1.4a), $X_x = Y_y$ and $X_y = -Y_x$. Thus $X_{xx} + X_{yy} = Y_{xy} - Y_{xy} = 0$; similarly, $Y_{xx} + Y_{yy} = 0$. Comparing these results with equation (1.4b), we see that $Z_x = Z_y = 0$, so $Z = \delta$, a constant. It follows from (1.4c) that $R = 0$. Equations (1.4a) then imply $X = X(y)$, $Y = Y(x)$ with $X'(y) = -Y'(x)$. This last equation implies $X' = -Y' = \gamma \in \mathcal{C}$. Thus, the general solution of equations (1.4) is

$$X = \alpha + \gamma y, \qquad Y = \beta - \gamma x, \qquad Z = \delta, \qquad R = 0, \tag{1.5}$$

and the symmetry operator L takes the form

$$L = (\alpha + \gamma y)\partial_x + (\beta - \gamma x)\partial_y + \delta. \tag{1.6}$$

Clearly the symmetry algebra $\mathcal{G}$ is four dimensional with basis

$$P_1 = \partial_x, \quad P_2 = \partial_y, \quad M = y\partial_x - x\partial_y, \quad E = 1, \tag{1.7}$$

obtained by setting $\alpha = 1$, $\beta = \gamma = \delta = 0$ for P_1; $\beta = 1$ and $\alpha = \gamma = \delta = 0$ for P_2; and so on. The commutation relations for this basis are easily verified to be

$$[P_1, P_2] = 0, \quad [M, P_1] = P_2, \quad [M, P_2] = -P_1 \tag{1.8}$$

and $[E, L] = 0$ for all $L \in \mathcal{G}$. The symmetry operator E is of no interest to us, so we will ignore it and concentrate on the three-dimensional Lie algebra with basis $\{P_1, P_2, M\}$ and commutation relations (1.8). Furthermore, for reasons that will become clear shortly we will restrict our attention to the *real* Lie algebra $\mathcal{E}(2)$ generated by $\{P_1, P_2, M\}$, that is, the Lie algebra consisting of all elements $\alpha P_1 + \beta P_2 + \gamma M$ where α, β, γ belong to the field of real numbers R. Here, $\mathcal{E}(2)$ is isomorphic to the Lie algebra of the Euclidean group in the plane $E(2)$. To show this we consider the well-known realization of $E(2)$ as a group of 3×3 matrices. The elements of $E(2)$ are

$$g(\theta, a, b) = \begin{bmatrix} \cos\theta & -\sin\theta & 0 \\ \sin\theta & \cos\theta & 0 \\ a & b & 1 \end{bmatrix}, \quad \begin{matrix} a, b \in R, \\ 0 \leqslant \theta < 2\pi \ (\mathrm{mod}\ 2\pi), \end{matrix} \tag{1.9}$$

and the group product is given by matrix multiplication,

$$g(\theta, a, b)\, g(\theta', a', b') = g(\theta + \theta', a\cos\theta' + b\sin\theta'$$

$$+ a', -a\sin\theta' + b\cos\theta' + b'). \tag{1.10}$$

$E(2)$ acts as a transformation group in the plane. Indeed, the group element $g(\theta, a, b)$ maps the point $\mathbf{x} = (x, y)$ in R^2 to the point

$$\mathbf{x}g = (x\cos\theta + y\sin\theta + a, -x\sin\theta + y\cos\theta + b). \tag{1.11}$$

It is easy to check that $\mathbf{x}(g_1 g_2) = (\mathbf{x}g_1)g_2$ for all $\mathbf{x} \in R^2$ and $g_1, g_2 \in E(2)$ and that $\mathbf{x}g(0,0,0) = \mathbf{x}$ where $g(0,0,0)$ is the identity element of $E(2)$. Geometrically, g corresponds to a rotation about the origin $(0,0)$ through the angle θ in a clockwise direction, followed by the translation (a,b).

Computing the Lie algebra of the matrix group $E(2)$ in the usual way

ISBN-0-201-13503-5

(see Appendix A), we find that a basis for the Lie algebra is given by the matrices

$$M = \begin{bmatrix} 0 & -1 & 0 \\ 1 & 0 & 0 \\ 0 & 0 & 0 \end{bmatrix}, \quad P_1 = \begin{bmatrix} 0 & 0 & 0 \\ 0 & 0 & 0 \\ 1 & 0 & 0 \end{bmatrix}, \quad P_2 = \begin{bmatrix} 0 & 0 & 0 \\ 0 & 0 & 0 \\ 0 & 1 & 0 \end{bmatrix} \quad (1.12)$$

with commutation relations identical to (1.8). (Here, the commutator $[A,B]$ of two $n \times n$ matrices is the *matrix commutator* $[A,B] = AB - BA$.) It follows that the symmetry algebra $\mathcal{E}(2)$ is isomorphic to the Lie algebra of $E(2)$.

We can construct a general group element (1.9) from the Lie algebra elements (1.12) through use of the matrix exponential. Indeed, it is straightforward to show that

$$g(\theta, a, b) = \exp(\theta M)\exp(aP_1 + bP_2) \quad (1.13)$$

where

$$\exp(A) = \sum_{k=0}^{\infty} (k!)^{-1} A^k, \quad A^0 = E_n, \quad (1.14)$$

for any $n \times n$ matrix A. Here E_n is the $n \times n$ identity matrix.

Using standard results from Lie theory (see Appendix A), we can extend the action of $\mathcal{E}(2)$ on $\mathcal{F}$ given by expressions (1.7) to a local representation $\mathbf{T}$ of $E(2)$ on $\mathcal{F}$. Indeed from Theorem A.3 we obtain the operators $\mathbf{T}(g)$ where

$$\mathbf{T}(g(0,a,0))\Phi(\mathbf{x}) = \exp(aP_1)\Phi(\mathbf{x}) = \Phi(x + a, y),$$

$$\mathbf{T}(g(0,0,b))\Phi(\mathbf{x}) = \exp(bP_2)\Phi(\mathbf{x}) = \Phi(x, y + b), \quad (1.15)$$

$$\mathbf{T}(g(\theta,0,0))\Phi(\mathbf{x}) = \exp(\theta M)\Phi(\mathbf{x}) = \Phi(x\cos\theta + y\sin\theta, -x\sin\theta + y\cos\theta)$$

and $\Phi \in \mathcal{F}$. In analogy with (1.13) the general operator $\mathbf{T}(g)$ is defined by

$$\mathbf{T}(g(\theta,a,b))\Phi(\mathbf{x}) = \exp(\theta M)\exp(aP_1)\exp(bP_2)\Phi(\mathbf{x})$$

$$= \Phi(\mathbf{x}g) \quad (1.16)$$

where $\mathbf{x}g$ is given by (1.11). Thus the action (1.11) of $E(2)$ as a transformation group is exactly that induced by the Lie derivatives (1.7). (Recall that if L is a Lie derivative, we have by definition

$$\exp(aL)\Phi(\mathbf{x}) = \sum_{k=0}^{\infty} \frac{a^k}{k!} L^k \Phi(\mathbf{x}), \quad \Phi \in \mathcal{F}; \quad (1.17)$$

see (A.8).)

ISBN-0-201-13503-5

It is a consequence of the fundamental results of Lie theory that the operators $T(g)$ satisfy the group homomorphism property

$$T(gg') = T(g)T(g'), \qquad g, g' \in E(2), \qquad (1.18)$$

although the dubious reader can verify this directly. The results (1.16), (1.18) have to be interpreted with some care because for a pair $x \in \mathcal{D}$, $g \in E(2)$, the element xg may not lie in $\mathcal{D}$, so that $\Phi(xg)$ is undefined. However, for fixed $x \in \mathcal{D}$ the element xg will lie in $\mathcal{D}$ as long as g is in a suitably small neighborhood of the identity element $g(0,0,0)$ in $E(2)$. Thus (1.16) and (1.18) have only local validity.

If L is a first-order symmetry operator of the Helmholtz equation, that is, L maps solutions into solutions, then also L^k maps solutions into solutions for each $k = 2, 3, 4, \ldots$. Furthermore, from (1.17) we see that the operator $\exp(aL)$ also maps solutions into solutions. Since the operators $T(g)$ are composed of products of terms of the form $\exp(aL)$, $L \in \mathcal{E}(2)$, we can conclude that if $\Psi(x)$ is an analytic solution of $Q\Psi = 0$, then $\Psi'(x) = T(g)\Psi(x) = \Psi(xg)$ is also an analytic solution, with domain the open set consisting of all $x \in R^2$ such that $xg \in \mathcal{D}$. (If $\mathcal{D} = R^2$, then the operators $T(g)$ are defined globally and there is no domain problem.) Based on these comments, we call $E(2)$ the *symmetry group* of the equation $Q\Psi = 0$.

It is now easy to see why we limit ourselves to the real Lie algebra with basis P_1, P_2, M. The exponential of an element of the complex Lie algebra, say iP_1 where $i = \sqrt{-1}$, is a symmetry of the Helmholtz equation. However, a straightforward application of Lie theory yields $\exp(iP_1)\Phi(x) = \Phi(x + i,y)$ and this is undefined for $\Phi \in \mathcal{F}$ because Φ is defined only for real x and y. Thus we limit ourselves to the Lie algebra whose elements have exponentials with the simple interpretation (1.16).

In analogy to our computation of the first-order symmetry operators for the Helmholtz equation, we can determine the second-order symmetry operators. We say that the second-order operator

$$S = A_{11}\partial_{xx} + A_{12}\partial_{xy} + A_{22}\partial_{yy} + B_1\partial_x + B_2\partial_y + C, \qquad A_{jk}, B_j, C \in \mathcal{F}, \quad (1.19)$$

is a *symmetry operator* for (0.1) provided

$$[S, Q] = U(x)Q \qquad (1.20)$$

where

$$U = H_1(x)\partial_x + H_2(x)\partial_y + J(x), \qquad H_j, J \in \mathcal{F}, \qquad (1.21)$$

is a first-order differential operator. (Here U may vary with S.) We consider a first-order symmetry operator L as a special second-order symmetry. When $S = L$, equation (1.20) holds with $H_1 = H_2 = 0$. We allow

ISBN-0-201-13503-5

U to be a first-order operator because the commutator of two second-order operators is an operator of order $\leqslant 3$.

The following result is proved exactly as is Theorem 1.1.

THEOREM 1.3. *A second-order symmetry operator S maps solutions of* (0.1) *into solutions; that is, if* $\Psi \in \mathcal{F}_0$, *then* $S\Psi \in \mathcal{F}_0$.

Furthermore, it is not difficult to show that if an operator S of the form (1.19) maps solutions Ψ of $Q\Psi = 0$ into solutions, then S satisfies the commutation relation (1.20) for some U of the form (1.21).

Let $\mathbb{S}$ be the vector space of all second-order symmetry operators S. Clearly $\mathbb{S}$ contains the first-order symmetry algebra $\mathcal{G}$. However, $\mathbb{S}$ is not a Lie algebra under the usual bracket operation because the commutator $[S, S']$ of two second-order symmetries is in general a third-order operator, hence not an element of $\mathbb{S}$. (Note that $[S, S']$ still maps solutions into solutions.)

Among the elements of $\mathbb{S}$ are all operators of the form RQ where R is any element of $\mathcal{F}$. Indeed $S = RQ$ satisfies (1.20) with $U = [R, Q]$, a first-order differential operator. We can check directly that RQ maps solutions Ψ of $Q\Psi = 0$ into other solutions. Indeed $(RQ)\Psi = R(Q\Psi) = 0$, so Ψ is mapped to the solution 0. It follows that the operators RQ are symmetries of a trivial sort; they act as the zero operator on the solution space $\mathcal{F}_0$.

The set of all trivial symmetries $\mathfrak{q} = \{RQ : R \in \mathcal{F}\}$ forms a subspace of $\mathbb{S}$ and each element of $\mathfrak{q}$ acts as the zero operator on $\mathcal{F}_0$. We will henceforth ignore $\mathfrak{q}$ and concentrate our attention on the factor space $\mathbb{S}/\mathfrak{q}$ of nontrivial symmetries. Thus we will regard two symmetries S, S' in $\mathbb{S}$ as identical ($S \equiv S'$) if $S' = S + RQ$ for some $R \in \mathcal{F}$. If S is given by (1.19), then $S \equiv S'$ where $S' = S - A_{22}Q$, so that the coefficient of ∂_{yy} in the expression for S' is zero. Thus every symmetry S is equivalent to a symmetry S' whose coefficient of ∂_{yy} is zero. (Note that the operators S and S' agree on the solution space $\mathcal{F}_0$.) Furthermore, two operators S_1, S_2 whose coefficients of ∂_{yy} are zero agree on $\mathcal{F}_0$ if and only if their remaining coefficients are identical.

The computation of all nontrivial symmetries is straightforward. We substitute expressions (1.19) (with $A_{22} = 0$) and (1.21) into (1.20) and equate coefficients of the various partial derivatives with respect to x and y on both sides of the resulting relation. The equations obtained are analogous to (1.3) and (1.4) but somewhat more complicated. Here we present only the results of the computation.

$\mathbb{S}/\mathfrak{q}$ is a nine-dimensional complex vector space with basis

$$
\begin{aligned}
&\text{(a)} \quad P_1, P_2, M, E, \\
&\text{(b)} \quad P_1^2, P_1 P_2, M^2, \{M, P_1\}, \{M, P_2\}.
\end{aligned}
\qquad (1.22)
$$

Here $\{A,B\}=AB+BA$ for operators A,B on $\mathcal{F}$. Note that if A and B are first-order symmetries, then the products AB and BA are second-order symmetries. The results (1.22) show that the Helmholtz equation admits no nontrivial symmetries other than these; that is, all second-order symmetries are quadratic polynomials in the elements of $\mathcal{G}$. (In fact, it can be shown that the nontrivial symmetry operators of any order are polynomials in the elements of $\mathcal{G}$, but we shall not need this.) In general, if $Q\Psi=0$ is a second-order partial differential equation whose nontrivial second-order symmetries are all quadratic polynomials in the elements of the first-order symmetry algebra $\mathcal{G}$, we call such an equation *class I*. If there exists a nontrivial second-order symmetry that is not expressible as a quadratic polynomial in the first-order symmetries, the equation is called *class II*. From (1.22) we conclude that the Helmholtz equation is class I.

A few comments are in order concerning the symbol $\{\cdot,\cdot\}$. Consider the second-order symmetry MP_1. Note that

$$MP_1 = \tfrac{1}{2}(MP_1 + P_1 M) + \tfrac{1}{2}(MP_1 - P_1 M) = \tfrac{1}{2}\{M,P_1\} + \tfrac{1}{2}[M,P_1].$$

Thus we have expressed MP_1 as the sum of the truly second-order (nonfirst-order) operator $\tfrac{1}{2}\{M,P_1\}$ and the first-order operator $\tfrac{1}{2}[M,P_1] = \tfrac{1}{2}P_2$. Similarly, any product AB of elements of $\mathcal{E}(2)$ can be written uniquely as the sum of a symmetrized purely second-order part $\tfrac{1}{2}\{A,B\}$ and a commutator $\tfrac{1}{2}[A,B]$ that belongs to $\mathcal{E}(2)$. In (1.22a) we have listed a basis for the first-order operators in $\mathcal{S}/\mathfrak{q}$, while in (1.22b) we have listed a basis for the subspace of purely second-order operators.

For another perspective on the five-dimensional space spanned by the basis (1.22b), consider the space $\mathcal{E}(2)^{(2)}$ of second-order symmetrized operators from $\mathcal{E}(2)$. This space is six-dimensional with a basis consisting of the five operators listed in (1.22b) plus the operator P_2^2. However, on $\mathcal{F}_0$ the operator $P_1^2 + P_2^2 \in \mathcal{E}(2)^{(2)}$ agrees with the first-order operator $-\omega^2$, that is, multiplication by the constant $-\omega^2$. Thus to characterize those elements of $\mathcal{E}(2)^{(2)}$ which act on $\mathcal{F}_0$ in distinct ways, we pass to the factor space $\mathcal{E}(2)^{(2)}/\{P_1^2 + P_2^2\}$, where $\{P_1^2 + P_2^2\}$ is the subspace of $\mathcal{E}(2)^{(2)}$ consisting of all constant multiples $a(P_1^2 + P_2^2), a \in R$. This makes sense because two operators S_1, S_2 in $\mathcal{E}(2)^{(2)}$ such that $S_1 - S_2 = a(P_1^2 + P_2^2)$ have the same eigenfunctions in $\mathcal{F}_0$ with corresponding eigenvalues differing by $a\omega^2$.

Up to now we have considered $\mathcal{S}/\mathfrak{q}$ as the space of all *complex* linear combinations of the basis operators (1.22). However, for purposes of describing the relationship between symmetry and separation of variables for the real Helmholtz equation we shall find that it is sufficient to consider only *real* linear combinations of the basis operators (1.22). Rather than introduce a new symbol to denote this real nine-dimensional vector

ISBN-0-201-13503-5

space, we shall retain the symbol $S/\mathfrak{q}$ but we shall henceforth consider this vector space to be defined over R rather than $\mathcal{C}$.

With this interpretation we see that the five-dimensional subspace of purely second-order operators in $S/\mathfrak{q}$ is isomorphic to $\mathcal{E}(2)^{(2)}/\{P_1^2+P_2^2\}$. That is, we can identify the purely second-order symmetries of the Helmholtz equation with the purely second-order elements in the universal enveloping algebra of $\mathcal{E}(2)$ modulo the center of the enveloping algebra. This point of view will be useful for the orbit analysis that we carry out in Section 1.2.

1.2 Separation of Variables for the Helmholtz Equation

The method of separation of variables for solving partial differential equations, although easy to illustrate for certain important examples, proves surprisingly subtle and difficult to describe in general. For this reason we begin with the simplest cases and then gradually consider cases of greater and greater complexity. At present we content ourselves with the vague assertion that separation of variables is a method for finding solutions of a second-order partial differential equation in n variables by reduction of this equation to a system of n (at most) second-order ordinary differential equations.

Let us begin by searching for solutions of (0.1) in the form $\Psi(x,y)=X(x)Y(y)$. Then the Helmholtz equation becomes

$$X''Y+XY''+\omega^2XY=0 \tag{2.1}$$

where a prime denotes differentiation. This equation can be written

$$\frac{X''}{X}=-\frac{Y''}{Y}-\omega^2 \tag{2.2}$$

where the left-hand side is a function of x alone and the right-hand side is a function of y alone. (Thus the Cartesian coordinates x, y have been separated in (2.2).) This is possible only if both sides of the equation are equal to a constant $-k^2$, called the *separation constant*. Thus equation (2.2) is equivalent to the pair of ordinary differential equations

$$X''(x)+k^2X(x)=0, \qquad Y''(y)+(\omega^2-k^2)Y(y)=0. \tag{2.3}$$

A basis of solutions for the x equation is $X_1=e^{ikx}$, $X_2=e^{-ikx}$ for $k\neq0$, while $Y_1=\exp(i(\omega^2-k^2)^{1/2}y)$, $Y_2=\exp(-i(\omega^2-k^2)^{1/2}y)$ is a basis for the y equation if $\omega^2-k^2\neq0$. Thus we can find solutions $\Psi(x,y)$ of (0.1) in the

ISBN-0-201-13503-5

form

$$\Psi_k(\mathbf{x}) = \sum_{j,l=1}^{2} A_{jl} X_j(x) Y_l(y) \tag{2.4}$$

where the complex constants A_{jl} are arbitrary. Although the Ψ_k are very special solutions of (0.1), it can be shown that essentially any solution of the Helmholtz equation can be represented as a sum or integral (with respect to k) of these special solutions.

Note that the separated solution $\Psi_k = X_1 Y_1 = \exp\{i[kx + (\omega^2 - k^2)^{1/2} y]\}$ is a simultaneous eigenvector of the commuting operators $P_1 = \partial_x$ and $P_2 = \partial_y$:

$$P_1 \Psi_k = ik\Psi_k, \qquad P_2 \Psi_k = i(\omega^2 - k^2)^{1/2}\Psi_k \tag{2.5}$$

with similar remarks for the other separated solutions $X_j Y_l$. Thus, we can characterize the separated solutions in Cartesian coordinates by saying that they are common eigenfunctions of the symmetry operators $P_1, P_2 \in \mathscr{E}(2)$ in $\mathscr{F}_0$.

For our next example we pass to polar coordinates r, θ:

$$x = r\cos\theta, \qquad y = r\sin\theta, \qquad 0 \leqslant r, \qquad 0 \leqslant \theta < 2\pi \ (\mathrm{mod}\, 2\pi). \tag{2.6}$$

In these coordinates the Helmholtz equation becomes

$$\left(\partial_{rr} + \frac{1}{r}\partial_r + \frac{1}{r^2}\partial_{\theta\theta} + \omega^2\right)\Psi(r,\theta) = 0. \tag{2.7}$$

We look for solutions of the form $\Psi = R(r)\Theta(\theta)$. Substituting this expression into (2.7) and rearranging terms, we obtain

$$(r^2 R'' + rR' + r^2\omega^2)R^{-1} = -\Theta''\Theta^{-1}. \tag{2.8}$$

Since the left-hand side of (2.8) is a function of r alone, while the right-hand side is a function of θ alone, both sides of this equation must be equal to a constant k^2. Thus (2.8) is equivalent to the two ordinary differential equations

$$\Theta''(\theta) + k^2\Theta(\theta) = 0, \qquad r^2 R''(r) + rR'(r) + (r^2\omega^2 - k^2)R = 0. \tag{2.9}$$

The first equation has solutions $\Theta = e^{\pm ik\theta}$ while the second, Bessel's equation, admits the solutions $R = J_{\pm k}(\omega r)$ where $J_\nu(z)$ is a Bessel function (see equation (B.14)). Note that the separated solution $\Psi_k = J_k(\omega r)e^{ik\theta}$ is an eigenvector of the operator $M \in \mathscr{E}(2)$. Indeed, in polar coordinates $M =$

ISBN-0-201-13503-5

$-\partial_\theta$, so $\Psi_k \in \mathcal{F}_0$ is a solution of the Helmholtz equation which also satisfies

$$M\Psi_k = -ik\Psi_k. \qquad (2.10)$$

Similar remarks hold for the other separated solutions in polar coordinates.

In each of our examples we have seen that the separated solutions were characterized as eigenfunctions of an element of the symmetry algebra $\mathcal{E}(2)$ with the separation constant k playing the part of an eigenvalue.

Given an arbitrary operator L in $\mathcal{E}(2)$, can we find a separable coordinate system $\{u,v\}$ for the Helmholtz equation such that the separated solutions are eigenfunctions of L? The following argument shows that this is the case. If L is a nonzero operator of the form

$$L = A(\mathbf{x})\partial_x + B(\mathbf{x})\partial_y, \qquad A, B \in \mathcal{F},$$

it is always possible to find new coordinates u, v valid in some neighborhood of any fixed point $\mathbf{x}_0 \in \mathcal{D}$ such that $L = \partial_u$. See [51, p. 95] or [34] for the straightforward construction of the (nonunique) coordinates u, v. (Here and in all further discussions of real coordinates we require that the new coordinates $u(\mathbf{x})$, $v(\mathbf{x})$ be real analytic functions of x and y with real analytic inverses. The new coordinate system may be defined only in a neighborhood of some point $\mathbf{x}_0$ and need not cover the entire x–y plane.)

Expressing the Laplace operator Δ_2 in the u, v coordinates, we find

$$Q = \Delta_2 + \omega^2 = B_{11}\partial_{uu} + B_{12}\partial_{uv} + B_{22}\partial_{vv} + C_1\partial_u + C_2\partial_v + \omega^2 \qquad (2.11)$$

where the functions B_{ij}, C_j are analytic in a neighborhood of $\mathbf{x}_0$. Assuming that L is a symmetry operator, that is, $L \in \mathcal{E}(2)$, we have $[L, Q] = 0$. Substituting $L = \partial_u$ and (2.11) into this expression and evaluating the commutator, we find that the functions B_{ij}, C_j are independent of u. The equation $Q\Psi = 0$ will admit separated solutions of the form $\Psi_k = e^{iku}V(v)$ where V satisfies the ordinary differential equation

$$B_{22}V'' + (ikB_{12} + C_2)V' + (-k^2B_{11} + ikC_1 + \omega^2)V = 0. \qquad (2.12)$$

The solutions Ψ_k are characterized by the eigenvalue equation

$$L\Psi_k = ik\Psi_k. \qquad (2.13)$$

Thus we have found separated solutions of (0.1) satisfying (2.13) and have achieved a separation of variables in the sense that the factors of the separated solution each satisfy an ordinary differential equation.

Note however, that for each $g \in E(2)$ the function $\mathbf{T}(g)\Psi_k$, where $\mathbf{T}(g)$ is given by (1.16), is a solution of the Helmholtz equation that also satisfies the eigenvalue equation

$$L^g(\mathbf{T}(g)\Psi_k) = ik\mathbf{T}(g)\Psi_k \tag{2.14}$$

if Ψ_k satisfies (2.13). Here,

$$L^g = \mathbf{T}(g)L\mathbf{T}(g^{-1}) \tag{2.15}$$

is a symmetry operator since it is a product of three operators that commute with Q. Moreover, L^g is a first-order differential operator. Indeed, if

$$L = A(\mathbf{x})\partial_x + B(\mathbf{x})\partial_y, \tag{2.16a}$$

a direct computation yields

$$L^g = A(\mathbf{x}')\partial_{x'} + B(\mathbf{x}')\partial_{y'} \tag{2.16b}$$

where $(x',y') = \mathbf{x}' = \mathbf{x}g$. (We can express L^g in terms of the original variables x,y using the chain rule.) Thus $L^g \in \mathcal{E}(2)$. If Ψ_k is a separated solution in the variables $\mathbf{u} = (u,v)$, then $\mathbf{T}(g)\Psi_k$ is a separated solution in the variables $\mathbf{u}' = \mathbf{u}g$ obtained by a Euclidean transformation of coordinates in the x–y plane. Since the separable coordinates $\mathbf{u}$ and $\mathbf{u}'$ can be mapped into one another by a transformation from the symmetry group $E(2)$, we regard these coordinate systems as *equivalent*. Note also that the coordinates and eigenfunctions associated with the operator L are the same as the coordinates and eigenfunctions associated with the operator cL where c is a nonzero real constant. Thus the operator $L \in \mathcal{E}(2)$ and all operators cL^g, $g \in E(2)$, lead to equivalent separable coordinates.

The action $L \to L^g$ of $E(2)$ on $\mathcal{E}(2)$, the *adjoint representation*, decomposes $\mathcal{E}(2)$ into *orbits* of one-dimensional subspaces. We say that $K \in \mathcal{E}(2)$ lies on the same orbit as L if $K = cL^g$ for some nonzero $c \in R$ and some $g \in E(2)$. (Note that $L^{gg'} = (L^{g'})^g$.)

From these comments, to find all possible inequivalent separable coordinate systems and associated separable solutions obtainable from expressions (2.12), (2.13), we should decompose $\mathcal{E}(2)$ into orbits under the action of $E(2)$. This orbit analysis can be carried out explicitly using expressions (2.16). Another useful expression is

$$\exp(aK)L\exp(-aK) = L + a[K,L] + \frac{a^2}{2}[K,[K,L]]$$

$$+ \cdots + \frac{a^n}{n!}[K,[\cdots,[K,L]]\cdots] + \cdots = e^{a\operatorname{Ad}K}(L) \tag{2.17}$$

where K is a Lie derivative from the symmetry algebra $\mathcal{E}(2)$, $a \in R$, and

ISBN-0-201-13503-5

$\text{Ad} K$ is the linear operator on $\mathcal{E}(2)$ defined by $\text{Ad} K(L) = [K, L]$. See [46, p. 66] for a proof.

We will now determine the adjoint action of $E(2)$ on the basis P_1, P_2, M for $\mathcal{E}(2)$. If $g_1 = \exp(aP_1)$ (a translation), we have

$$P_1^{g_1} = P_1, \qquad P_2^{g_1} = P_2, \qquad M^{g_1} = M - aP_2; \qquad (2.18)$$

if $g_2 = \exp(bP_2)$ (a translation), we have

$$P_1^{g_2} = P_1, \qquad P_2^{g_2} = P_2, \qquad M^{g_2} = M + bP_1; \qquad (2.19)$$

and if $g_3 = \exp(\alpha M)$ (a rotation), we have

$$P_1^{g_3} = \cos(\alpha)P_1 + \sin(\alpha)P_2, \qquad P_2^{g_3} = -\sin(\alpha)P_1 + \cos(\alpha)P_2, \qquad M^{g_3} = M.$$
$$(2.20)$$

Let

$$L = c_1 P_1 + c_2 P_2 + c_3 M \in \mathcal{E}(2) \qquad (2.21)$$

with $c_3 \neq 0$. It follows from (2.18) and (2.19) that $L^{g_1 g_2} = c_3 M$ if we choose a and b such that $a = c_2/c_3$, $b = -c_1/c_3$. Thus, L lies on the same orbit as M. On the other hand, if $c_3 = 0$ and $c_1^2 + c_2^2 > 0$, it is easy to see that g_3 can be chosen such that $L^{g_3} = (c_1^2 + c_2^2)^{1/2} P_2$. Thus, L lies on the same orbit as P_2.

In conclusion, there are only two orbits in $\mathcal{E}(2)$ under the adjoint action of $E(2)$; namely, the orbits containing the representatives

$$M, \quad P_2, \qquad (2.22)$$

respectively. A nonzero $L \in \mathcal{E}(2)$ given by (2.21) lies on the first or second orbit, depending on whether $c_3 \neq 0$ or $c_3 = 0$. Thus the polar and Cartesian coordinate systems are the only separable systems for the Helmholtz equation that arise via expressions (2.12), (2.13). These systems are called *subgroup coordinates* because they correspond to diagonalization of generators for a rotation subgroup and a translation subgroup of $E(2)$.

We will now explicitly compute the separable coordinate systems for

$$\left(\partial_{xx} + \partial_{yy} + \omega^2\right)\Psi = 0 \qquad (2.23)$$

and show that there exist more than just those which are subgroup coordinates. Let $\{u, v\}$ be a separable system. Then $x = x(u, v)$, $y = y(u, v)$; also, $u = u(x, y)$, $v = v(x, y)$, and the Jacobian $J = v_x u_y - u_x v_y = (y_u x_v -$

ISBN-0-201-13503-5

$x_u y_v)^{-1}$ is nonzero. Writing (2.23) in the $\{u,v\}$ system, we obtain

$$\{(u_x^2+u_y^2)\partial_{uu}+(u_{xx}+u_{yy})\partial_u+2(u_xv_x+u_yv_y)\partial_{uv}$$
$$+(v_x^2+v_y^2)\partial_{vv}+(v_{xx}+v_{yy})\partial_v+\omega^2\}\Psi=0. \qquad (2.24)$$

We now require that (2.24) separate in the $\{u,v\}$ system. It is convenient to consider two cases, depending on whether the coefficient of ∂_{uv} is zero or nonzero.

Case I. $u_xv_x+u_yv_y=0$

By interchanging u and v if necessary and using the fact that $J\neq0$, we can assume there exists a nonzero function $\mathfrak{R}$ such that $v_y=\mathfrak{R}u_x$, $v_x=-\mathfrak{R}u_y$. Due to the occurrence of the term ω^2 in (2.24), in order for variables to separate it is necessary that

$$u_x^2+u_y^2=\frac{\mathfrak{U}(u)}{\mathfrak{U}_1(u)+\mathfrak{V}_1(v)}, \qquad v_x^2+v_y^2=\frac{\mathfrak{V}(v)}{\mathfrak{U}_1(u)+\mathfrak{V}_1(v)} \qquad (2.25)$$

where $\mathfrak{U}$, $\mathfrak{U}_1+\mathfrak{V}_1$, and $\mathfrak{V}$ are nonzero functions. Furthermore, since $v_x^2+v_y^2=\mathfrak{R}^2(u_x^2+u_y^2)$, it follows that $\mathfrak{R}^2=\mathfrak{V}/\mathfrak{U}$, hence that $\mathfrak{R}$ is the quotient of a function of u and a function of v.

Suppose $\tilde{u}=\tilde{u}(u)$, $\tilde{v}=\tilde{v}(v)$ are real analytic functions of u and v, respectively, with $d\tilde{u}/du\neq0$, $d\tilde{v}/dv\neq0$. Then $\{\tilde{u},\tilde{v}\}$ defines a coordinate system which, for obvious reasons, we consider as equivalent to the original $\{u,v\}$ system. If in (2.24) we had adopted coordinates $\{\tilde{u},\tilde{v}\}$ where

$$\frac{d\tilde{u}}{du}=\mathfrak{U}^{-1/2}, \qquad \frac{d\tilde{v}}{dv}=\mathfrak{V}^{-1/2},$$

then equations (2.25) would hold for $\{\tilde{u},\tilde{v}\}$ in place of $\{u,v\}$, where also the numerators $\tilde{\mathfrak{U}}=1$, $\tilde{\mathfrak{V}}=1$ on the right-hand sides. Thus without loss of generality we can drop the tildes and assume that the $\{u,v\}$ system satisfies (2.25) with $\mathfrak{U}(u)\equiv\mathfrak{V}(v)\equiv1$. It then follows that we can assume $\mathfrak{R}\equiv1$, so

$$u_x=v_y, \qquad u_y=-v_x, \qquad (2.26)$$

and the functions u,v satisfy the Cauchy–Riemann equations [1, p. 39]. This means that if we define the complex variables z,w by

$$z=x+iy, \qquad w=u+iv, \qquad (2.27)$$

then $w=f(z)$ where f is a complex analytic function. Furthermore, equa-

ISBN-0-201-13503-5

tions (2.25) become $|dw/dz|^2 = (\mathscr{U}_1(u) + \mathscr{V}_1(v))^{-1}$ or

$$\left|\frac{dz}{dw}\right|^2 = \mathscr{U}_1(u) + \mathscr{V}_1(v), \tag{2.28}$$

or, finally,

$$\partial_{uv}\left(\left|\frac{dz}{dw}\right|^2\right) = 0. \tag{2.29}$$

We now follow the argument presented in Chapter 5.1 of [98]. Instead of the variables u, v, we write (2.29) in terms of the variables w and $\bar{w} = u - iv$. Using the facts that $\partial_{uv} = i\partial_{ww} - i\partial_{\bar{w}\bar{w}}$ and $|dz/dw|^2 = dz/dw\,d\bar{z}/d\bar{w}$, where the first factor is a function of w alone while the second is a function of $\bar{w}$ alone, we find

$$\left(\frac{d\bar{z}}{d\bar{w}}\right)\frac{d^2}{dw^2}\left(\frac{dz}{dw}\right) = \left(\frac{dz}{dw}\right)\frac{d^2}{d\bar{w}^2}\left(\frac{d\bar{z}}{d\bar{w}}\right),$$

which implies

$$\left(\frac{dz}{dw}\right)^{-1}\frac{d^2}{dw^2}\left(\frac{dz}{dw}\right) = \left(\frac{d\bar{z}}{d\bar{w}}\right)^{-1}\frac{d^2}{d\bar{w}^2}\left(\frac{d\bar{z}}{d\bar{w}}\right)$$

where the left-hand side depends on w alone and the right-hand side depends on $\bar{w}$ alone. Thus, there exists a complex constant λ such that

$$\frac{d^2}{dw^2}\left(\frac{dz}{dw}\right) = \lambda\frac{dz}{dw}, \qquad \frac{d^2}{d\bar{w}^2}\left(\frac{d\bar{z}}{d\bar{w}}\right) = \lambda\frac{d\bar{z}}{d\bar{w}}. \tag{2.30}$$

The solutions of these third-order differential equations yield the separable solutions of the Helmholtz equation corresponding to case I.

Before solving these equations we consider

Case II. $u_x v_x + u_y v_y \neq 0$

Here the only way for variables to separate is for (by replacing one of the variables, say u, by a function of itself if necessary) all the coefficients of the partial derivatives $\partial_{uu}, \partial_u, \partial_{uv}, \partial_{vv}, \partial_v$ in (2.24) to be functions of v alone. Then, substituting $\Psi(u, v) = e^{iku}\Phi(v)$, we see that the u dependence in (2.24) factors out and we are left with a second-order (k-dependent) ordinary differential equation for Φ. Clearly the operator ∂_u is a symmetry operator for the Helmholtz equation, so we are led to the type of separation of variables described by expressions (2.12) and (2.13). As we saw

ISBN-0-201-13503-5

earlier, modulo a Euclidean transformation we can take either $\partial_u = P_2$ or $\partial_u = M$ where P_2, M are given by (1.7).

In the first case we have a coordinate system $\{u,v\}$ related to the $\{x,y\}$ system by

$$\partial_y = u_y \partial_u + v_y \partial_v = \partial_u.$$

Thus $u_y = 1$, $v_y = 0$, and $v(x,y)$ depends on x alone. Replacing v by a new variable $\tilde{v}(v)$ if necessary, we can assume $v = x$. Integrating the equation $u_y = 1$, we obtain the separable system

$$u = y + h(x), \qquad v = x \qquad (2.31)$$

where $h'(x) \neq 0$ in order to satisfy the condition for a case II solution. Note that these coordinates are nonorthogonal; that is, the curves $u = $ constant are not perpendicular to the curves $v = $ constant with respect to the usual Euclidean inner product. Moreover, by constructing the separated solutions of (0.1) corresponding to system (2.31), the reader can easily verify that these solutions differ only slightly from the separated solutions in Cartesian coordinates.

Similarly, if $\partial_u = M$, we can choose the separable system in the form

$$u = \theta + h(r), \qquad v = r, \qquad h'(r) \neq 0, \qquad (2.32)$$

where r, θ are polar coordinates. Again these coordinates are nonorthogonal and differ only slightly from the separated solutions in polar coordinates.

Note that there are an infinite number of case II separable coordinate systems corresponding to a single symmetry operator L, but the systems are essentially the same. Some writers do not consider these systems to be truly separable and reserve the appellation "separable" for the systems of case I. From the group-theoretic point of view, however, I see no reason to exclude such systems, uninteresting though they may be, from the family of separable solutions.

Before returning to our detailed examination of the case I systems it is worthwhile remarking that these systems are all orthogonal. Indeed, as the reader can verify, the orthogonality follows immediately from the relation $u_x v_x + u_y v_y = 0$.

Now we solve equations (2.30) in the special case where $\lambda = 0$. The solution for dz/dw is

$$\frac{dz}{dw} = \beta + \gamma w, \qquad \beta, \gamma \in \mathcal{C}. \qquad (2.33)$$

ISBN-0-201-13503-5

If $\gamma = 0$, $\beta = c + id$, we find that $z = \beta w + \alpha$ or

$$x = a + cu - dv, \qquad y = b + du + cv, \qquad \alpha = a + ib, \qquad (2.34)$$

where $a, b, c, d \in R$. It is easy to see that the coordinates u, v are obtained from the Cartesian coordinates x, y by a Euclidean transformation and a dilatation $x, y \rightarrow (c^2 + d^2)^{1/2} x$, $(c^2 + d^2)^{1/2} y$. Since we are identifying coordinate systems $\{u, v\}$ and $\{h(u), k(v)\}$, and since we consider systems obtainable from one another by Euclidean transformations as equivalent, we see that all systems given by (2.34) are equivalent to Cartesian coordinates.

If $\gamma \neq 0$ in (2.33) we obtain the solution

$$z = (\gamma/2)w^2 + \beta w + \alpha, \qquad \alpha, \beta, \gamma \in \mathbb{C}.$$

However, just as in the previous case, by choosing appropriate dilatations and Euclidean transformations we can show that this system is equivalent to a system for which $\gamma = 1$, $\beta = \alpha = 0$. Thus,

$$x = \tfrac{1}{2}(u^2 - v^2), \qquad y = uv. \qquad (2.35)$$

The u, v coordinates are called *parabolic* because the coordinate lines $u = [(x^2 + y^2)^{1/2} + x]^{1/2} = \text{constant}$ and $v = \pm[(x^2 + y^2)^{1/2} - x]^{1/2} = \text{constant}$ form two orthogonal families of confocal parabolas. (By convention we restrict u to positive values and allow v to range over both positive and negative values. Since the inverse function $w = (2z)^{1/2}$ is not well defined in the entire x–y plane, we introduce a cut on the negative x axis. See [98, section 5.1] for more details.)

Substituting expressions (2.35) into (2.24), we obtain

$$\partial_{uu}\Psi + \partial_{vv}\Psi + (u^2 + v^2)\omega^2\Psi = 0 \qquad (2.36)$$

in which variables clearly separate. Indeed, setting $\Psi = U(u)V(v)$, we find

$$U'' + (\omega^2 u^2 - k^2)U = 0, \qquad V'' + (\omega^2 v^2 + k^2)V = 0 \qquad (2.37)$$

where k^2 is the separation constant. The two separated equations are slightly different forms of the parabolic cylinder equation and the separated solutions Ψ_k are products of parabolic cylinder functions (see Appendix B).

Multiplying the first of the separated equations (2.37) by $v^2 V$ and the second by $u^2 U$, and subtracting the second equation from the first, we

ISBN-0-201-13503-5

obtain the eigenvalue equation

$$(u^2 + v^2)^{-1}(v^2 \partial_{uu} - u^2 \partial_{vv})\Psi_k = k^2 \Psi_k.$$

Since the solutions Ψ_k of the Helmholtz equation are mapped into other solutions, namely, $k^2\Psi_k$, by the operator S on the left-hand side of this expression, we might expect that $S \in \mathbb{S}/\mathfrak{q}$; that is, S is a second-order symmetry operator. Indeed, a straightforward computation yields

$$\{M, P_2\}\Psi_k = k^2 \Psi_k, \tag{2.38}$$

so $S = \{M, P_2\}$. Thus we have characterized the separable solutions in parabolic coordinates as eigenfunctions of the symmetry operator $\{M, P_2\}$. The separation constant k^2 is the eigenvalue of this operator. Similarly, the separable solutions in Cartesian coordinates are eigenfunctions of the symmetry operator P_2^2.

Next we solve equations (2.30) for $\lambda \neq 0$. Since the second of these equations is the complex conjugate of the first, it follows that λ is real. Furthermore, by performing a dilatation of coordinates if necessary, we can assume $\lambda = 1$. The solution for dz/dw is

$$\frac{dz}{dw} = \alpha e^w - \beta e^{-w}, \qquad \alpha, \beta \in \mathbb{C},$$

so

$$z = \alpha e^w + \beta e^{-w} + \gamma, \qquad \gamma \in \mathbb{C}. \tag{2.39}$$

By translation and rotation of coordinates in the x–y plane if necessary, we can assume $\gamma = 0$ and $\alpha \geqslant 0$. If $\beta = 0$, $\alpha > 0$, we set $r = \alpha e^u$, $\theta = v$, to obtain the *polar* coordinate system

$$x = r\cos\theta, \qquad y = r\sin\theta, \qquad r \geqslant 0, \qquad 0 \leqslant \theta < 2\pi. \tag{2.40}$$

(Clearly, if the coordinates $\{u, v\}$ permit separable solutions of the Helmholtz equation, so do the coordinates $\{r, \theta\}$.)

If $\alpha\beta \neq 0$, we can rotate our coordinate system in the x–y plane to achieve $\alpha\beta > 0$. Thus

$$2\alpha = \exp(a - b + i\varphi), \qquad 2\beta = \exp(a + b - i\varphi)$$

and setting $d = e^a$, $\xi = u - b$, $\eta = v + \varphi$, we obtain the separable *elliptic coordinates* $\{\xi, \eta\}$ where

$$x = d\cosh\xi\cos\eta, \qquad y = d\sinh\xi\sin\eta. \tag{2.41}$$

ISBN-0-201-13503-5

(We could set the positive constant d equal to 1, but we retain d to agree with convention.) The coordinate curves $\xi = \text{constant}$ and $\eta = \text{constant}$ are

$$\frac{x^2}{d^2\cosh^2\xi} + \frac{y^2}{d^2\sin^2\xi} = 1, \qquad \frac{x^2}{d^2\cos^2\eta} - \frac{y^2}{d^2\sin^2\eta} = 1,$$

confocal ellipses and hyperbolas, respectively, with foci at $(x,y) = (\pm d, 0)$. We can sweep out every point in the x–y plane by allowing ξ, η to vary in the range $\xi \geqslant 0, 0 \leqslant \eta < 2\pi$. Substituting expressions (2.41) into (2.24), we obtain

$$\partial_{\xi\xi}\Psi + \partial_{\eta\eta}\Psi + d^2\omega^2(\cosh^2\xi - \cos^2\eta)\Psi = 0. \tag{2.42}$$

Then, setting $\Psi = U(\xi)V(\eta)$, we find the separated equations

$$U'' + (d^2w^2\cosh^2\xi + k^2)U = 0, \qquad V'' - (d^2w^2\cos^2\eta + k^2)V = 0 \tag{2.43}$$

where k^2 is the separation constant. These equations are variants of Mathieu's equation and the separated solutions are products of Mathieu functions (see Appendix B).

Multiplying the first equation (2.43) by $V\cos^2\eta$, the second by $U\cosh^2\xi$, and adding, we obtain the eigenvalue equation

$$(\cosh^2\xi - \cos^2\eta)^{-1}(\cos^2\eta\,\partial_{\xi\xi} + \cosh^2\xi\,\partial_{\eta\eta})\Psi_k = k^2\Psi_k$$

where $\Psi_k = UV$. In analogy with the corresponding results for Cartesian and parabolic coordinates, we expect that the operator S on the left-hand side of this equation belongs to $\mathcal{S}/\mathfrak{q}$. A straightforward (but tedious) computation yields the result

$$(M^2 + d^2P_1^2)\Psi_k = k^2\Psi_k, \tag{2.44}$$

verifying our conjecture. Thus the separated solutions in elliptic coordinates are eigenfunctions of the symmetry operator $M^2 + d^2P_1^2$ (or $M^2 - d^2P_2^2$) and the separation constant k^2 is the corresponding eigenvalue. A similar but much easier computation shows that the separated solutions in polar coordinates are characterized as eigenfunctions of the symmetry operator M^2.

We have now associated each of the four orthogonal coordinate systems that permit variable separation in the Helmholtz equation with a second-order symmetry operator. Moreover, we have seen that the separable systems actually decompose into equivalence classes under the action of $E(2)$. Indeed, if Ψ_k is a separated solution in the variables $\mathbf{u} = (u_1, u_2)$, then $\mathbf{T}(g)\Psi_k$ is a separated solution in the variables $\mathbf{u}' = \mathbf{u}g$ obtained by a

ISBN-0-201-13503-5

Euclidean transformation of coordinates. We regard $\mathbf{u}$ and $\mathbf{u}'$ as equivalent systems. Let us examine this equivalence on the operator level. Suppose $S \in \mathbb{S}/\mathfrak{q}$ is the second-order symmetry operator associated with the coordinates $\mathbf{u}$, $S\Psi_k = k^2\Psi_k$. Then $S^g = \mathbf{T}(g)S\mathbf{T}(g)^{-1}$ is the corresponding operator associated with $\Psi_k^g = \mathbf{T}(g)\Psi_k$. Indeed, $S^g\Psi_k^g = k^2\Psi_k^g$. Furthermore, it is easy to show that S^g is a second-order symmetry operator. In particular, if $S = L_1 L_2$ where the L_j are first-order symmetries, then $S^g = L_1^g L_2^g$ where L_j^g is given by (2.15).

Since $S^{gg'} = (S^g)^{g'}$, $g, g' \in E(2)$, the action of $E(2)$ on $\mathbb{S}$, called the *adjoint representation*, decomposes $\mathbb{S}$ into orbits of one-dimensional subspaces. We say that S lies on the same orbit as S' if $S = c(S')^g$ for some nonzero $c \in R$ and some $g \in E(2)$. Now $P_1^2 + P_2^2$ commutes with all elements of $\mathbb{E}(2)$, hence with all operators $\mathbf{T}(g)$. Thus $R(\mathbf{x})(P_1^2 + P_2^2) = W \in \mathfrak{q}$ satisfies $W^g = R(\mathbf{x}g)(P_1^2 + P_2^2) \in \mathfrak{q}$ for $R \in \mathcal{F}$. This shows that $E(2)$ acts on $\mathbb{S}/\mathfrak{q}$ via the adjoint representation and decomposes this space into orbits.

Let $\mathbb{S}^{(2)}$ be the space of purely second-order elements in $\mathbb{S}/\mathfrak{q}$, that is, the five-dimensional vector space of symmetric second-order operators with basis (1.22b). (We can also include the operator P_2^2 in this space provided we recognize that $P_2^2 \equiv -P_1^2$. This is because $P_1^2 + P_2^2$ corresponds to the zero operator in $\mathbb{S}^{(2)}/\mathfrak{q}$.) From the comments above we see that $E(2)$ acts on $\mathbb{S}^{(2)}$ via the adjoint representation and decomposes $\mathbb{S}^{(2)}$ into orbits of one-dimensional subspaces. In particular, $\{L_1, L_2\}^g = \{L_1^g, L_2^g\}$, so symmetric second-order operators are mapped into symmetric second-order operators.

We have seen that each of the four orthogonal separable coordinate systems for the Helmholtz equation is associated with a symmetry operator S of the form

$$S = aP_1^2 + bP_1 P_2 + cP_2^2 + dM^2 + e\{M, P_1\} + f\{M, P_2\}, \qquad a, \dots, f \in R,$$

hence with $\hat{S} \in \mathbb{S}^{(2)}$ where

$$\hat{S} = (a - c)P_1^2 + bP_1 P_2 + dM^2 + e\{M, P_1\} + f\{M, P_2\}. \qquad (2.45)$$

If one of the coordinate systems $\{u, v\}$ is subjected to a Euclidean transformation g, it is transformed into another (equivalent) separable orthogonal system which is associated with the symmetry operator S^g. Furthermore, if two coordinate systems are associated with operators S, S', respectively, where $\hat{S} = (\hat{S}')$, it follows that the separated eigenfunction Ψ_k, $S\Psi_k = k^2\Psi_k$, of S satisfies the equation $S'\Psi_k = (k^2 + (a - a')\omega^2)\Psi_k$. (Note: $(P_1^2 + P_2^2)\Psi_k = -\omega^2\Psi_k$.) Thus S and S' have the same separated eigenfunctions but the spectrum of S' is translated by the fixed amount $(a - a')\omega^2$ from the spectrum of S. Clearly, the separated coordinates for S and S' are identical. From these comments and our earlier computations

ISBN-0-201-13503-5

we see that each orthogonal separable coordinate system should be associated with a one-dimensional subspace of operators $\{c\hat{S}\}$ in $\mathbb{S}^{(2)}$ (since $c\hat{S}$ and $\hat{S}$ correspond to the same coordinates for $c \neq 0$ in R). Furthermore, the collection of all coordinate systems equivalent to a given system under the action of $E(2)$ should be associated with an orbit in $\mathbb{S}^{(2)}$.

These remarks lead us to the problem of determining the orbit structure of $\mathbb{S}^{(2)}$. Suppose the first-order symmetries L_1, L_2, L_3 form a basis for $\mathcal{E}(2)$. For each $g \in E(2)$ we can find a 3×3 matrix G such that

$$L_j^g = \sum_{k=1}^{3} G_{kj} L_k, \qquad j = 1, 2, 3. \tag{2.46}$$

It follows easily that

$$\{L_{j_1}, L_{j_2}\}^g = \{L_{j_1}^g, L_{j_2}^g\} = \sum_{k_1, k_2 = 1}^{3} G_{k_1 j_1} G_{k_2 j_2} \{L_{k_1}, L_{k_2}\}, \tag{2.47}$$

an element of $\mathbb{S}$. (Note that $L_j^2 = \frac{1}{2}\{L_j, L_j\}$.) We can now use (2.46) and (2.18)–(2.20) to find a representative on each $\mathbb{S}^{(2)}$ orbit.

Any $\hat{S} \in \mathbb{S}^{(2)}$ can be expressed uniquely in the form (2.45) with $c = 0$. Suppose $d \neq 0$. Then applying translations (2.18), (2.19), we can transform $\hat{S}$ to

$$a' P_1^2 + b' P_1 P_2 + c' P_2^2 + d M^2,$$

that is, we can choose the translations such that the coefficients of $\{M, P_1\}$ and $\{M, P_2\}$ are zero. Next, by an appropriate rotation (2.20) we can diagonalize the quadratic form in $P_j P_k$ to obtain

$$a'' P_1^2 + c'' P_2^2 + d M^2 \equiv (a'' - c'') P_1^2 + d M^2.$$

There are two possibilities: If $a'' = c''$, then $\hat{S}$ is on the same orbit as M^2, whereas if $a'' \neq c''$, then $\hat{S}$ is on the same orbit as $M^2 + r^2 P_1^2, r > 0$.

Next suppose $d = 0$ while $e^2 + f^2 > 0$. Applying an appropriate rotation (2.20), we can assume $e = 0$, $f \neq 0$. Furthermore, by choosing appropriate translations (2.18), (2.19) we can transform $\hat{S}$ to $c\{M, P_2\}$. Thus $\hat{S}$ lies on the same orbit as $\{M, P_2\}$.

Finally, suppose $d = e = f = 0$, $a^2 + b^2 > 0$. Then by an appropriate rotation we can diagonalize the quadratic form $a P_1^2 + b P_1 P_2$ to obtain

$$a' P_1^2 + b' P_2^2 \equiv (a' - b') P_1^2,$$

so $\hat{S}$ lies on the same orbit as P_1^2 (or P_2^2).

ISBN-0-201-13503-5

Table 1. Operators and Coordinates for $(\Delta_2+\omega^2)\Psi=0$

Operator S	Coordinates	Separated solutions
1 P_2^2	Cartesian x, y	Product of exponential functions
2 M^2	Polar $x=r\cos\theta, y=r\sin\theta$	Bessel times exponential function
3 $\{M,P_2\}$	Parabolic $x=\frac{1}{2}(\xi^2-\eta^2), y=\xi\eta$	Product of parabolic cylinder functions
4 $M^2+d^2P_1^2$	Elliptic $x=d\cosh\alpha\cos\beta,$ $y=d\sinh\alpha\sin\beta$	Product of Mathieu functions

We have shown that $\mathbb{S}^{(2)}$ contains exactly four orbits with orbit representatives

$$M^2, \qquad M^2+r^2P_1^2, \qquad \{M,P_2\}, \qquad P_1^2. \qquad (2.48)$$

It follows that there is a one-to-one correspondence between the orthogonal separable coordinate systems for the Helmholtz equation and orbits in $\mathbb{S}^{(2)}$.

1.3 Expansion Formulas Relating Separable Solutions

We now describe a method for exploiting the relationship between separable solutions of the Helmholtz equation and orbits of symmetry operators to derive properties of the separable solutions. This method involves use of the Fourier transform to introduce a Hilbert space structure on the solution space of the Helmholtz equation.

Let $\Psi(x,y)$ be a solution of $(\Delta_2+\omega^2)\Psi=0$ and suppose

$$\Psi(x,y)=\iint_{-\infty}^{\infty}\exp\left[i(\omega_1 x+\omega_2 y)\right]\tilde{h}\,(\omega_1,\omega_2)\,d\omega_1\,d\omega_2$$

where $\tilde{h}$ is the Fourier transform of Ψ. Proceeding formally we have

$$(\Delta_2+\omega^2)\Psi=\iint(\omega^2-\omega_1^2-\omega_2^2)\exp\left[i(\omega_1 x+\omega_2 y)\right]\tilde{h}\,(\omega_1,\omega_2)\,d\omega_1\,d\omega_2=0$$

provided $\tilde{h}(\omega_1,\omega_2)=\omega^{-1}\delta(\omega-s)h(\varphi)$ where $\delta(r)$ is the Dirac delta function, s and φ are polar coordinates in the $\omega_1-\omega_2$ plane ($\omega_1=s\cos\varphi$, $\omega_2=s\sin\varphi$), and $d\omega_1\,d\omega_2=s\,ds\,d\varphi$.

Carrying out the integration in s, we find

$$\Psi(x,y)=\int_{-\pi}^{\pi}\exp\left[i\omega(x\cos\varphi+y\sin\varphi)\right]h(\varphi)\,d\varphi=I(h). \qquad (3.1)$$

The elements $g(\theta,a,b)$ of $E(2)$ act on the solutions Ψ of the Helmholtz

ISBN-0-201-13503-5

equation via the operators $\mathbf{T}(g)$ defined by expressions (1.11) and (1.16). Using (3.1) we find

$$\mathbf{T}(g)\Psi(x,y) = \int_{-\pi}^{\pi} \exp\big[\, i\omega(x\cos(\varphi+\theta)+y\sin(\varphi+\theta)$$

$$+ a\cos\varphi + b\sin\varphi)\big]h(\varphi)\,d\varphi$$

$$= \int_{-\pi}^{\pi} \exp\big[\, i\omega(x\cos\varphi+y\sin\varphi)\big]\mathbf{T}(g)h(\varphi)\,d\varphi$$

where

$$\mathbf{T}(g)h(\varphi) = \exp\big[\, i\omega(a\cos(\varphi-\theta)+b\sin(\varphi-\theta))\big]h(\varphi-\theta). \qquad (3.2)$$

(We are requiring $h(\varphi)=h(\varphi+2\pi)$, so that the integration in (3.1) is over the unit circle.)

Thus the operators $\mathbf{T}(g)$ acting on Ψ induce operators (which we also call $\mathbf{T}(g)$) acting on h and given by (3.2). It is easy to see that the operators (3.2) satisfy the group homomorphism property $\mathbf{T}(g_1 g_2)=\mathbf{T}(g_1)\mathbf{T}(g_2)$.

Now we restrict the functions h to the Hilbert space $L_2(S_1)$ of Lebesgue square-integrable functions on the unit circle $S_1 : \omega_1^2+\omega_2^2=1$, that is, $s=1$, $-\pi \leqslant \varphi < \pi \pmod{2\pi}$. Thus, we consider the space of all measurable functions h such that $\int_{-\pi}^{\pi}|h(\varphi)|^2\,d\varphi < \infty$. (The reader who is not familiar with Lebesgue integration may replace the Lebesgue integral by the Riemann integral in the following. Very little of practical importance is lost thereby.) On $L_2(S_1)$ we have the inner product

$$\langle h_1,h_2\rangle = \int_{-\pi}^{\pi} h_1(\varphi)\bar{h}_2(\varphi)\,d\varphi, \qquad h_j \in L_2(S_1). \qquad (3.3)$$

(See [69] and [112] for a thorough discussion of Hilbert space concepts.)

Any $h \in L_2(S_1)$ can be extended to a function defined on the real line by imposition of the periodicity condition $h(\varphi)=h(\varphi+2\pi)$. Adopting this convention, we see that the operators $\mathbf{T}(g)$ given by (3.2) are well defined on $L_2(S_1)$ and map this Hilbert space into itself. Moreover, a simple computation yields

$$\langle \mathbf{T}(g)h_1, \mathbf{T}(g)h_2\rangle = \langle h_1, h_2\rangle$$

for all $h_j \in L_2(S_1)$, that is, the operators $\mathbf{T}(g)$ are *unitary*. (See [112] for a discussion of unitary operators on a Hilbert space.) Thus we have constructed a unitary representation (ω) of $E(2)$ on $L_2(S_1)$, labeled by the positive constant ω. It is shown in [75] or [128] that this representation is actually irreducible; that is, there exists no proper closed subspace of $L_2(S_1)$ that is invariant under all the operators $\mathbf{T}(g)$, but this fact need not concern us here.

ISBN-0-201-13503-5

Proceeding in analogy to (3.1), (3.2) and employing integration by parts, we can compute the Lie algebra operators P_1, P_2, M on $L_2(S_1)$ induced by the operators P_1, P_2, M, (1.7), acting on solutions of the Helmholtz equation. The results are

$$P_1 = i\omega\cos\varphi, \qquad P_2 = i\omega\sin\varphi, \qquad M = -d/d\varphi. \tag{3.4}$$

Clearly, these operators satisfy the commutation relations (1.8) and form a basis for $\mathcal{E}(2)$. In strict analogy with (1.16), the new Lie algebra operators are related to the $\mathbf{T}(g)$, (3.2), by

$$\mathbf{T}(g) = \exp(\theta M)\exp(aP_1 + bP_2).$$

Since the $\mathbf{T}(g)$ are unitary, the new Lie algebra operators are skew-Hermitian; that is,

$$\langle Lh_1, h_2 \rangle = -\langle h_1, Lh_2 \rangle, \qquad h_j \in L_2(S_1) \tag{3.5}$$

for $L = c_1 P_1 + c_2 P_2 + c_3 M$, $c_k \in R$, as can be directly verified. Actually we have to be careful about the domain of definition of L. The P_j make sense when applied to any $h \in L_2(S_1)$, but M is meaningful only when applied to a differentiable function. To be definite we will define all of our Lie algebra operators L on the subspace $\mathcal{D}$ of $L_2(S_1)$ consisting of all functions $h(\varphi)$ that are infinitely differentiable and vanish in neighborhoods of $-\pi$ and π. For $h_j \in \mathcal{D}$ expression (3.5) is easy to check.

We now restrict ourselves to the space $\mathcal{H}$ consisting of solutions Ψ of the Helmholtz equation such that $\Psi = I(h)$, (3.1), for some $h \in L_2(S_1)$. Here $\mathcal{H}$ is a Hilbert space with inner product

$$(\Psi_1, \Psi_2) \equiv \langle h_1, h_2 \rangle, \qquad \Psi_j = I(h_j). \tag{3.6}$$

Note that each function $\Psi(x,y)$ in $\mathcal{H}$ can be considered as an inner product

$$\Psi(x,y) = I(h) = \langle h, H(x,y,\cdot)\rangle,$$
$$H(x,y,\varphi) = \exp\left[-i\omega(x\cos\varphi + y\sin\varphi)\right] \in L_2(S_1). \tag{3.7}$$

One can check that no nonzero $h \in L_2(S_1)$ (nonzero means $\|h\|^2 = \langle h, h \rangle > 0$) has the property that $\Psi = I(h)$ is the identically zero solution of the Helmholtz equation. Thus, distinct functions h_1, h_2 in $L_2(S_1)$ define distinct solutions Ψ_1, Ψ_2 of the Helmholtz equation. It follows that our formal definition (3.6) for the Hilbert space structure of $\mathcal{H}$ is meaningful. Also, the operators $\mathbf{T}(g)$ on $\mathcal{H}$, given by (1.16), are unitary and the Lie algebra operators (1.7) on $\mathcal{H}$ are skew-Hermitian.

ISBN-0-201-13503-5

Finally, the linear transformation I is a one-to-one mapping of $L_2(S_1)$ onto $\mathcal{H}$ that preserves inner product, that is, I is a *unitary transformation* from $L_2(S_1)$ to $\mathcal{H}$. Due to the existence of this invertible mapping we can transform problems involving $\mathcal{H}$ to problems involving $L_2(S_1)$. In particular, we have shown that the representation of the symmetry group $E(2)$ on $\mathcal{H}$ defined by operators (1.16) is unitary equivalent via I to the unitary representation of $E(2)$ on $L_2(S_1)$ defined by (3.2). Since for many purposes the second Hilbert space is much simpler than the first, this observation will be very useful in the computations to follow.

We have already seen that the separable solutions of the Helmholtz equation corresponding to an orthogonal separable coordinate system $\{u,v\}$ are characterized as eigenfunctions of an operator $S \in \mathbb{S}^{(2)}$ and S is a symmetric second-order polynomial in the elements of $\mathcal{E}(2)$. In an obvious manner we can define S as an operator on the domain $D = I(\mathcal{D})$ in $\mathcal{H}$ and equivalently, on the domain $\mathcal{D}$ in $L_2(S_1)$. Moreover, S is symmetric on this domain. (Recall that an operator A defined on a dense domain D of a Hilbert space $\mathcal{H}$ is *symmetric* if $(A\Psi_1, \Psi_2) = (\Psi_1, A\Psi_2)$ for all $\Psi_1, \Psi_2 \in D$, where $(\cdot, \cdot)$ is the inner product on $\mathcal{H}$.) Indeed, if $S = \{L_1, L_2\} = L_1 L_2 + L_2 L_1, L_j \in \mathcal{E}(2)$, we have

$$(S\Psi_1, \Psi_2) = (L_1 L_2 \Psi_1, \Psi_2) + (L_2 L_1 \Psi_1, \Psi_2)$$
$$= -(L_2 \Psi_1, L_1 \Psi_2) - (L_1 \Psi_1, L_2 \Psi_2)$$
$$= (\Psi_1, S\Psi_2)$$

since the L_j are skew-Hermitian and map D into itself. It follows from these comments that each operator S characterizing a separable coordinate system can be defined as a symmetric operator on D.

Symmetric operators have many pleasant properties, not the least of which are the facts that their eigenvectors corresponding to distinct eigenvalues are orthogonal and that their eigenvalues are real. Indeed, if $S\Psi_j = \lambda_j \Psi_j$, $j = 1, 2$, where the Ψ_j are nonzero elements of D and $\lambda_j \in \mathbb{C}$, then

$$\lambda_1(\Psi_1, \Psi_2) = (S\Psi_1, \Psi_2) = (\Psi_1, S\Psi_2) = \bar{\lambda}_2(\Psi_1, \Psi_2). \qquad (3.8)$$

Setting $\Psi_1 = \Psi_2$ and comparing the left- and right-hand sides of (3.8), we see that $\lambda_1 = \bar{\lambda}_1$, so λ_1 (and hence λ_2) is real. Furthermore, if $\Psi_1 \neq \Psi_2$ and $\lambda_1 \neq \lambda_2$, a comparison of the two sides of (3.8) yields $(\Psi_1, \Psi_2) = 0$; that is, Ψ_1 is *orthogonal* to Ψ_2.

It follows from these remarks and the well-known Gram–Schmidt orthogonalization process that S will have a countable set of mutually orthogonal eigenvectors $\{\Psi_j\}$ where each eigenvector can be normalized to have length one: $(\Psi_j, \Psi_j) = 1$, $S\Psi_j = \lambda_j \Psi_j$, $(\Psi_j, \Psi_l) = 0$ for $j \neq l$.

ISBN-0-201-13503-5

Now suppose $\Psi \in \mathcal{H}$ can be expressed as an infinite linear combination of our orthonormal sequence of eigenvectors.

$$\Psi = \sum_{j=0}^{\infty} a_j \Psi_j, \qquad a_j \in \mathcal{C}. \tag{3.9}$$

Using the orthonormality relations $(\Psi_j, \Psi_l) = \delta_{jl}$, we find Parseval's equality

$$(\Psi, \Psi) = \sum_{j,l=0}^{\infty} a_j \bar{a}_l (\Psi_j, \Psi_l) = \sum_{j=0}^{\infty} a_j \bar{a}_j. \tag{3.10}$$

Furthermore, the coefficients a_j are uniquely determined:

$$(\Psi, \Psi_l) = \sum_{j=0}^{\infty} a_j (\Psi_j, \Psi_l) = a_l. \tag{3.11}$$

The question now arises whether the orthonormal (ON) set $\{\Psi_j\}$ is actually an ON *basis* for $\mathcal{H}$; that is, whether an arbitrary $\Psi \in \mathcal{H}$ can be expanded in a sum (3.9). In general, this is not the case, since the domain D is usually too restricted to provide sufficient eigenvectors Ψ_j to form a basis. However, there is a well-developed theory of extensions of symmetric operators in which this difficulty is largely overcome. (One says that a symmetric operator S' is an *extension* of S if the domain of S' properly contains the domain of S and the two operators agree on their common domain.) It is not our intention to go into the details of this theory, which is contained in many textbooks (e.g., [33, 112]). A principal result of the theory is that many symmetric operators (those with equal deficiency indices) can be extended to special symmetric operators called *self-adjoint* (not unique unless the deficiency indices are $(0,0)$), which do have the property that their eigenfunctions form a basis for $\mathcal{H}$. However, one must pay a price for this by considering not only eigenvectors of the self-adjoint extension but also generalized eigenvectors. (We shall give some examples of generalized eigenvectors shortly.)

Most of the symmetric operators that appear in the separation of variables context developed in this book can be extended to self-adjoint operators via the standard theory. For the few operators that cannot be so extended we can still find bases of eigenfunctions, but by a process that is somewhat arbitrary and nonunique.

We now consider the relationship between the foregoing theory and separation of variables in a purely formal context. Let S be the symmetric operator on D corresponding to a separable coordinate system for the Helmholtz equation. We shall soon see that each such operator can be extended to a self-adjoint operator S' defined on a domain $D' \supseteq D$. Then

ISBN-0-201-13503-5

every solution $\Psi \in \mathcal{H}$ of the Helmholtz equation can be expanded uniquely in terms of the eigenfunctions of S'. However, the eigenfunctions of S' are necessarily separable solutions of the Helmholtz equation in the coordinate system corresponding to S. Moreover, the eigenvalues of S' are just values of the separation constant. Our Hilbert space formalism allows us to expand rather arbitrary solutions of the Helmholtz equation in terms of an eigenbasis of separable solutions. In practice the computations required to obtain these expansions will be carried out not in $\mathcal{H}$, but in the more convenient space $L_2(S_1)$.

We now compute the spectral decomposition for each of the four operators S listed in Table 1.

Orbit 2. $S = M^2$

To compute the spectral decomposition for M^2 it is sufficient to compute the decomposition for iM, $i = \sqrt{-1}$, and square the resulting eigenvalues. (Note that iM is symmetric on the domain $\mathcal{D}$, since M is skew-Hermitian.) The operator M on $L_2(S_1)$ takes the form $M = -d/d\varphi$, so the eigenvalue equation $iMf_\lambda = \lambda f_\lambda$ becomes

$$-i \frac{df_\lambda(\varphi)}{d\varphi} = \lambda f_\lambda(\varphi). \tag{3.12}$$

This equation has no nonzero solutions for $f_\lambda \in \mathcal{D}$. However, if iM is extended to the domain $\mathcal{D}'$ of all functions $f \in L_2(S_1)$ whose first derivative exists and is continuous at every point of S_1, it is easy to check that iM is symmetric on $\mathcal{D}'$ and has the normalized eigenfunctions $\{ f_n^{(2)}(\varphi) \}$ where

$$f_n^{(2)}(\varphi) = \frac{e^{in\varphi}}{(2\pi)^{1/2}}, \qquad iMf_n^{(2)} = nf_n^{(2)}, \qquad n = 0, \pm 1, \pm 2, \dots. \tag{3.13}$$

(Note that for $f \in \mathcal{D}'$ we have the periodicity condition $f(-\pi) = f(\pi)$ which implies that the solutions $f_\lambda(\varphi) = ce^{i\lambda\varphi}$, $c \neq 0$, of (3.12) belong to $\mathcal{D}'$ if and only if $\lambda = n$.) We can verify directly that

$$\langle f_n^{(2)}, f_m^{(2)} \rangle = \frac{1}{2\pi} \int_{-\pi}^{\pi} e^{i(n-m)\varphi} d\varphi = \delta_{nm} = \begin{cases} 0 & \text{if } n \neq m, \\ 1 & \text{if } n = m. \end{cases} \tag{3.14}$$

It is well known from the theory of Fourier series that the ON set $\{ f_n^{(2)} \}$ is actually an ON basis for $L_2(S_1)$ [101]. Thus, any $f \in L_2(S_1)$ can be expanded uniquely in the form

$$f(\varphi) \sim \sum_{n=-\infty}^{\infty} c_n f_n^{(2)}(\varphi), \qquad c_n = \langle f, f_n^{(2)} \rangle \tag{3.15}$$

ISBN-0-201-13503-5

where $\sim$ denotes that the sum converges to f in the Hilbert space sense,

$$\lim_{m\to\infty}\left\langle f-\sum_{n=-m}^{m}c_nf_n^{(2)},f-\sum_{n=-m}^{m}c_nf_n^{(2)}\right\rangle=0$$

and not necessarily pointwise.

(Actually iM is not self-adjoint on $\mathcal{D}'$ but it can easily be shown that the further extension of iM to the domain where it is self-adjoint does not lead to new eigenvalues or eigenvectors. In the future we shall not always discuss these domain problems, but shall proceed directly to list the appropriate basis of eigenfunctions.)

The operator $iM'=i(y\,\partial_x-x\,\partial_y)$ on $\mathcal{H}$ which corresponds to $iM=-i\,d/d\varphi$ on $L_2(S_1)$ is given by $M'=IMI^{-1}$ where the unitary transformation I is defined by (3.1). It follows immediately that iM' is unitarily equivalent to iM, hence iM' has the same spectrum as iM and the ON basis $\{f_n^{(2)}\}$ of eigenfunctions of iM is mapped by I to the ON basis $\{\Psi_n^{(2)}\}$,

$$\Psi_n^{(2)}=I\big(f_n^{(2)}\big),\qquad n=0,\pm1,\ldots,\tag{3.16}$$

of eigenfunctions of iM'. Thus,

$$\Psi_n^{(2)}(r,\theta)=I\Big[\exp(in\varphi)/(2\pi)^{1/2}\Big]$$

$$=(2\pi)^{-1/2}\int_{-\pi}^{\pi}\exp\big[i\omega r\cos(\varphi-\theta)\big]\,\exp(in\varphi)\,d\varphi\tag{3.17}$$

where we have switched to polar coordinates. The change of variable $\alpha=\varphi-\theta$ yields

$$\Psi_n^{(2)}(r,\theta)=\exp(in\theta)R_n(r),$$

$$R_n(r)=(2\pi)^{-1/2}\int_{-\pi}^{\pi}\exp(i\omega r\cos\alpha)\exp(in\alpha)\,d\alpha.\tag{3.18}$$

Although the integral for $R_n(r)$ is well known, we shall derive it from first principles to demonstrate the connection between Lie theory and separation of variables. Since $\Psi_n^{(2)}$ is an eigenfunction of iM' with eigenvalue n, it follows that $\Psi_n^{(2)}$ separates in polar coordinates and $R_n(r)$ satisfies Bessel's equation (2.9) with $k=n$. Since $R_n(0)$ is finite, $R_n(r)=c_nJ_n(\omega r)$ where c_n is a constant (see Appendix B, Section 5). To compute c_n we use the fact that the coefficient of r^n in the expansion (B.14) for $J_n(\omega r)$ is $(\omega/2)^n/n!$. Expanding $\exp(i\omega r\cos\alpha)$ in a power series in r, we see that the coefficient

ISBN-0-201-13503-5

of r^n in the integral expression (3.18) for $R_n(r)$ is

$$\frac{(i\omega)^n}{n!(2\pi)^{1/2}}\int_{-\pi}^{\pi}\cos^n\alpha\exp(in\alpha)\,d\alpha=\frac{(2\pi)^{1/2}}{n!}\left(\frac{i\omega}{2}\right)^n.$$

(This follows easily from the relation $\cos\alpha=(e^{i\alpha}+e^{-i\alpha})/2$ and the orthogonality relations (3.14).) Thus,

$$\Psi_n^{(2)}(r,\theta)=i^n(2\pi)^{1/2}J_n(\omega r)\exp(in\theta).\qquad(3.19)$$

Note also that (3.17) implies $\Psi_{-n}^{(2)}$ is the nth Fourier coefficient of $\exp[i\omega r\cos(\varphi-\theta)]$,

$$\exp\big[i\omega r\cos(\varphi-\theta)\big]=\sum_{n=-\infty}^{\infty}\Psi_n^{(2)}(r,\theta)\bar{f}_n^{(2)}(\varphi)$$

$$=\sum_{n=-\infty}^{\infty}i^n J_n(\omega r)\exp\big[in(\theta-\varphi)\big].\qquad(3.20)$$

Orbit 1. $S=P_2^2$

As in the previous case it is sufficient to compute the spectral decomposition for the symmetric operator iP_2. Here $iP_2=-\omega\sin\varphi$ and this operator is defined and self-adjoint on the full Hilbert space $L_2(S_1)$. Clearly iP_2 has no eigenvalues or eigenfunctions in the usual sense, since if $iP_2 f=\lambda f$, then $f(\varphi)=0$ except for at most two values φ_1,φ_2 where $\lambda=-\omega\sin\varphi_j$. Thus $\langle f,f\rangle=\int_{-\pi}^{\pi}|f(\varphi)|^2\,d\varphi=0$ and f is not an eigenfunction. However, iP_2 does have generalized eigenfunctions $f_\alpha^{(1)}(\varphi)=\delta(\varphi-\alpha)$, $-\pi\leqslant\alpha<\pi$, where $\delta(\varphi-\alpha)=\delta_\alpha(\varphi)$ is the *Dirac delta function* (not a function at all) defined formally by

$$\langle h,\delta_\alpha\rangle=\int_{-\pi}^{\pi}h(\varphi)\delta(\varphi-\alpha)\,d\varphi=h(\alpha)\qquad(3.21)$$

where h is any infinitely differentiable function belonging to $L_2(S_1)$. (See [42] or [77] for precise definitions of generalized eigenfunctions and discussions of their relationship to spectral theory.) Thus we have the formal relations

$$iP_2 f_\alpha^{(1)}=-\omega\sin\alpha f_\alpha^{(1)},\qquad-\pi\leqslant\alpha<\pi,$$
$$\langle f_\alpha^{(1)},f_{\alpha'}^{(1)}\rangle=\delta(\alpha-\alpha').\qquad(3.22)$$

ISBN-0-201-13503-5

As α ranges over $[-\pi, \pi)$, the generalized eigenvalues range over the interval $(-\omega, \omega]$ covering almost every point twice. We say that the spectrum of iP_2 is *continuous* and covers the interval $(-\omega, \omega]$ with *multiplicity two*. A general $h \in L_2(S_1)$ can be expanded in the eigenbasis $\{ f_\alpha^{(1)} \}$ as an integral

$$h = \int_{-\pi}^{\pi} c_\alpha f_\alpha^{(1)} \, d\alpha, \qquad c_\alpha = \langle h, f_\alpha^{(1)} \rangle = h(\alpha). \tag{3.23}$$

The Parseval equality becomes

$$\langle h, h \rangle = \int_{-\pi}^{\pi} c_\alpha \bar{c}_\alpha \, d\alpha = \int_{-\pi}^{\pi} |h(\alpha)|^2 \, d\alpha.$$

(These formulas seem so trivial because we have obtained iP_2 in a model where its spectral decomposition is obvious.)

The corresponding operator iP_2 on $\mathcal{H}$ is given by $iP_2 = i \partial_y$ and is unitary equivalent to the operator (3.22) on $L_2(S_1)$. The corresponding basis of generalized eigenfunctions $\{ \Psi_\alpha^{(1)} \}$ is

$$\Psi_\alpha^{(1)}(x, y) = I\left(f_\alpha^{(1)} \right) = \int_{-\pi}^{\pi} \exp\left[i\omega(x \cos\varphi + y \sin\varphi) \right] \delta(\varphi - \alpha) \, d\varphi$$

$$= \exp\left[i\omega(x \cos\alpha + y \sin\alpha) \right], \qquad -\pi \leqslant \alpha < \pi. \tag{3.24}$$

The unitary equivalence gives

$$iP_2 \Psi_\alpha^{(1)} = -\omega \sin\alpha \, \Psi_\alpha^{(1)}, \qquad \left(\Psi_\alpha^{(1)}, \Psi_{\alpha'}^{(1)} \right) = \delta(\alpha - \alpha'), \tag{3.25}$$

and the expansion theorem

$$\Psi(x, y) = \int_{-\pi}^{\pi} c_\alpha \Psi_\alpha^{(1)}(x, y) \, d\alpha = \int_{-\pi}^{\pi} c_\alpha \exp\left[i\omega(x \cos\alpha + y \sin\alpha) \right] d\alpha,$$

$$c_\alpha = \left(\Psi, \Psi_\alpha^{(1)} \right) = \langle h, f_\alpha^{(1)} \rangle = h(\alpha) \tag{3.26}$$

where $\Psi \in \mathcal{H}$ corresponds to $h \in L_2(S_1)$, reduces to (3.1).

Orbit 3. $S = \{ M, P_2 \}$

Here $\{ M, P_2 \} = -2i\omega \sin\varphi(d/d\varphi) - i\omega \cos\varphi$ on the domain of infinitely differentiable functions in $L_2(S_1)$ that vanish in neighborhoods of $\varphi = 0$ and $\pm \pi$. It is straightforward to verify that S is symmetric on this domain (and essentially self-adjoint). To determine the self-adjoint extension, we

ISBN-0-201-13503-5

define a unitary mapping U from $L_2(S_1)$ onto $L_2(R) \oplus L_2(R)$ by

$$Uf(\nu) = \mathbf{F}(\nu) = \begin{pmatrix} F_+(\nu) \\ F_-(\nu) \end{pmatrix} = |\sin\varphi|^{1/2} \begin{pmatrix} f_+(\cos\varphi) \\ f_-(\cos\varphi) \end{pmatrix}, \qquad \cos\varphi = \tanh\nu.$$

(3.27)

Thus to each $f \in L_2(S_1)$ we associate a column two-vector with components $F_\pm(\nu) = |\sin\varphi|^{1/2} f_\pm(\cos\varphi) \in L_2(R)$. Here,

$$f_-(\cos\varphi) = f(\varphi), \qquad -\pi \leqslant \varphi < 0,$$
$$f_+(\cos\varphi) = f(\varphi), \qquad 0 < \varphi \leqslant \pi,$$

and $L_2(R)$ is the Hilbert space of Lebesgue square-integrable functions $F(x)$ on the real line R, $\int_{-\infty}^{\infty}|F(x)|^2\,dx < \infty$, with inner product $\langle F, G \rangle' = \int_{-\infty}^{\infty} F(x)\overline{G}(x)\,dx$. (We are considering only functions that vanish near $\varphi = 0$ because the coefficient of the first derivative term in the expression for S vanishes at $\varphi = 0$; that is, S has a singularity at $\varphi = 0$.)

$$\mathfrak{L} = L_2(R) \oplus L_2(R)$$

is the Hilbert space of vector-valued Lebesgue square-integrable functions

$$\mathbf{F}(\nu): \int_{-\infty}^{\infty} \left(|F_+(\nu)|^2 + |F_-(\nu)|^2 \right) d\nu < \infty$$

and inner product

$$\langle \mathbf{F}, \mathbf{G} \rangle = \int_{-\infty}^{\infty} \left(F_+(\nu)\overline{G}_+(\nu) + F_-(\nu)\overline{G}_-(\nu) \right) d\nu, \qquad \mathbf{F}, \mathbf{G} \in \mathfrak{L}.$$

It is straightforward to check that

$$\langle f_1, f_2 \rangle = \langle \mathbf{F}_1, \mathbf{F}_2 \rangle$$

for $\mathbf{F}_j = Uf_j$, so U is indeed a unitary transformation from $L_2(S_1)$ onto $\mathfrak{L}$. Furthermore, the operator USU^{-1} on $\mathfrak{L}$, which we also denote S, takes the form $S\mathbf{F}(\nu) = 2i\omega(d/d\nu)\mathbf{F}(\nu)$. (At this point it is apparent that S is unitary equivalent to two copies of the momentum operator in the sense of quantum theory.) To make the spectrum of S even more obvious, we take the vector Fourier transform

$$\mathscr{F}(\lambda) = \begin{pmatrix} \mathscr{F}_+(\lambda) \\ \mathscr{F}_-(\lambda) \end{pmatrix} = (2\pi)^{-1/2} \int_{-\infty}^{\infty} \mathbf{F}(\nu) e^{i\nu\lambda}\,d\nu$$

(3.28)

ISBN-0-201-13503-5

with inverse

$$\mathbf{F}(v) = (2\pi)^{-1/2} \int_{-\infty}^{\infty} \mathcal{F}(\lambda) e^{-iv\lambda} d\lambda. \qquad (3.29)$$

(For a discussion of the Fourier transform see [112].) Then introducing the inner product

$$(\mathcal{F}, \mathcal{G}) = \int_{-\infty}^{\infty} \left(\mathcal{F}_+(\lambda) \bar{\mathcal{G}}_+(\lambda) + \mathcal{F}_-(\lambda) \bar{\mathcal{G}}_-(\lambda) \right) d\lambda, \qquad (3.30)$$

we obtain a Hilbert space $\mathcal{L}'$ of vector-valued functions $\mathcal{F}$ such that

$$(\mathcal{F}, \mathcal{G}) = \langle \mathbf{F}, \mathbf{G} \rangle = \langle f, g \rangle,$$

and on $\mathcal{L}'$, S takes the form

$$S\mathcal{F}(\lambda) = 2\lambda \omega \mathcal{F}(\lambda).$$

We conclude that $S = \{M, P_2\}$ can be extended to a unique self-adjoint operator with continuous spectrum of multiplicity two covering the real axis. The generalized eigenfunctions of S in $\mathcal{L}'$ are

$$\mathcal{F}_\mu^+(\lambda) = \begin{pmatrix} \delta(\lambda - \mu) \\ 0 \end{pmatrix}, \qquad \mathcal{F}_\mu^-(\lambda) = \begin{pmatrix} 0 \\ \delta(\lambda - \mu) \end{pmatrix}.$$

Mapping back to $L_2(S_1)$, we obtain generalized eigenfunctions

$$f_{\mu+}^{(3)}(\varphi) = \begin{cases} (2\pi)^{-1/2}(1+\cos\varphi)^{-i\mu/2-\frac{1}{4}}(1-\cos\varphi)^{i\mu/2-\frac{1}{4}}, & 0 < \varphi \leqslant \pi, \\ 0, & -\pi \leqslant \varphi < 0, \end{cases}$$

$$f_{\mu-}^{(3)}(\varphi) = f_{\mu+}^{(3)}(-\varphi), \qquad \{M, P_2\} f_{\mu\pm}^{(3)} = 2\mu\omega f_{\mu\pm}^{(3)}, \qquad -\infty < \mu < \infty,$$

$$\langle f_{\mu\pm}^{(3)}, f_{\mu'\pm}^{(3)} \rangle = \delta(\mu - \mu'), \qquad \langle f_{\mu\pm}^{(3)}, f_{\mu'\mp}^{(3)} \rangle = 0. \qquad (3.31)$$

The solution of the Helmholtz equation corresponding to $f_{\mu+}^{(3)}$ is

$$\Psi_{\mu+}^{(3)}(x,y) = I\left(f_{\mu+}^{(3)}\right) = \int_0^\pi \exp\left[i\omega(x\cos\varphi + y\sin\varphi) \right] f_{\mu+}^{(3)}(\varphi) \, d\varphi$$

$$= \frac{1}{\sqrt{\pi}} \int_0^\infty \frac{t^{i\mu-\frac{1}{2}}}{(1+t^2)^{1/2}} \exp\left\{ i\omega\left[x\left(\frac{1-t^2}{1+t^2}\right) + \frac{2yt}{1+t^2} \right] \right\} dt,$$

$$\cos\varphi = (t^{-1} - t)/(t^{-1} + t). \qquad (3.32)$$

ISBN-0-201-13503-5

Since $\Psi^{(3)}_{\mu+}$ is an eigenfunction of $\{M, P_2\}$ in $\mathcal{H}$ with eigenvalue $2\mu\omega$, we know that it must separate in parabolic coordinates

$$x = \tfrac{1}{2}(\xi^2 - \eta^2), \qquad y = \xi\eta,$$

or, more precisely, it must be expressible as a sum of at most four linearly independent terms, each of which separates. The four terms are multiples of $U_j(\xi)V_l(\eta), j, l = 1, 2$, where the U_j and the V_l form bases of solutions for the parabolic cylinder equations (2.37) with $k^2 = 2\mu\omega$. Thus, the integral is determined to within four constants and these constants can be fixed by evaluation of the integral for various special cases (e.g., $x = 0$). An explicit computation yields

$$\Psi^{(3)}_{\mu+}(\xi, \eta) = \left[\sqrt{2}\, \cos(i\mu\pi) \right]^{-1} \left[D_{i\mu - \frac{1}{2}}(\sigma\xi) D_{-i\mu - \frac{1}{2}}(\sigma\eta) \right.$$

$$\left. + D_{i\mu - \frac{1}{2}}(-\sigma\xi) D_{-i\mu - \frac{1}{2}}(-\sigma\eta) \right] \qquad (3.33)$$

where $\sigma = \exp(i\pi/4)(2\omega)^{1/2}$ and $D_\nu(x)$ is a parabolic cylinder function (B.9). Furthermore,

$$\Psi^{(3)}_{\mu-}(\xi, \eta) = \Psi^{(3)}_{\mu+}(\xi, -\eta). \qquad (3.34)$$

Orbit 4. $S = M^2 + d^2 P_1^2$

In this case $S = (d^2/d\varphi^2) - d^2\omega^2 \cos^2\varphi$ on the domain $\mathcal{D} \subset L_2(S_1)$. The eigenvalue equation $Sf = \lambda f$ can be rewritten

$$\frac{d^2 f}{d\varphi^2} + (a - 2q\cos 2\varphi)f = 0, \qquad a = -\lambda - \frac{d^2\omega^2}{2}, \qquad q = \frac{d^2\omega^2}{4}. \qquad (3.35)$$

Equation (3.35) is Mathieu's equation (B.25). Now S has no eigenfunctions on $\mathcal{D}$ but S can be extended uniquely to a symmetric operator on the space of twice continuously differentiable functions f on S_1. Then the eigenvalue problem for S becomes a regular Sturm–Liouville problem [72, 101], and as seen from Appendix B, Section 8, there is a countable infinity of discrete eigenvalues λ_n decreasing to $-\infty$, each eigenvalue of multiplicity one. The corresponding normalized eigenvectors are

$$f^{(4)}_{nc}(\varphi) = \pi^{-1/2} ce_n(\varphi, q), \qquad n = 0, 1, 2, \ldots, \qquad (3.36)$$

$$f^{(4)}_{ns}(\varphi) = \pi^{-1/2} se_n(\varphi, q), \qquad n = 1, 2,.$$

ISBN-0-201-13503-5

The $\{f_{nc}^{(4)}, f_{ns}^{(4)}\}$ form an ON basis for $L_2(S_1)$.

$$\langle f_{nt}^{(4)}, f_{mt}^{(4)} \rangle = \delta_{mn}, \qquad t = s, c, \qquad \langle f_{nc}^{(4)}, f_{ms}^{(4)} \rangle = 0. \qquad (3.37)$$

The solution of the Helmholtz equation corresponding to $f_{nc}^{(4)}$ is

$$\Psi_{nc}^{(4)}(x, y) = I\left(f_{nc}^{(4)}\right) = \pi^{-1/2} \int_{-\pi}^{\pi} \exp\left[i\omega(x\cos\varphi + y\sin\varphi)\right] ce_n(\varphi, q)\, d\varphi. \qquad (3.38)$$

To evaluate this integral we use the fact that $\Psi_{nc}^{(4)}$ is an eigenfunction of $M^2 + d^2 P_1^2$ in $\mathcal{H}$. Thus it must be a sum of at most four terms, each of which separates in elliptic coordinates

$$x = d\cosh\alpha\cos\beta, \qquad y = d\sinh\alpha\sin\beta.$$

Furthermore, comparing (2.43) and (B.25), we find, with $k^2 = \lambda = -a - d^2\omega^2/2$, that $\Psi_{nc}^{(4)}$ satisfies Mathieu's equation in the variable β. Inspection of the integral (3.38) shows that $\Psi_{nc}^{(4)}$ is periodic in β with period 2π. Hence,

$$\Psi_{nc}^{(4)}(\alpha, \beta) = U(\alpha) ce_n(\beta, q) \qquad (3.39)$$

where $U(\alpha)$ satisfies the *modified Mathieu equation*

$$\frac{d^2 U}{d\alpha^2} + (-a + 2q\cosh 2\alpha) U = 0, \qquad (3.40)$$

obtained from Mathieu's equation (B.25) by setting $x = i\alpha$. Depending on whether $a = a_n$ or $a = b_n$ in (3.40), this equation has solutions $Ce_n(\alpha, q) = ce_n(i\alpha, q)$ or $Se_n(\alpha, q) = i se_n(i\alpha, q)$ (see (B.26)), which are either even or odd functions of α and are characterized by this symmetry property. An inspection of the integral (3.38) shows that it is even in α. Hence,

$$\Psi_{nc}^{(4)}(\alpha, \beta) = C_n Ce_n(\alpha, q) ce_n(\beta, q), \qquad n = 0, 1, \ldots, \qquad (3.41a)$$

where C_n is a constant to be determined from the integral. A similar argument yields

$$\Psi_{ns}^{(4)}(\alpha, \beta) = S_n Se_n(\alpha, q) se_n(\beta, q), \qquad n = 1, 2, \ldots, \qquad (3.41b)$$

where S_n is a constant. Note that (3.38) can now be interpreted as an integral representation for a product of a Mathieu and a modified Mathieu function. (The Ce_n and Se_n are called *modified Mathieu functions* of the *first kind*.)

We have now determined the spectra and eigenbases for the four operators that characterize separation of variables in (0.1), and by a simple

ISBN-0-201-13503-5

extension, the spectra of all operators in $\mathbb{S}^{(2)}$. Indeed, if two operators in this space are unitary equivalent under the adjoint action of $E(2)$, they have the same spectra. If one operator is a real multiple of another, then the spectrum of the first operator is the same real multiple of the spectrum of the second operator. Thus when we computed the spectrum of one of the four operators above, we were essentially computing the spectrum for every operator on the same orbit as the given operator.

In practical problems involving solutions of the Helmholtz equation it is of great interest to obtain formulas yielding the expansion of a separable basis function $\Psi_n^{(j)}$ as a sum or integral over basis functions $\Psi_m^{(l)}$. More generally, we often want to apply a Euclidean transformation to $\Psi_n^{(j)}$ and then expand in terms of the $\Psi_m^{(l)}$ basis. Since $\mathcal{H}$ is a Hilbert space, we have

$$\mathbf{T}(g)\Psi_n^{(j)} = \sum_m \left(\mathbf{T}(g)\Psi_n^{(j)}, \Psi_m^{(l)}\right)\Psi_m^{(l)} \tag{3.42}$$

where the sum is replaced by an integral if $\Psi_m^{(l)}$ corresponds to continuous spectrum. However, by definition

$$\left(\mathbf{T}(g)\Psi_n^{(j)}, \Psi_m^{(l)}\right) = \langle \mathbf{T}(g)f_n^{(j)}, f_m^{(l)}\rangle, \tag{3.43}$$

so we can compute the expansion coefficients in $L_2(S_1)$ rather than $\mathcal{H}$. This simplification makes the problem tractable. Moreover, since the operators $\mathbf{T}(g)$ that appear in (3.43) define a unitary representation of $E(2)$, we can move them from the left-hand to the right-hand side of the inner product or decompose them in any way that simplifies the computation of the integral in $L_2(S_1)$; for example,

$$\langle \mathbf{T}(g)f_n^{(j)}, f_m^{(l)}\rangle = \langle f_n^{(j)}, \mathbf{T}(g^{-1})f_m^{(l)}\rangle.$$

In discussing expansions of the form (3.42) it is useful to introduce new terminology (which will be adopted in the remainder of this book.). For $g = g(0,0,0)$, so that $\mathbf{T}(g)$ is the identity operator, the expansion coefficients $\langle f_n^{(j)}, f_m^{(l)}\rangle$ are called *overlap functions*. For $j = l$ and general $g \in E(2)$ formula (3.42) is called an *addition theorem* for the basis $\{\Psi_n^{(j)}\}$ and the coefficients $T_{mn}^{(j)}(g) = \langle \mathbf{T}(g)f_n^{(j)}, f_m^{(j)}\rangle$ are the *matrix elements* of the operator $\mathbf{T}(g)$ in the j basis. Since $\mathbf{T}(gg') = \mathbf{T}(g)\mathbf{T}(g')$, it follows immediately that the matrix elements satisfy the identities

$$T_{mn}^{(j)}(gg') = \sum_k T_{mk}^{(j)}(g)T_{kn}^{(j)}(g'). \tag{3.44}$$

Moreover, the fact that $\mathbf{T}(g)$ is unitary implies

$$T_{mn}^{(j)}(g^{-1}) = \bar{T}_{nm}^{(j)}(g). \tag{3.45}$$

Finally, for $g(0,0,0) = e$ we have

$$T_{mn}^{(j)}(e) = \delta_{mn}. \tag{3.46}$$

If $j \neq l$ in (3.42) and g is arbitrary, we call the expansion coefficients (3.43) *mixed-basis matrix elements*.

We now proceed to the computation of some expansions of special interest. From (3.22) we have

$$\langle f_n^{(j)}, f_\alpha^{(1)} \rangle = \int_{-\pi}^{\pi} f_n^{(j)}(\varphi) \delta(\varphi - \alpha) d\varphi = f_n^{(j)}(\alpha), \qquad j = 2, 3, 4,$$

so

$$\Psi_n^{(j)}(x,y) = \int_{-\pi}^{\pi} f_n^{(j)}(\alpha) \Psi_\alpha^{(1)}(x,y) d\alpha \tag{3.47}$$

where $\Psi_\alpha^{(1)}$ is the so-called *plane wave* solution (3.24) of the Helmholtz equation. Note that this plane wave expansion is just expression (3.1) specialized to the eigenbasis $\{\Psi_n^{(j)}\}$.

Furthermore, we have the expansion

$$\Psi_\alpha^{(1)}(x,y) = \sum_n \langle f_\alpha^{(1)}, f_n^{(j)} \rangle \Psi_n^{(j)} = \sum_n \bar{f}_n^{(j)}(\alpha) \Psi_n^{(j)} \tag{3.48}$$

where the sum is replaced by an integral if $\{\Psi_n^{(j)}\}$ is a continuum eigenbasis. Now we have an expansion of a plane wave solution in terms of another eigenbasis. For example, the Bessel function solutions (3.19), $j = 2$, called *cylindrical wave* solutions, yield the identity (3.20) when substituted into (3.48). For $j = 3$ we find

$$\exp\left[i\omega(x\cos\alpha + y\sin\alpha) \right] = \frac{(\sin\alpha)^{-1/2}}{(2\pi)^{1/2}} \int_{-\infty}^{\infty} \left(\cot\frac{\alpha}{2} \right)^{i\mu} \Psi_{\mu+}^{(3)}(x,y) d\mu,$$

$$0 < \alpha < \pi, \quad (3.49)$$

with a similar result for $-\pi < \alpha < 0$. A tedious but straightforward integra-

ISBN-0-201-13503-5

tion yields the results

$$\langle f_n^{(2)}, f_{\mu+}^{(3)} \rangle = (2\pi)^{-1} \int_0^\pi e^{in\varphi} (1+\cos\varphi)^{i\mu/2-\frac{1}{4}} (1-\cos\varphi)^{-i\mu/2-\frac{1}{4}} d\varphi$$

$$= \frac{\exp[\pi/2(i/2-\mu)]}{\pi\sqrt{2}} \Gamma(\tfrac{1}{2}-n) \left[\frac{(-1)^n \Gamma(\tfrac{1}{2}+i\mu)}{\Gamma(\tfrac{1}{2}+i\mu-n)} {}_2F_1 \left[\begin{matrix} \tfrac{1}{2}+i\mu, \tfrac{1}{2}-n \\ 1+i\mu-n \end{matrix} \middle| -1 \right] \right.$$

$$\left. - i\frac{\Gamma(\tfrac{1}{2}-i\mu)}{\Gamma(1-i\mu-n)} {}_2F_1 \left[\begin{matrix} \tfrac{1}{2}-i\mu, \tfrac{1}{2}-n \\ 1-i\mu-n \end{matrix} \middle| -1 \right] \right]$$

$$\langle f_n^{(2)}, f_{\mu-}^{(3)} \rangle = \langle f_{-n}^{(2)}, f_{\mu+}^{(3)} \rangle, \qquad (3.50)$$

giving the overlaps between the Bessel and parabolic cylinder bases. Thus we have the expansion

$$(2\pi)^{1/2} i^n J_n(\omega r) e^{in\theta} = \int_{-\infty}^\infty \left[\langle f_n^{(2)}, f_{\mu+}^{(3)} \rangle \Psi_{\mu+}^{(3)}(x,y) \right.$$

$$\left. + \langle f_n^{(2)}, f_{\mu-}^{(3)} \rangle \Psi_{\mu-}^{(3)}(x,y) \right] d\mu \qquad (3.51)$$

of a cylindrical wave in terms of parabolic cylindrical waves. The overlaps between the Mathieu basis ($j=4$) and the Bessel basis are especially easy to compute because the Mathieu functions $ce_n(\varphi,q)$, $se_n(\varphi,q)$ are defined by their coefficients A_m^n, B_m^n with respect to a Fourier series expansion on the unit circle. For example, from (B.26)

$$ce_{2n+1}(\varphi,q) = \frac{1}{2} \sum_{m=0}^\infty A_{2m+1}^{(2n+1)} \{ \exp[i(2m+1)\varphi] + \exp[-i(2m+1)\varphi] \}.$$

Since $f_{2n+1,c}^{(4)}(\varphi) = \pi^{-1/2} ce_{2n+1}(\varphi,q)$, we have

$$\langle f_{2n+1,c}^{(4)}, f_p^{(2)} \rangle = \begin{cases} A_{2m+1}^{(2n+1)} & \text{if } |p|=2m+1, \\ 0 & \text{otherwise.} \end{cases} \qquad (3.52)$$

The remaining overlaps are just as simple. With these overlaps the reader can easily express a cylindrical wave solution as an infinite sum of Mathieu solutions (3.41), (3.42), or a Mathieu solution as an infinite sum of cylindrical waves. (See also [108].)

Next we present an example of an addition theorem. The matrix elements of the operator $T(g)$, (3.2), with respect to the Bessel basis $\{f_n^{(2)}\}$,

ISBN-0-201-13503-5

(3.13), are

$$\mathbf{T}_{mn}^{(2)}(\theta,a,b)=\langle\mathbf{T}(g)f_n^{(2)},f_m^{(2)}\rangle$$

$$=(2\pi)^{-1}\int_{-\pi}^{\pi}\exp\big[\,i\omega(a\cos(\varphi-\theta)+b\sin(\varphi-\theta))-in\theta$$

$$+i(n-m)\varphi\big]\,d\varphi,\ \ g=g(\theta,a,b). \tag{3.53}$$

Setting $a=r\cos\alpha$, $b=r\sin\alpha$, and introducing the new variable $\beta=\varphi-\theta-\alpha$, we find

$$T_{mn}^{(2)}(\theta,a,b)=(2\pi)^{-1}\exp\big[\,i(n-m)\alpha-im\theta\,\big]$$

$$\int_{-\pi}^{\pi}\exp\big[\,i\omega r\cos\beta+i(n-m)\,\beta\,\big]\,d\beta.$$

Comparing this integral with (3.18), (3.19), we finally obtain

$$T_{mn}^{(2)}[\,\theta,r,\alpha\,]=i^{n-m}\exp\big[\,i(n-m)\alpha-im\theta\,\big]J_{n-m}(\omega r), \tag{3.54}$$

$$a=r\cos\alpha,\ b=r\sin\alpha,\ n,m=0,\pm1,\pm2,\cdots.$$

Substituting these matrix elements into (3.44)–(3.46), we obtain a series of identities for Bessel functions. (See [82, 124, 128] for more complete discussions of these results.) The addition theorem for the Bessel basis turns out to be a special case of (3.44) with $j=2$, since

$$\Psi_n^{(2)}(r,\theta)=(2\pi)^{1/2}T_{0n}^{(2)}[\,0,r,\theta\,].$$

As a final example, note that the plane wave (3.7), (3.24)

$$\overline{H}(x,y,\varphi)=\Psi_\varphi^{(1)}(x,y)=\exp\big[\,i\omega(x\cos\varphi+y\sin\varphi)\big],$$

considered as a function of φ, belongs to $L_2(S_1)$. Indeed, a direct computation yields

$$\langle\Psi^{(1)}(\mathbf{x}),\Psi^{(1)}(\mathbf{x}')\rangle=2\pi J_0\big\{\omega\big[(x-x')^2+(y-y')^2\big]^{1/2}\big\}. \tag{3.55}$$

On the other hand,

$$\langle\Psi^{(1)}(\mathbf{x}),\Psi^{(2)}(\mathbf{x}')\rangle=\sum_n\langle\Psi^{(1)}(\mathbf{x}),f_n^{(j)}\rangle\langle f_n^{(j)},\Psi^{(1)}(\mathbf{x}')\rangle$$

$$=\sum_n\overline{\Psi}_n^{(j)}(-\mathbf{x})\Psi_n^{(j)}(-\mathbf{x}'), \tag{3.56}$$

as follows from the completeness property of the $\{f_n^{(j)}\}$ basis and (3.7). (Again, the sum on n may be replaced by an integral.) Comparing (3.55)

ISBN-0-201-13503-5

and (3.56), we obtain for each $j = 1, \ldots, 4$ a bilinear expansion of $J_0\{\omega[(x - x')^2 + (y - y')^2]^{1/2}\}$ in terms of separable solutions of the Helmholtz equation.

For additional examples of Helmholtz expansions see [94].

1.4 Separation of Variables for the Klein–Gordon Equation

The method of the preceding sections yields dramatically different results when applied to the Klein–Gordon equation in two-dimensional space-time,

$$\left(\partial_{tt} - \partial_{xx} + \omega^2\right)\Phi(t, x) = 0. \tag{4.1}$$

Here, x and t are real variables and ω is a positive constant. Since our method has been explained in great detail for the Helmholtz equation, we shall often present only the results and omit routine computations.

Discarding the trivial identity operator, we find that the symmetry algebra of (4.1) is three dimensional with basis

$$P_1 = \partial_t, \qquad P_2 = \partial_x, \qquad M = -t\,\partial_x - x\,\partial_t \tag{4.2}$$

and commutation relations

$$[P_1, P_2] = 0, \qquad [M, P_1] = P_2, \qquad [M, P_2] = P_1. \tag{4.3}$$

As the symmetry algebra of (4.1) we take the *real* Lie algebra $\mathscr{E}(1,1)$ with basis (4.2). In terms of the elements of $\mathscr{E}(1,1)$ (4.1) reads

$$\left(P_1^2 - P_2^2 + \omega^2\right)\Phi = 0. \tag{4.4}$$

Thus $P_1^2 - P_2^2 \equiv -\omega^2$ on the solution space of the Klein–Gordon equation. Furthermore, it is easy to verify that every $L \in \mathscr{E}(1,1)$ commutes with $P_1^2 - P_2^2$; that is, this operator lies in the center of the enveloping algebra of $\mathscr{E}(1,1)$.

The symmetry group of (4.1) is $E(1,1)$, isomorphic to the group of matrices

$$g(\theta, a, b) = \begin{bmatrix} \cosh\theta & -\sinh\theta & 0 \\ -\sinh\theta & \cosh\theta & 0 \\ a & b & 1 \end{bmatrix}, \qquad -\infty < \theta, a, b < \infty. \tag{4.5}$$

The group product is

$$g(\theta, a, b)\, g(\theta', a', b') = g(\theta + \theta', a\cosh\theta' - b\sinh\theta' + a',$$
$$-a\sinh\theta' + b\cosh\theta' + b'). \tag{4.6}$$

$E(1,1)$ acts as a transformation group in the t–x plane: the group element

$g(\theta,a,b)$ maps the point $\mathbf{x}=(t,x)$ to the point

$$\mathbf{x}g=(t\cosh\theta-x\sinh\theta+a,\,-t\sinh\theta+x\cosh\theta+b). \qquad (4.7)$$

(Note that $E(1,1)$ is also called the *Poincaré group* for two-dimensional space-time. It is a connected group of transformations that preserve the (special relativistic) space-time distance s between two points $\mathbf{x}$ and $\mathbf{x}'$.

$$s^2=(t-t')^2-(x-x')^2.$$

The reader can easily verify that the transformations (4.7) have this property.) It is straightforward to check that $\mathbf{x}(g_1 g_2)=(\mathbf{x}g_1)g_2$ for $g_1,g_2\in E(1,1)$ and that $\mathbf{x}g(0,0,0)=\mathbf{x}$.
   A basis for the Lie algebra of the matrix group $E(1,1)$ is given by the matrices

$$P_1=\begin{bmatrix}0&0&0\\0&0&0\\1&0&0\end{bmatrix},\qquad P_2=\begin{bmatrix}0&0&0\\0&0&0\\0&1&0\end{bmatrix},\qquad M=\begin{bmatrix}0&-1&0\\-1&0&0\\0&0&0\end{bmatrix} \quad (4.8)$$

with commutation relations identical to (4.3). The Lie algebra generators and the matrices (4.5) are related by (matrix exponential)

$$g(\theta,a,b)=\exp(\theta M)\exp(aP_1+bP_2). \qquad (4.9)$$

The corresponding action of $E(1,1)$ on the space $\mathcal{F}$ of analytic functions f in t,x is

$$\mathbf{T}(g)f(\mathbf{x})=\exp(\theta M)\exp(aP_1+bP_2)f(\mathbf{x})$$

where P_1,P_2,M are now given by (4.2). Exponentiating the Lie derivatives as in Appendix A, we obtain

$$\mathbf{T}(g)f(\mathbf{x})=f(\mathbf{x}g) \qquad (4.10)$$

where $\mathbf{x}g$ is defined by (4.7). As usual, it is easy to check that

$$\mathbf{T}(gg')=\mathbf{T}(g)\mathbf{T}(g'),\qquad g,g'\in E(1,1),\quad \mathbf{T}(g(0,0,0))=E, \quad (4.11)$$

with E the identity operator on $\mathcal{F}$.
   The space $\mathcal{S}$ of second-order symmetry operators can be computed in a manner analogous to that presented in Section 1.1. Factoring out the space $\mathfrak{q}$ of trivial symmetries $f(\mathbf{x})(\partial_{tt}-\partial_{xx}+\omega^2)$, $f\in\mathcal{F}$, we find that $\mathcal{S}/\mathfrak{q}$ is nine

ISBN-0-201-13503-5

dimensional with basis

$$\begin{aligned}
&\text{(a)} \quad P_1, P_2, M, E, \\
&\text{(b)} \quad P_1^2, P_1 P_2, M^2, \{M, P_1\}, \{M, P_2\}.
\end{aligned} \tag{4.12}$$

The first-order symmetries are listed in part (a) and the purely second-order symmetries in part (b). It follows that the Klein–Gordon equation is class I.

In analogy with expression (2.15) for the Helmholtz equation, $E(1,1)$ acts on the Lie algebra $\mathscr{E}(1,1)$ of symmetry operators via the *adjoint representation*

$$L \to L^g = \mathbf{T}(g) L \mathbf{T}(g^{-1}), \qquad L \in \mathscr{E}(1,1). \tag{4.13}$$

This action decomposes $\mathscr{E}(1,1)$ into orbits. To determine the orbit structure we first compute the adjoint action on the basis P_j, M. If $g_1 = \exp(aP_1)$, we have

$$P_1^{g_1} = P_1, \qquad P_2^{g_1} = P_2, \qquad M^{g_1} = M - aP_2;$$

if $g_2 = \exp(bP_2)$, then

$$P_1^{g_2} = P_1, \qquad P_2^{g_2} = P_2, \qquad M^{g_2} = M - bP_1; \tag{4.15}$$

and if $g_3 = \exp(\alpha M)$, then

$$P_1^{g_3} = \cosh(\alpha) P_1 + \sinh(\alpha) P_2,$$

$$P_2^{g_3} = \sinh(\alpha) P_1 + \cosh(\alpha) P_2, \qquad M^{g_3} = M. \tag{4.16}$$

Let $L = c_1 P_1 + c_2 P_2 + c_3 M \in \mathscr{E}(1,1)$, $L \neq 0$. A computation analogous to that carried out in Section 1.2 shows that if $c_3 \neq 0$, then L lies on the same orbit as M; if $c_3 = 0$, $c_1^2 > c_2^2$, L lies on the same orbit as P_1; if $c_3 = 0$, $c_1^2 < c_2^2$, L lies on the same orbit as P_2; if $c_3 = 0$, $c_1 = c_2$, L is on the orbit with $P_1 + P_2$; and if $c_3 = 0$, $c_1 = -c_2$, L is on the orbit with $P_1 - P_2$.

To simplify these results, note that the space and time inversion operators I_1, I_2,

$$I_1 \Phi(t, x) = \Phi(t, -x), \qquad I_2 \Phi(t, x) = \Phi(-t, x), \qquad \Phi \in \mathscr{F}, \tag{4.17}$$

map solutions of the Klein–Gordon equation into solutions and these symmetries cannot be written in the form (4.10); that is, the I_j cannot be obtained by exponentiation of the symmetry algebra but must be found by inspection. The I_j and the $\mathbf{T}(g)$, $g \in E(1,1)$, generate a larger symmetry

ISBN-0-201-13503-5

group $\tilde{E}(1,1)$, the *extended Poincaré group*. Here $\tilde{E}(1,1)$ is no longer connected and its connected component containing the identity is $E(1,1)$. The Lie symmetry algebra provides no information about the connected components bounded away from the identity.

Since $\tilde{E}(1,1)$ is a symmetry group, we do not wish to distinguish between coordinates that are related by an $\tilde{E}(1,1)$ transformation. Thus we need to compute the orbits in $\mathscr{E}(1,1)$ under the adjoint action of $\tilde{E}(1,1)$. This is easily accomplished. From (4.17) we find ($I_j^2 = E$)

$$P_1^{I_1} = I_1 P_1 I_1^{-1} = P_1, \qquad P_2^{I_1} = -P_2, \qquad M^{I_1} = -M, \qquad (4.18)$$

$$P_1^{I_2} = -P_1, \qquad P_2^{I_2} = P_2, \qquad M^{I_2} = -M. \qquad (4.19)$$

Thus, under the adjoint action of $\tilde{E}(1,1)$, $\mathscr{E}(1,1)$ decomposes into four orbits with orbit representatives

$$M, \quad P_1, \quad P_2, \quad P_1 + P_2. \qquad (4.20)$$

Since the last three operators commute pairwise, they can be simultaneously diagonalized.

Let $\mathbb{S}^{(2)}$ be the real five-dimensional vector space of purely second-order symmetries in $\mathbb{S}/\mathfrak{q}$. A basis for $\mathbb{S}^{(2)}$ is given by the operators (4.12b). Note that $P_1^2 = P_2^2$ on this space, since $P_1^2 - P_2^2$ corresponds to the zero operator. Again, $\tilde{E}(1,1)$ acts on $\mathbb{S}^{(2)}$ via the adjoint representation and decomposes this space into disjoint orbits. The classification of orbits was made by Kalnins in [55]. Here we present only the results. The following is a list of orbit representatives:

$$P_2^2, \ P_1 P_2, \quad (P_1+P_2)^2, \ M^2 \pm \alpha P_2^2, \quad M^2 - \alpha P_1 P_2, \ M^2 \pm \alpha(P_1+P_2)^2,$$
$$M^2, \{M,P_1\}, \{M,P_2\}, \quad \{M,P_1-P_2\}, \{M,P_1-P_2\} + \alpha(P_1+P_2)^2.$$

$$(4.21)$$

Here $\alpha > 0$, and there are 13 orbit types. When we relate these operators to separable coordinate systems we find that the parameter α plays a role exactly analogous to the parameter d for the elliptic coordinates $M^2 + d^2 P_1^2$ listed in Table 1. That is, α merely scales the coordinates, and coordinates differing only in their values of α are not essentially different. For that reason we henceforth set $\alpha = 1$.

In [55] Kalnins has derived all coordinate systems in which the Klein–Gordon equation separates and related these systems to the symmetry operators (4.21), exactly as we did in Section 1.2 for the Helmholtz equation. We present here an abbreviated form of his results.

ISBN-0-201-13503-5

System 1. P_2^2, P_1P_2, $(P_1+P_2)^2$

Kalnins finds three Cartesian-like coordinate systems corresponding to these three operators. However, we shall consider only the coordinate system in which all three operators are simultaneously diagonalized. This is the Cartesian system $\{t,x\}$ with separated solutions $\Phi(t,x)=\exp[i(\alpha t+\beta x)]$, $\alpha^2-\beta^2=\omega^2$.

System 2. $S_M=M^2$

This operator is related to polar coordinates

$$t = u\cosh v, \qquad x = u\sinh v, \qquad 0 \leqslant u < \infty, \qquad -\infty < v < \infty. \quad (4.22)$$

The separated equations are

$$\left(\frac{d^2}{du^2}+\frac{\tau^2}{u^2}+\omega^2\right)U=0, \qquad \left(\frac{d^2}{dv^2}+\tau^2\right)V=0, \qquad (4.23)$$

and the separable solutions take the form

$$\Phi_\tau = UV = \sqrt{u}\,J_\nu(\omega u)e^{i\tau v}, \qquad S_M\Phi_\tau = -\tau^2\Phi_\tau \qquad (4.24)$$

where $J_\nu(z)$ is a solution of Bessel's equation (B.16) with $\nu^2=\frac{1}{4}-\tau^2$. The parametrization (4.22) covers only that portion of the x–t plane given by $t\pm x>0$. Similar separable coordinates can be defined in the other three quadrants of the x–t plane.

System 3. $S_{D'}=\{M,P_2\}=u^2\partial_{vv}-v^2\partial_{uu}$

$$t=\tfrac{1}{2}(u^2+v^2), \qquad x=uv, \qquad -\infty<u<\infty, \qquad 0\leqslant v<\infty. \quad (4.25)$$

The separated equations are

$$\left(\frac{d^2}{du^2}+\omega^2u^2+\omega\lambda\right)U=0, \qquad \left(\frac{d^2}{dv^2}+\omega^2v^2+\omega\lambda\right)V=0, \qquad (4.26)$$

and the separable solutions take the form

$$\Phi_\lambda = UV = D_{(i\lambda-1)/2}\!\left(\varepsilon_1(1+i)\sqrt{\omega}\,u\right)D_{(i\lambda-1)/2}\!\left(\varepsilon_2(1+i)\sqrt{\omega}\,v\right) \quad (4.27)$$
$$S_{D'}\Phi_\lambda = -\omega\lambda\Phi_\lambda, \qquad \varepsilon_1,\varepsilon_2 = \pm1,$$

where $D_\nu(z)$ is a parabolic cylinder function (see (B.9iii)).

System 4. $S_D = \{M, P_1\} = u^2 \partial_{vv} - v^2 \partial_{uu}$

$$t = uv, \qquad x = \tfrac{1}{2}(u^2 + v^2), \qquad -\infty < u < \infty, \qquad 0 \leqslant v < \infty. \quad (4.28)$$

All equations are as in System 3 except that ω is replaced by $i\omega$ wherever it occurs. (Note that $\{M, P_1\}$ can be obtained from $\{M, P_2\}$ by making the interchange $x \leftrightarrow t$. Under this interchange the Klein–Gordon equation is mapped into a new equation of the same type with ω replaced by $i\omega$.)

System 5. $S_A = \{M, P_1 - P_2\} + (P_1 + P_2)^2 = (u + v)^{-1}(u \partial_{vv} - v \partial_{uu})$

The coordinates are

$$t = \tfrac{1}{2}(u - v)^2 + u + v, \qquad x = -\tfrac{1}{2}(u - v)^2 + u + v, \qquad -\infty < u, v < \infty,$$
$$(4.29)$$

and the separable solutions take the form

$$\Phi_\lambda = UV = \varphi_{\varepsilon_1}(u)\varphi_{\varepsilon_2}(v), \qquad \varepsilon_j = \pm 1,$$

$$S_A \Phi_\lambda = \lambda \Phi_\lambda, \qquad \varphi_\varepsilon(z) = \left(z + \frac{\lambda}{\omega}\right)^{1/2} J_{\varepsilon/3}\left[\frac{2\omega}{3}\left(z + \frac{\lambda}{\omega}\right)^{3/2}\right].$$

System 6. $S = M^2 - P_1 P_2 = 4(\sinh u - \sinh v)^{-1}(\sinh u \, \partial_{vv} - \sinh v \, \partial_{uu})$

The separable coordinates are

$$2t = \cosh\left[(u - v)/2\right] + \sinh\left[(u + v)/2\right],$$
$$2x = \cosh\left[(u - v)/2\right] - \sinh\left[(u + v)/2\right], \qquad -\infty < u, v < \infty, \quad (4.30)$$

and the separated equations become

$$\left(\frac{d^2}{dz^2} + \omega^2 \sinh z + \lambda\right)\varphi(z) = 0, \qquad z = u, v. \qquad (4.31)$$

This is a form of the modified Mathieu equation (3.40), so the separated solutions are products of Mathieu functions. Here $S\Phi_\lambda = 4\lambda\Phi_\lambda$.

System 7. $S_K = M^2 + (P_1 + P_2)^2 = (e^{2u} + e^{2v})^{-1}(e^{2v} \partial_{uu} + e^{2u} \partial_{vv})$

The separable coordinates are

$$t = \sinh(u - v) + \tfrac{1}{2}e^{u+v}, \qquad x = \sinh(u - v) - \tfrac{1}{2}e^{u+v}, \qquad -\infty < u, v < \infty,$$
$$(4.32)$$

ISBN-0-201-13503-5

and the separated equations become

$$\left(\frac{d^2}{du^2} + \omega^2 e^{2u} - \nu^2\right)U = 0, \qquad \left(\frac{d^2}{dv^2} - \omega^2 e^{2v} - \nu^2\right)V = 0. \qquad (4.33)$$

The separated solutions are

$$\Phi_{\nu^2} = J_{\pm\nu}(\omega e^u)J_{\pm\nu}(i\omega e^v), \qquad S_K\Phi_{\nu^2} = \nu^2\Phi_{\nu^2}. \qquad (4.34)$$

System 8. $S_B = M^2 - (P_1 + P_2)^2 = (e^{2v} - e^{2u})^{-1}(e^{2v}\partial_{uu} - e^{2u}\partial_{vv})$

$$t = \cosh(u-v) + \tfrac{1}{2}e^{u+v}, \quad x = \cosh(u-v) - \tfrac{1}{2}e^{u+v}, \qquad -\infty < u,v < \infty. \qquad (4.35)$$

These coordinates are very similar to the coordinates of type 7. The separated solutions take the form

$$\Phi_{\nu^2} = J_{\pm\nu}(\omega e^u)J_{\pm\nu}(\omega e^v), \qquad S_B\Phi_{\nu^2} = -\nu^2\Phi_{\nu^2}. \qquad (4.36)$$

As Kalnins points out, even by employing space and time inversion these coordinates cannot be made to cover the whole x–t plane.

System 9. $S = M^2 + P_2^2 = (\cosh^2 u + \sinh^2 v)^{-1}(\cosh^2 u\,\partial_{vv} + \sinh^2 v\,\partial_{uu})$

$$t = \sinh u \cosh v, \qquad x = \cosh u \sinh v, \qquad -\infty < u,v < \infty. \qquad (4.37)$$

The separated equations are

$$\left(\frac{d^2}{du^2} + \frac{\omega^2}{2}\cosh 2u + \lambda\right)U = 0, \qquad \left(\frac{d^2}{dv^2} - \frac{\omega^2}{2}\cosh 2v + \lambda\right)V = 0, \qquad (4.38)$$

variants of the modified Mathieu equation (3.40). The separated solutions are products of Mathieu functions and $S\Phi_\lambda = (\lambda - \omega^2/2)\Phi_\lambda$.

System 10. $L = M^2 - P_2^2 = (\sinh^2 a - \sinh^2 b)^{-1}(\sinh^2 a\,\partial_{bb} - \sinh^2 b\,\partial_{aa})$
$$= (\sin^2\alpha - \sin^2\beta)^{-1}(\sin^2\beta\,\partial_{\alpha\alpha} - \sin^2\alpha\,\partial_{\beta\beta})$$

There are two separable coordinate systems here, covering disjoint regions in the x–t plane.

(i) $t = \cosh a \cosh b$, $x = \sinh a \sinh b$, $\quad -\infty < a < \infty, 0 \leqslant b < \infty$.

$$(4.39)$$

The separated equations take the form

$$\left(\frac{d^2}{dz^2} + \frac{\omega^2}{2}\cosh 2z + \lambda - \frac{\omega^2}{2}\right)\varphi(z) = 0, \qquad z = a, b. \tag{4.40}$$

Comparing with (3.40), we see that the separated solutions are products of modified Mathieu functions. Here $L\Phi_\lambda = \lambda\Phi_\lambda$.

(ii) $t = \cos\alpha\cos\beta, \quad x = \sin\alpha\sin\beta, \qquad 0 < \alpha < 2\pi, \ 0 \leqslant \beta < \pi. \tag{4.41}$

The separated equations take the form

$$\left(\frac{d^2}{dz^2} - \frac{\omega^2}{2}\cos 2z + \lambda + \frac{\omega^2}{2}\right)\varphi(z) = 0, \qquad z = \alpha, \beta. \tag{4.42}$$

This is Mathieu's equation and the separated solutions are products of Mathieu functions. Here $L\Phi_\lambda = \lambda\Phi_\lambda$. Note that the coordinates (i) and (ii) do not cover the full x–t plane.

This completes the list of separable coordinate systems that correspond to orbits of second-order symmetry operators. As shown in [55], these systems are all orthogonal with respect to the Minkowski metric $ds^2 = dt^2 - dx^2$, that is, $ds^2 = g_{11}(u,v)\,du^2 + g_{22}(u,v)\,dv^2$ for every coordinate system $\{u,v\}$ (Systems 1–10). Furthermore, these are the only orthogonal separable coordinate systems for the Klein–Gordon equation.

There do exist nonorthogonal separable systems derived exactly as in Section 1.2 for the Helmholtz equation. Again we find that these systems always correspond to the diagonalization of a first-order symmetry operator and many distinct systems correspond to the same operator. We shall not consider these nonorthogonal systems further, but simply refer the reader to [55] for details.

Our results for the orthogonal systems are presented in Table 2.

Note that the simple relationship between separable coordinate systems and orbits of second-order symmetry operators that we established for the Helmholtz equation does not carry over in its entirety to the Klein–Gordon equation. First of all, for systems 5 and 10 the separable coordinates cannot be made to cover the full x–t plane. Second, we have shown that 12 of the 13 orbit types (4.21) in $\mathfrak{S}^{(2)}$ correspond to separable systems. However, this still leaves the orbit containing the operator $S_E = \{M, P_1 - P_2\}$, which we will show does *not* correspond to a separable system.

Thus the relationship between separable coordinate systems and orbits of symmetry operators is not one to one. In all cases that we know, a given separable system corresponds to a symmetry operator of order two or less. However, a given symmetry operator does not necessarily correspond to a

ISBN-0-201-13503-5

TABLE 2: Operators and Separable Coordinates for the Klein–Gordon Equation

Operators S	Coordinates	Separated solutions
1 $P_2^2, P_1P_2, (P_1+P_2)^2$	t, x	Product of exponentials
2 $S_M = M^2$	$t = u\cosh v,$ $x = u\sinh v$	Exponential times Bessel function
3 $\{M, P_2\}$	$t = \frac{1}{2}(u^2+v^2),$ $x = uv$	Product of parabolic cylinder functions
4 $S_D = \{M, P_1\}$	$t = uv,$ $x = \frac{1}{2}(u^2+v^2)$	Product of parabolic cylinder functions
5 $\{M, P_1 - P_2\}$ $+(P_1+P_2)^2$	$t = \frac{1}{2}(u-v)^2 + u + v,$ $x = -\frac{1}{2}(u-v)^2 + u + v$	Product of Airy functions
6 $M^2 - P_1P_2$	$t + x = \cosh[(u-v)/2],$ $t - x = \sinh[(u+v)/2]$	Product of Mathieu functions
7 $S_K = M^2 + (P_1+P_2)^2$	$t + x = 2\sinh(u-v),$ $t - x = e^{u+v}$	Product of Bessel functions
8 $S_B = M^2 - (P_1+P_2)^2$	$t + x = 2\cosh(u-v),$ $t - x = e^{u+v}$	Product of Bessel functions
9 $M^2 + P_2^2$	$t = \sinh u \cosh v,$ $x = \cosh u \sinh v$	Product of Mathieu functions
10 $M^2 - P_2^2$	$t = \cosh u \cosh v,$ $x = \sinh u \sinh v$ or $t = \cos u \cos v,$ $x = \sin u \sin v$	Product of Mathieu functions
11 $S_E = \{M, P_1 - P_2\}$		No separable solutions

separable system. An explanation of this fact is one of the principal problems remaining in the theory. We have distinguished those operators S in Table 2 that appear most frequently in the remainder of the book by a subscript (e.g., S_M).

1.5 Expansion Formulas for Solutions of the Klein–Gordon Equation

Our method for relating the various separable solutions of the Klein–Gordon equation is almost identical to the method applied to the Helmholtz equation in Section 1.3. Indeed, in analogy with (3.1) we consider solutions $\Phi(t, x)$ of equation (4.1) which can be represented in the form

$$\Phi(t, x) = \text{l.i.m.} \int_{-\infty}^{\infty} \exp\left[i\omega(t\cosh y + x\sinh y)\right] h(y)\, dy = I(h) \quad (5.1)$$

where h belongs to the Hilbert space $L_2(R)$ of functions square integrable with respect to Lebesgue measure on the real line: $\int_{-\infty}^{\infty} |h(y)|^2 dy < \infty$. The inner product on this space is

$$\langle h_1, h_2 \rangle = \int_{-\infty}^{\infty} h_1(y)\bar{h}_2(y)\,dy, \qquad h_j \in L_2(R). \tag{5.2}$$

(Note: Integral (5.1) may not converge pointwise for general $h \in L_2(R)$. We will explain the notation l.i.m. shortly.)

The action of $E(1,1)$ on solutions of the Klein–Gordon equation, defined by operators $\mathbf{T}(g)$, (4.7), (4.10), corresponds to the action

$$\mathbf{T}(g)h(y) = \exp[i\omega(a\cosh(y+\theta)+b\sinh(y+\theta))]h(y+\theta), \tag{5.3}$$

$$h \in L_2(R), \quad g(\theta,a,b) \in E(1,1),$$

of $E(1,1)$ on $L_2(R)$. That is, $\mathbf{T}(g)\Phi = I(\mathbf{T}(g)h)$ if $\Phi = I(h)$. The corresponding Lie algebra action on $L_2(R)$ is given by the operators

$$P_1 = i\omega\cosh y, \qquad P_2 = i\omega\sinh y, \qquad M = \partial_y, \tag{5.4}$$

which, of course, satisfy the commutation relations (4.3).

In distinction to the Helmholtz case, the functions Φ representable in the form (5.1) may not be true solutions of the Klein–Gordon equation since they may not have two continuous derivatives with respect to t and x. For example, if Φ is given by (5.1), then one expects $\partial_t\Phi = I(i\omega h\cosh y)$ by formal differentiation under the integral sign. However $h\cosh y$ may not be square integrable for a given square-integrable h, so the integral $I(i\omega h \cosh y)$ may not converge. Nevertheless, if we confine h to the dense subspace $\mathcal{D}$ consisting of all infinitely differentiable functions with compact support (i.e., which vanish outside some closed bounded interval on the real line) it is easy to check that the corresponding functions $\Phi = I(h)$, $h \in \mathcal{D}$, are true infinitely differentiable solutions of the Klein–Gordon equation. Also, the operators (5.4) are well-defined on $\mathcal{D}$.

Let $\mathcal{H}$ be the space of all $\Phi = I(h)$ with $h \in L_2(R)$. Here $\mathcal{H}$ is a Hilbert space with inner product

$$(\Phi_1, \Phi_2) \equiv \langle h_1, h_2 \rangle, \qquad \Phi_j = I(h_j). \tag{5.5}$$

(This definition makes sense because we can check that no nonzero $h \in L_2(R)$ is mapped by I to the zero solution of the Klein–Gordon equation.) We call the elements of $\mathcal{H}$ *weak* solutions of the Klein–Gordon equation. Since $\mathcal{D}$ is dense in $L_2(R)$, it follows easily that any weak solution Φ is the (Hilbert space) limit $\Phi_j \to \Phi$ of a sequence of true infinitely differentiable solutions.

In this case we can derive explicit integral expressions for the inner product on $\mathcal{H}$. Indeed for $\Phi_j = I(h_j)$ with $h_j \in \mathcal{D}$, we can easily verify the

ISBN-0-201-13503-5

expressions

$$(\Phi_1,\Phi_2)= i\int_{-\infty}^{\infty}\Phi_1\partial_t\overline{\Phi}_2\,dx = -i\int_{-\infty}^{\infty}(\partial_t\Phi_1)\overline{\Phi}_2\,dx \qquad (5.5')$$
$$\scriptstyle t=t_0 \qquad\qquad\qquad t=t_0$$

where the integrals are independent of the constant t_0. (These expressions are well known in quantum field theory (e.g., [118]). We have simply constructed the Hilbert space of positive energy solutions of the Klein–Gordon equation.) However, the integrals (5.5') make sense only when the functions Φ_j are differentiable with respect to t. Furthermore, we can make precise sense of (5.1) by regarding Φ as the limit in the Hilbert space $\mathcal{H}$ as $n\to\infty$ of the integrals over the finite intervals $[-n,n]$ (limit in the mean).

It is easy to check that the operators $T(g)$, (5.3), are unitary on $L_2(R)$ and define a unitary (even irreducible) representation of $E(1,1)$. The unitary transformation I, which maps $L_2(R)$ onto $\mathcal{H}$, maps the $T(g)$ to the operators $IT(g)I^{-1}$, (4.10), on $\mathcal{H}$. Finally, the Lie algebra operators (5.4) on the domain $\mathcal{D}$ are skew-Hermitian. This implies that the elements of $\mathcal{S}^{(2)}$ formed from the operators (5.4) and defined on $\mathcal{D}$ are symmetric. Thus we can associate each of the formal operators S listed in Table 2 with a symmetric operator on $\mathcal{D}$ and then attempt to extend the domain to obtain a self-adjoint operator whose spectral resolution is computable. We can finally map the results to $\mathcal{H}$ using the transformation I.

Note that expression (5.1) can be interpreted as an inner product:

$$\Phi(t,x)=I(h)=\langle h, H(\cdot,t,x)\rangle,$$

$$(5.6)$$

$$H(y,t,x)=\exp\left[-i\omega\left(\overline{t}\cosh y+\overline{x}\sinh y\right)\right].$$

As a function of y, H does not belong to $L_2(R)$ for real t, x, so this interpretation is not strictly correct. However, if we allow t and x to become complex such that $\mathrm{Im}(t\pm x)>0$, then $H(y,t,x)\in L_2(R)$ and the interpretation becomes valid. Moreover, $\Phi(t,x)$ satisfies $(\partial_{tt}-\partial_{xx}+\omega^2)\Phi(t,x)=0$ for complex t and x. All of the separable coordinate systems listed in Table 2 can be analytically continued into the region $\mathrm{Im}(t\pm x)>0$ and yield separation of variables for the complex Klein–Gordon equation. Furthermore, the integrals $I(h)$ where h belongs to one of the eigenbases corresponding to some operator S in Table 2 will converge absolutely due to the exponential damping factor contained in the function H. In the following list of eigenbases we shall confine ourselves to the case in which t, x are real, but the results are ordinarily obtained by assuming that $\mathrm{Im}(t\pm x)>0$ and then justifying the limit to the case of t, x real.

System 1. $S = P_2^2,\ P_1 P_2,\ (P_1 + P_2)^2$

To compute the simultaneous eigenfunctions of these commuting operators it is sufficient to obtain the simultaneous eigenfunctions of iP_1 and iP_2. A basis of generalized eigenfunctions for $L_2(R)$ is $\{f_\lambda^c(y)\}$:

$$f_\lambda^c(y) = \delta(y-\lambda) \quad (-\infty < \lambda < \infty), \qquad iP_1 f_\lambda^c = -\omega \cosh\lambda f_\lambda^c,$$
$$iP_2 f_\lambda^c = -\omega \sinh\lambda f_\lambda^c, \qquad \langle f_\lambda^c, f_{\lambda'}^c \rangle = \delta(\lambda - \lambda'). \tag{5.7}$$

Here, the superscript C stands for Cartesian. The corresponding basis of generalized eigenfunctions $\{\Phi_\lambda^c\}$ in $\mathcal{H}$ is

$$\Phi_\lambda^c(t,x) = I(f_\lambda^c) = \exp[i\omega(t\cosh\lambda + x\sinh\lambda)],$$
$$(\Phi_\lambda^c, \Phi_{\lambda'}^c) = \delta(\lambda - \lambda').$$

System 2. $S_M = M^2$

Here, it is sufficient to diagonalize the symmetric operator $iM = i(d/dy)$ on $L_2(R)$. The spectral resolution of this operator corresponds exactly to the theory of the Fourier transform on $L_2(R)$ (see [33, 112]). Here, we merely give the results. The symmetric operator iM on $\mathcal{D}$ has a unique self-adjoint extension, which we also call iM. A basis of generalized eigenfunctions for $L_2(R)$ is $\{f_\lambda^M\}$,

$$f_\lambda^M(y) = \frac{e^{i\lambda y}}{(2\pi)^{1/2}} \quad (-\infty < \lambda < \infty), \qquad iM f_\lambda^M = -\lambda f_\lambda^M,$$
$$\langle f_\lambda^M, f_\mu^M \rangle = \delta(\lambda - \mu). \tag{5.9}$$

Thus, any $h \in L_2(R)$ can be expanded uniquely in the form

$$h(y) = (2\pi)^{1/2} \int_{-\infty}^{\infty} \tilde{h}(\lambda) e^{i\lambda y}\, d\lambda \tag{5.10}$$

where

$$\tilde{h}(\lambda) = \langle h, f_\lambda^M \rangle = (2\pi)^{-1/2} \int_{-\infty}^{\infty} h(y) e^{-i\lambda y}\, dy, \tag{5.11}$$

$$\int_{-\infty}^{\infty} |\tilde{h}(\lambda)|^2\, d\lambda = \int_{-\infty}^{\infty} |h(y)|^2\, dy. \tag{5.12}$$

Note that the expansion coefficient $\tilde{h}(\lambda)$, (5.11), is just the *Fourier transform* of h, and (5.12) is Parseval's equality. It follows that iM has continuous spectrum covering the whole real axis.

ISBN-0-201-13503-5

The corresponding basis of generalized eigenfunctions $\{\Phi_\lambda^M\}$ in $\mathfrak{K}$ is

$$\Phi_\lambda^M(t,x)=(2\pi)^{-1/2}\int_{-\infty}^{\infty}\exp\left[i\omega(t\cosh y+x\sinh y)+i\lambda y\right]dy$$

$$=\frac{\sqrt{2}}{\pi}e^{-i\lambda v}K_{i\lambda}(i\omega u) \tag{5.13}$$

where

$$t=u\cosh v,\qquad x=u\sinh v$$

and

$$K_\nu(z)=\frac{\pi}{2\sin\nu\pi}\left(e^{i\nu\pi/2}J_{-\nu}(ze^{i\pi/2})-e^{-i\nu\pi/2}J_\nu(ze^{i\pi/2})\right) \tag{5.14}$$

is a *Macdonald function*. Note that our evaluation of (5.13) is correct only for $t\pm x>0$. For the other quadrants of the x–t plane we can find similar parametrizations.

System 3. $\{M,P_2\}$

On the domain $\mathfrak{D}$, $\{M,P_2\}=2i\omega\sinh y(d/dy)+i\omega\cosh y$. We can show that this operator has nonequal deficiency indices; hence, it cannot be extended to a self-adjoint operator on some dense subspace of $L_2(R)$. It is possible to make $\{M,P_2\}$ self-adjoint by extending the Hilbert space $L_2(R)$ itself, but this procedure is highly nonunique. Since orbit 4 also involves parabolic cylinder functions and is much simpler to study than this example, we shall say no more concerning orbit 3.

System 4. $S_D=\{M,P_1\}$

On $\mathfrak{D}$, $S_D=2i\omega\cosh y(d/dy)+i\omega\sinh y$. To determine the spectral resolution of S_D we shall subject $L_2(R)$ to a unitary transformation and change of variable such that the transformed $S_D=2i\omega(d/du)$ where u is the new variable. In this form the spectrum of S_D will be easy to compute.

The mapping $V:L_2(R)\to L_2(\mu,R)$, defined by

$$Vh(y)=(\cosh y)^{1/2}h(y),\qquad h\in L_2(R),$$

ISBN-0-201-13503-5

is a unitary transformation of $L_2(R)$ onto the Hilbert space $L_2(\mu,R)$ of functions on R, square integrable with respect to the measure $d\mu(y)=dy/\cosh y$ and inner product

$$\langle f_1,f_2\rangle'=\int_{-\infty}^{\infty}f_1(y)\bar{f_2}(y)d\mu(y).$$

Thus,

$$\langle h_1, h_2 \rangle = \langle V h_1, V h_2 \rangle',$$

and the symmetric operator $V S_D V^{-1}$ on $L_2(\mu, R)$, which we also call S_D, becomes $S_D = 2i\omega \cosh y(d/dy)$. Now we make the change of variable $e^y = \tan(u/2)$, $0 < u < \pi$. Then the inner product on $L_2(\mu, R) \equiv L_2[0, \pi]$ becomes

$$\langle f_1, f_2 \rangle' = \int_0^\pi f_1(u) \bar{f}_2(u) \, du, \qquad f_j(y) \equiv f_j(u),$$

and $S_D = 2i\omega(d/du)$. The subspace $\mathcal{D}$ of $L_2(R)$ maps onto the subspace $\mathcal{D}' \subset L_2[0, \pi]$ consisting of infinitely differentiable functions (in u) on the interval $[0, \pi]$ which vanish near the endpoints. The theory of self-adjoint extensions of S_D on $\mathcal{D}'$ is well known [33, 2]. There are an infinite number of extensions, designated by the parameter α, $0 \leqslant \alpha < 2$. For fixed α we extend $\mathcal{D}'$ to the space $\mathcal{D}'_\alpha$ of all continuous functions f on $[0, \pi]$ that are continuously differentiable on $(0, \pi)$ and such that $f(0) = e^{i\alpha\pi} f(\pi)$. It is easy to check that S_D is symmetric on $\mathcal{D}'_\alpha$ and has an ON basis of eigenfunctions

$$f_n(u) = \pi^{-1/2} \exp(-i\lambda u / 2\omega), \qquad \lambda = 2\omega(\alpha + 2n), \qquad n = 0, \pm 1, \pm 2, \ldots.$$

Mapping back to $L_2(R)$, we see that the operator S_D^α, α fixed, has eigenfunctions

$$f_n^{D,\alpha}(y) = (2/\pi)^{1/2} e^{y/2} (1 - ie^y)^{\alpha + 2n - \frac{1}{2}} (1 + ie^y)^{-\alpha - 2n - \frac{1}{2}},$$

$$n = 0, \pm 1, \ldots,$$

$$S_D^\alpha f_n^{D,\alpha} = \lambda f_n^{D,\alpha}, \qquad \langle f_n^{D,\alpha}, f_{n'}^{D,\alpha} \rangle = \delta_{nn'}. \tag{5.15}$$

Moreover, the $\{f_\lambda^{D,\alpha}\}$ form an ON basis for $L_2(R)$.
Eigenfunctions of S_D^α and $S_D^{\alpha'}$ are related by

$$f_m^{D,\alpha'} = \sum_{n=-\infty}^{\infty} \langle f_m^{D,\alpha'}, f_n^{D,\alpha} \rangle f_n^{D,\alpha},$$

$$\langle f_m^{D,\alpha'}, f_n^{D,\alpha} \rangle = \frac{e^{i\beta\pi} - 1}{i\beta\pi}, \qquad \beta = \alpha' - \alpha + 2(m - n),$$

as can be obtained from a simple computation in $L_2[0, \pi]$.

Although the basis functions $f_n^{D,\alpha}$ do not lie in the domains of P_1 or P_2, they do lie in the domain of M. An explicit computation yields

$$M f_n^{D,\alpha} = - \sum_{m=-\infty}^{\infty} \frac{2i}{\pi} \frac{\alpha + m + n}{4(n-m)^2 - 1} f_m^{D,\alpha}. \tag{5.16}$$

ISBN-0-201-13503-5

To compute the corresponding basis of generalized eigenfunctions $\{\Phi_n^{D,\alpha}\}$ in $\mathcal{H}$,

$$\Phi_n^{D,\alpha}(t,x) = I\left(f_n^{D,\alpha}\right),$$

we use the fact that variables separate in the integral if

$$t = uv, \qquad x = \tfrac{1}{2}(u^2 + v^2).$$

The result is

$$\Phi_n^{D,\alpha}(u,v) = 2\exp\left[\,3i\pi(\alpha+2n)/2\,\right] D_{-\alpha-2n-\frac{1}{2}}$$

$$\times\left((2\omega)^{1/2}u\right) D_{\alpha+2n-\frac{1}{2}}\left(i(2\omega)^{1/2}v\right). \qquad (5.17)$$

System 5. $\{M, P_1 - P_2\} + (P_1 + P_2)^2$

This operator on $\mathcal{D}$ has deficiency indices $(0,1)$, hence no self-adjoint extension in the usual sense. A self-adjoint extension (nonunique) obtained by augmenting the Hilbert space $L_2(R)$ to $L_2(R) \oplus L_2(R)$ can be found in [57].

Systems 6, 9, and 10

We will not treat the Mathieu bases here, although they are not particularly difficult. They occur rather infrequently in applications.

System 7. $S_K = M^2 + (P_1 + P_2)^2$

On $\mathcal{D}$, $S_K = d^2/dy^2 - \omega^2 e^{2y}$ and this operator has a unique self-adjoint extension (its closure). The spectral resolution of S_K is not elementary but it can be obtained from the known form of the Lebedev integral transform [125, p. 96]. There is a basis of generalized eigenfunctions

$$f_\lambda^K(y) = \pi^{-1}(2\lambda\sinh\lambda)^{1/2} K_{i\lambda}(\omega e^y), \qquad 0<\lambda<\infty, \qquad (5.18)$$
$$S_K f_\lambda^K = -\lambda^2 f_\lambda^K, \qquad \langle f_\lambda^K, f_{\lambda'}^K\rangle = \delta(\lambda-\lambda').$$

Here $K_\nu(z)$ is a Macdonald function (5.14). The corresponding basis of generalized eigenfunctions $\{\Phi_\lambda^K\}$ in $\mathcal{H}$ is given by

$$\Phi_\lambda^K(u,v) = 2\pi^{-1}(\lambda\sinh\lambda)^{1/2} K_{i\lambda}(\omega u) K_{i\lambda}(-i\omega v),$$
$$t = (u^2 - u^2v^2 - v^2)/2uv, \quad x = (u^2 + u^2v^2 - v^2)/2uv, \quad |u/v|>1, \qquad (5.19)$$

with similar results in other regions of the x–t plane.

ISBN-0-201-13503-5

System 8. $S_B = M^2 - (P_1 + P_2)^2$

On $\mathfrak{D}$, $S_B = d^2/dy^2 + \omega^2 e^{2y}$. This operator has a family of self-adjoint extensions $S_B^\alpha, 0 < \alpha \leqslant 2$ (see [125, pp. 93–95]). Here S_B^α has both discrete and continuous spectrum. The discrete eigenfunctions are

$$f_n^{B,\alpha}(y) = \left[2(\alpha+2n)\right]^{1/2} J_{\alpha+2n}(\omega e^y), \qquad n=0,1,2,\ldots,$$

$$\langle f_n^{B,\alpha}, f_m^{B,\alpha}\rangle = \delta_{n,m}, \qquad S_B^\alpha f_n^{B,\alpha} = (2n+\alpha)^2 f_n^{B,\alpha}. \tag{5.20}$$

Restricting ourselves to the case $\alpha=2$ for simplicity, we find that the operator S_B^2 also has continuous spectrum on the half-line $\lambda<0$ with generalized eigenfunctions

$$\tilde{f}_\lambda^B(y) = \tfrac{1}{2}\left[\sinh(\pi\sqrt{-\lambda})\right]^{-1/2}(J_{i\sqrt{-\lambda}}(\omega e^y) + J_{-i\sqrt{-\lambda}}(\omega e^y)),$$

$$\langle \tilde{f}_\lambda^B, \tilde{f}_{\lambda'}^B\rangle = \delta(\lambda-\lambda'), \qquad \langle f_n^{B,2}, \tilde{f}_\lambda^B\rangle = 0, \qquad S_B^2 \tilde{f}_\lambda^B = \lambda \tilde{f}_\lambda^B. \tag{5.21}$$

The functions $\{f_n^{B,2}, \tilde{f}_\lambda^B\}$ form a basis for $L_2(R)$.
The corresponding basis functions in $\mathfrak{H}$ are

$$\Phi_n^{B,\alpha} = 2\left[2(\alpha+2n)\right]^{1/2} J_{\alpha+2n}(\omega u) K_{\alpha+2n}(-i\omega v), \qquad |u/v|<1,$$

$$\tilde{\Phi}_\lambda^B = \left[\sinh(\pi\sqrt{-\lambda})\right]^{-1/2}(J_{i\sqrt{-\lambda}}(\omega u) + J_{-i\sqrt{-\lambda}}(\omega u)) K_{i\sqrt{-\lambda}}(-i\omega v),$$

$$t=(u^2+u^2v^2+v^2)/2uv, \qquad x=(u^2-u^2v^2+v^2)/2uv, \qquad v>u>0, \tag{5.22}$$

with similar expressions in other regions of the x–t plane.

System 11. $S_E = \{M, P_1 - P_2\}$

On $\mathfrak{D}$, $S_E = i\omega(2e^{-y}(d/dy) - e^{-y})$. Formal solutions of the equation $S_E f = \lambda f$ are $f(y) = ce^{y/2}\exp(-i\lambda e^y/2\omega)$. However, the deficiency indices of S_E are $(0,1)$ so S_E has no natural self-adjoint extensions. As shown in [57], one can obtain a (nonunique) self-adjoint operator from S_E by extending the Hilbert space to $L_2(R)\oplus L_2(R)$. The self-adjoint operator has continuous spectrum covering the real axis and a basis of generalized eigenfunctions for $L_2(R)\oplus L_2(R)$ whose restrictions to the original space $L_2(R)$ are

$$f_\lambda^E(y) = \frac{e^{y/2}}{(4\pi\omega)^{1/2}} \exp\left(\frac{-i\lambda e^y}{2\omega}\right), \qquad S_E f_\lambda^E = \lambda f_\lambda^E, \qquad -\infty<\lambda<\infty. \tag{5.23}$$

ISBN-0-201-13503-5

Any $h \in L_2(R)$ can be expanded uniquely in terms of the eigenbasis:

$$h(y) \sim \int_{-\infty}^{\infty} c(\lambda) f_\lambda^{E}(y) d\lambda, \qquad c(\lambda) = \langle h, f_\lambda^{E} \rangle. \qquad (5.24)$$

However, the $\{ f_\lambda^{E} \}$ do not satisfy orthogonality relations on $L_2(R)$.

Note: As shown in [57] the extensions of $L_2(R)$ to $L_2(R) \oplus L_2(R)$ which are used to obtain self-adjoint operators corresponding to orbits 5 and 11 are rather natural. Up to now we have restricted ourselves to the Hilbert space $L_2(R)$ corresponding to positive energy solutions of the Klein–Gordon equation. In relativistic quantum theory one frequently uses the Hilbert space $L_2(R) \oplus L_2(R)$, corresponding to both positive and negative energy solutions [118]. The larger space is invariant under the extended Poincaré group, including space and time inversion, but the smaller one is not. It is on this larger Hilbert space that the operators corresponding to all orbits can be made self-adjoint.

Passing to $\mathcal{H}$, we find

$$\Phi_\lambda^{E}(t, x) = I(f_\lambda^{E}) = \left[-2i\omega^2(t + x) - 2i\lambda \right]^{-1/2}$$

$$\times \exp\left\{ -\left[\omega^2(x^2 - t^2) - \lambda(x - t)/\omega \right]^{1/2} \right\} \qquad (5.25)$$

with similar functional expressions for other ranges of t, x, and λ. We can see by inspection that it is impossible to find coordinates u, v such that $\Phi_\lambda^{E}(t, x) = \sum_{j=1}^{4} U_\lambda^{(j)}(u) V_\lambda^{(j)}(v)$; that is, such that variables separate. The functions Φ_λ^{E} still form a basis for $\mathcal{H}$ consisting of eigenfunctions of S_E but these solutions are no longer separable.

Next we list some overlap functions, computed in the $L_2(R)$ model, which allow us to relate different bases. Most of the overlaps for the Klein–Gordon equation are rather complicated and we give here only the more tractable expressions. More details can be found in [57, 94].

The overlaps between an arbitrary basis $\{ f_\mu^{G} \}$ and the Cartesian basis $\{ f_\lambda^{c}(y) = \delta(y - \lambda) \}$ are especially easy to compute. Indeed,

$$\langle f_\mu^{G}, f_\lambda^{c} \rangle = f_\mu^{G}(\lambda) \qquad (5.26)$$

and the expansion of Φ_μ^{G} in the Cartesian basis is

$$\Phi_\mu^{G}(t, x) = I(f_\mu^{G}) = \int_{-\infty}^{\infty} f_\mu^{G}(\lambda) \Phi_\lambda^{c}(t, x) d\lambda.$$

ISBN-0-201-13503-5

The overlaps between the M and the (D,α) bases are

$$\langle f_n^{D,\alpha}, f_\lambda^M \rangle = \exp\left[\frac{\pi}{2}\left(\lambda+\frac{i}{2}\right)\right]\frac{\Gamma(\frac{1}{2}-\delta)}{\pi}\left[\frac{\Gamma(\frac{1}{2}-i\lambda)}{\Gamma(1-\delta-i\lambda)}\,{}_2F_1\left(\begin{array}{c}\frac{1}{2}-\delta,\frac{1}{2}-i\lambda\\1-\delta-i\lambda\end{array}\middle|-1\right)\right.$$

$$\left.+\exp\left[-i\pi\left(\delta+\tfrac{1}{2}\right)\right]\frac{\Gamma(\frac{1}{2}+i\lambda)}{\Gamma(1-\delta+i\lambda)}\,{}_2F_1\left(\begin{array}{c}\frac{1}{2}-\delta,\frac{1}{2}+i\lambda\\1-\delta+i\lambda\end{array}\middle|-1\right)\right],$$

$$\langle f_{-n}^{D,\alpha}, f_\lambda^M \rangle = \langle \overline{f_n^{D,\alpha}, f_{-\lambda}^M} \rangle, \qquad \delta=\alpha+2n<0. \tag{5.27}$$

We can expand products of Bessel functions as integrals of a single Bessel function through use of the overlaps

$$\langle f_n^{B,\alpha}, f_\lambda^M \rangle = \frac{1}{2}\left(\frac{\alpha+2n}{\pi}\right)^{1/2}\left(\frac{\omega}{2}\right)^{i\lambda}\frac{\Gamma((\alpha/2)+n-i\lambda/2)}{\Gamma(1+n+[\alpha+i\lambda]/2)}, \tag{5.28}$$

and as integrals over products of Macdonald functions through use of

$$\langle f_n^{B,\alpha}, f_\mu^K \rangle = \left[(\alpha+2n)\,\mu\sinh(\mu\pi)\right]^{1/2}\frac{\Gamma(n+[\alpha+i\mu]/2)\Gamma(n+[\alpha-i\mu]/2)}{2\pi\Gamma(1+\alpha+2n)}$$

$$\times{}_2F_1\left[\begin{array}{c}n+\dfrac{\alpha+i\mu}{2},n+\dfrac{\alpha-i\mu}{2}\\1+\alpha+2n\end{array}\middle|-1\right]. \tag{5.29}$$

Some matrix elements of the operators $T(g)$ with respect to the D basis are computed in [94], but they are complicated. In [128], however, Vilenkin has computed the matrix elements in the M basis and obtained relatively simple results.

The matrix elements in the M basis are

$$\langle T(g)f_\lambda^M, f_\mu^M \rangle = T_{\mu\lambda}(\theta,a,b)$$

$$=(2\pi)^{-1}\exp(i\mu\theta)\int_{-\infty}^{\infty}\exp\left[i\omega(a\cosh y+b\sinh y)\right.$$

$$\left.+iy(\lambda-\mu)\right]dy, \qquad g=g(\theta,a,b),\ -\infty<\lambda,\mu<\infty, \tag{5.30}$$

ISBN-0-201-13503-5

where g is given by (4.5). The addition theorem for these matrix elements reads

$$T_{\mu\lambda}(g_1 g_2) = \int_{-\infty}^{\infty} T_{\mu\nu}(g_1) T_{\nu\lambda}(g_2)\, dv. \qquad (5.31)$$

Special cases are

$$T_{\mu\lambda}(\theta,0,0) = e^{i\mu\theta}\delta(\mu-\lambda),$$

$$T_{\mu\lambda}(0,a,0) = (i/2)\exp\left[(\lambda-\mu)\pi/2\right] H^{(1)}_{i(\mu-\lambda)}(\omega a),$$

$$T_{\mu\lambda}(\theta,-a,0) = (-i/2)\exp\left[(\mu-\lambda)\pi/2\right] H^{(2)}_{i(\mu-\lambda)}(\omega a),$$

$$T_{\mu\lambda}(0,0,b) = \pi^{-1}\exp\left[(\mu-\lambda)\pi/2\right] K_{i(\mu-\lambda)}(\omega b),$$

$$T_{\mu\lambda}(0,0,-b) = \pi^{-1}\exp\left[(\lambda-\mu)\pi/2\right] K_{i(\lambda-\mu)}(\omega b),$$

$$T_{\mu\lambda}(0,a,a) = (2\pi)^{-1}\exp\left[(\mu-\lambda)\pi/2\right]\Gamma(i\lambda-i\mu)(\omega a)^{i(\mu-\lambda)},$$

$$T_{\mu\lambda}(0,-a,-a) = (2\pi)^{-1}\exp\left[(\lambda-\mu)\pi/2\right]\Gamma(i\lambda-i\mu)(\omega a)^{i(\mu-\lambda)},$$

$$T_{\mu\lambda}(0,-a,a) = (2\pi)^{-1}\exp\left[(\mu-\lambda)\pi/2\right]\Gamma(i\mu-i\lambda)(\omega a)^{i(\lambda-\mu)},$$

$$T_{\mu\lambda}(0,a,-a) = (2\pi)^{-1}\exp\left[(\lambda-\mu)\pi/2\right]\Gamma(i\mu-i\lambda)(\omega a)^{i(\lambda-\mu)}, \qquad (5.32)$$

$$a>0, \qquad b>0.$$

Here the $H_\nu^{(j)}(x)$ are Hankel functions, expressible in terms of Macdonald functions by

$$H_\nu^{(1)}(x) = \frac{2}{i\pi}\exp\left(\frac{-\nu\pi i}{2}\right) K_\nu\left[\exp\left(\frac{-\pi i}{2}\right)x\right],$$

$$H_\nu^{(2)}(x) = \frac{-2}{i\pi}\exp\left(\frac{\nu\pi i}{2}\right) K_\nu\left[\exp\left(\frac{\pi i}{2}\right)x\right]. \qquad (5.33)$$

Let us note that the addition theorem (5.31) has to be interpreted with great care because the integrals in (5.30) and (5.31) never converge absolutely. Indeed, some of the matrix elements are generalized functions. This problem plagued our whole treatment of the Klein–Gordon equation, since for most of the separable eigenbases the integral (5.1) did not converge absolutely. The origin of the difficulty was the fact that the kernel $H(y,t,x)$ of this integral transform, (5.6), was not an element of $L_2(R)$. To get around this problem we allowed t and x to take complex values such that $\mathrm{Im}(t\pm x)>0$. Then $H(\cdot,t,x)\in L_2(R)$ and the evaluation of the integrals was simplified. The resulting solutions of the Helmholtz equation $\Phi(t,x)=I(h)$ no longer belonged to $\mathcal{H}$ but all expansion theorems that we derived were valid for these solutions, with convergence in

ISBN-0-201-13503-5

the pointwise sense. Vilenkin [128] applies a similar idea to make sense of
(5.31). He allows ω to be a complex number such that $\operatorname{Im}\omega>0$. Then the
integral (5.30) for the matrix elements $T_{\mu\lambda}(\theta,a,b)$ converges absolutely for
all $g(\theta,a,b)$ such that $a\pm b>0$. Also, one can easily show that the product
$g_1 g_2=g'$ of two such group elements has the same property $a'\pm b'>0$. For
this set of group elements (5.31) can be proved rigorously and the matrix
elements can be calculated, with results similar to (5.32). Vilenkin obtains
the corresponding identities for real ω by a careful passage to the limit
from the $\operatorname{Im}\omega>0$ case rather than by trying to justify (5.31) directly.
(Similarly, in this section we have obtained eigensolutions $\Phi(t,x)$ of the
Klein–Gordon equation for real t,x by passing to the limit from the case
$\operatorname{Im}(t\pm x)>0$.) Nonetheless, (5.31) gives correct results.

1.6 The Complex Helmholtz Equation

Here we study the Helmholtz equation

$$\left(\partial_{xx}+\partial_{yy}+\omega^2\right)\Psi(x,y)=0 \qquad (6.1)$$

where now x and y are complex variables, ω is a nonzero complex
constant, and Ψ is analytic in x,y. The real Helmholtz and Klein–Gordon
equations can be considered as different real forms of the complex equation (6.1). If x,y, and ω are real, (6.1) becomes the real Helmholtz
equation, while if $x=t'$, $y=ix'$, and ω is real, then (6.1) becomes the
Klein–Gordon equation

$$\left(\partial_{t't'}-\partial_{x'x'}+\omega^2\right)\Psi(t',x')=0. \qquad (6.2)$$

The symmetry algebra of (6.1) is the *complex* Lie algebra $\mathcal{E}(2)^c$ with
basis

$$P_1=\partial_x, \qquad P_2=\partial_y, \qquad M=y\,\partial_x-x\,\partial_y \qquad (6.3)$$

and commutation relations

$$[P_1,P_2]=0, \qquad [M,P_1]=P_2, \qquad [M,P_2]=-P_1. \qquad (6.4)$$

Note that $\mathcal{E}(2)$, the symmetry algebra of the real Helmholtz equation, is
the *real* Lie algebra with basis (6.3). Furthermore, the real Lie algebra with
basis

$$\mathcal{P}_1=P_1, \qquad \mathcal{P}_2=iP_2, \qquad \mathcal{M}=iM \qquad (6.5)$$

and commutation relations

$$[\mathcal{P}_1,\mathcal{P}_2]=0, \qquad [\mathcal{M},\mathcal{P}_1]=\mathcal{P}_2, \qquad [\mathcal{M},\mathcal{P}_2]=\mathcal{P}_1 \qquad (6.6)$$

ISBN-0-201-13503-5

can be identified with $\mathcal{E}(1,1)$. We say that $\mathcal{E}(2)$ and $\mathcal{E}(1,1)$ are *real forms* of $\mathcal{E}(2)^c$.

The symmetry group of (6.1), obtained by exponentiating the symmetry algebra, is $E(2)^c$, the complex Euclidean group. Just as for $E(2)$ we can realize $E(2)^c$ as a group of 3×3 matrices:

$$g(\theta, a, b) = \begin{bmatrix} \cos\theta & -\sin\theta & 0 \\ \sin\theta & \cos\theta & 0 \\ a & b & 1 \end{bmatrix} \tag{6.7}$$

where now θ, a, b are complex numbers. The group product is given by expression (1.10) and the action $\mathbf{x}g$ of $E(2)^c$ as a transformation group by (1.11). Just as in Section 1.1 we have

$$g(\theta, a, b) = \exp(\theta M) \exp(a P_1 + b P_2), \tag{6.8}$$

and there is a local representation of $E(2)^c$ on the space $\mathcal{F}$ of complex analytic functions of x and y, defined by operators $\mathbf{T}(g)$.

$$\mathbf{T}(g)\Phi(\mathbf{x}) = \Phi(\mathbf{x}g), \qquad g \in E(2)^c. \tag{6.9}$$

The determination of the space $\mathcal{S}$ of second-order symmetry operators is almost identical to that presented in Section 1.1. Factoring out the space $\mathfrak{q}$ of trivial symmetries $f(\mathbf{x})(\partial_{xx} + \partial_{yy} + \omega^2)$, $f \in \mathcal{F}$, we find that $\mathcal{S}/\mathfrak{q}$ is a *complex* nine-dimensional vector space with basis

(a) $P_1, P_2, M, E,$

(b) $P_1^2, P_1 P_2, M^2, \{M, P_1\}, \{M, P_2\}.$ $\tag{6.10}$

(Recall that $E = 1$.) The first-order symmetries are listed in part (a) and the purely second-order symmetries in part (b). Again we see that our equation is class I: the second-order symmetries are polynomials in the first-order symmetries.

In strict analogy with expression (2.15), $E(2)^c$ acts on the algebra $\mathcal{E}(2)^c$ of symmetry operators via the adjoint representation

$$L \rightarrow L^g = \mathbf{T}(g)L\mathbf{T}(g^{-1}), \qquad L \in \mathcal{E}(2)^c, \qquad g \in E(2)^c,$$

and this action decomposes $\mathcal{E}(2)^c$ into orbits. The adjoint action on the basis P_1, P_2, M is given by (2.18)–(2.20) where now the parameters a, b, α are arbitrary complex numbers.

We shall find it useful to introduce two symmetries of the complex Helmholtz equation which cannot be obtained by exponentiation of the

ISBN-0-201-13503-5

symmetry algebra. These are the space inversion operators R_j.

$$R_1\Phi(x,y)=\Phi(-x,y), \qquad R_2\Phi(x,y)=\Phi(x,-y), \qquad \Phi\in\mathcal{F}. \quad (6.11)$$

The R_j and the $T(g)$, $g\in E(2)^c$, generate a larger group $\tilde{E}(2)^c$ of symmetries, whose connected component containing the identity is $E(2)^c$. The adjoint action of the R_j on $\mathcal{E}(2)^c$ is given by $(R_j^2=E)$:

$$P_1^{R_1}=R_1P_1R_1^{-1}=-P_1, \qquad P_2^{R_1}=P_2, \qquad M^{R_1}=-M, \qquad (6.12)$$

$$P_1^{R_2}=P_1, \qquad P_2^{R_2}=-P_2, \qquad M^{R_2}=-M. \qquad (6.13)$$

A computation analogous to those given in Sections 1.2 and 1.4 shows that under the adjoint action of $\tilde{E}(2)^c$, $\mathcal{E}(2)^c$ decomposes into three orbits with orbit representatives

$$M, P_1, P_1+iP_2. \qquad (6.14)$$

Since the last two operators commute, they can be simultaneously diagonalized.

Let $\mathcal{S}^{(2)}$ be the complex five-dimensional vector space of purely second-order symmetries in $\mathcal{S}/q$. The operators (6.10b) form a basis for this space. Note that $P_1^2=-P_2^2$ on $\mathcal{S}^{(2)}$ since $P_1^2+P_2^2$ corresponds to the zero operator. The group $\tilde{E}(2)^c$ acts on $\mathcal{S}^{(2)}$ via the adjoint representation and decomposes this space into disjoint orbits. A straightforward computation yields exactly eight orbit types, with orbit representatives

$$P_1^2, \quad (P_1+iP_2)^2, \quad M^2, \quad M^2-a^2P_2^2 \ (a\neq0), \quad \{M,P_2\},$$

$$(6.15)$$

$$M^2+(P_1+iP_2)^2, \quad \{M,P_1+iP_2\}+(P_1-iP_2)^2, \quad \{M,P_1+iP_2\}.$$

The problem of separation of variables for the complex Helmholtz equation can be formulated in a manner analogous to the problem for the real Helmholtz equation in Section 1.2. Now, however, a new set of coordinates $\{u,v\}$ must range over an open set in the space $\mathbb{C}\times\mathbb{C}$ of ordered pairs of complex numbers and $u(\mathbf{x})$, $v(\mathbf{x})$ must be complex analytic functions of x and y, with nonzero Jacobian. We regard two separable systems $\{u,v\}$, $\{u',v'\}$ as equivalent if one system can be mapped onto the other by an element of $\tilde{E}(2)^c$.

We have seen that the real Helmholtz and Klein–Gordon equations are real forms of the complex Helmholtz equation. Furthermore, we can easily

ISBN-0-201-13503-5

check that each of the real separable coordinate systems for these real
equations can be uniquely extended to complex analytic separable coordinates for the complex equation. Thus we already have a considerable
number of separable coordinate systems for (6.1). The corresponding
separated solutions can all be uniquely extended to analytic solutions of
(6.1) and each of these separated solutions is an eigenfunction of an
operator in $S^{(2)}$. (To determine the operator that is diagonalized we use
(6.3), (6.4) to translate the contents of Table 1 to $S^{(2)}$ and we use (6.5), (6.6)
to translate the contents of Table 2.) Finally, a tedious computation shows
that there are no nontrivial orthogonal separable systems other than those
we have already found.

However, the relation between the separable coordinate systems in
Table 1 and 2 and the distinct separable systems for (6.1) is not one to one.
This is because systems that are real inequivalent under $E(2)$ and $\tilde{E}(1,1)$
may be complex equivalent under the larger group $\tilde{E}(2)^c$. Put another way,
several distinct orbits of $S^{(2)}$ with respect to $E(2)$ or $\tilde{E}(1,1)$ may belong to
the same $\tilde{E}(2)^c$ orbit.

The relationships between $\tilde{E}(2)^c$ orbits of operators in $S^{(2)}$ and complex
separable coordinates are presented in Table 3.

Table 3. Operators and Coordinates for the Complex Helmholtz Equation

Operator S	Complex coordinates	Separated solutions
1 $P_1^2, (P_1 + iP_2)^2$	x, y	Product of exponentials
2 M^2	$x = r\cos\theta,$ $y = r\sin\theta$	Exponential times Bessel function
3 $M^2 - a^2P_2^2,$ $a \neq 0$	$x = a\cosh\alpha\cos\beta,$ $y = a\sinh\alpha\sin\beta$	Product of Mathieu functions
4 $\{M, P_2\}$	$x = \frac{1}{2}(\xi^2 - \eta^2),$ $y = \xi\eta$	Product of parabolic cylinder functions
5 $M^2 + (P_1 + iP_2)^2$	$x = (u^2 + u^2v^2 - v^2)/2uv,$ $y = i(u^2 - u^2v^2 + v^2)/2uv$	Product of Bessel functions
6 $\{M, P_1 + iP_2\}$ $+ (P_1 - iP_2)^2$	$x = -\frac{1}{4}(w - z)^2 + \frac{1}{2}(w + z),$ $iy = -\frac{1}{4}(w - z)^2 - \frac{1}{2}(w + z)$	Product of Airy functions
7 $\{M, P_1 + iP_2\}$		No separable solutions

The commuting operators P_1^2, $(P_1 + iP_2)^2$ belong to distinct orbits but
correspond to the same coordinates. Table 4 shows which distinct separable solutions of the real Helmholtz and Klein–Gordon equations correspond to equivalent solutions of the complex Helmholtz equation. Note
that the three Mathieu function systems for the Klein–Gordon equation
and the Mathieu function system for the real Helmholtz equation are all
equivalent under $E(2)^c$ and correspond to the single system 3. There are
similar but less dramatic equivalences for the other coordinate systems.

ISBN-0-201-13503-5

Table 4. Equivalence Classes of Separable Solutions

Operator S	Equivalent real Helmholtz solutions (Table 1)	Equivalent real Klein–Gordon solutions (Table 2)
1 $P_1^2, (P_1+iP_2)^2$	1	1
2 M^2	2	2
3 $M^2-a^2P_2^2$	4	6, 9, 10
4 $\{M,P_2\}$	3	3, 4
5 $M^2+(P_1+iP_2)^2$		7, 8
6 $\{M,P_1+iP_2\}$ $+(P_1-iP_2)^2$		5
7 $\{M,P_1+iP_2\}$		11

1.7 Weisner's Method for the Complex Helmholtz Equation

Although the solution space of (6.1) does not appear to admit a Hilbert space structure, one can employ a method due to Louis Weisner to derive identities for the separated solutions [82, 134]. Weisner's method focuses on the solutions of orbit 2 with representative M^2. On introducing complex polar coordinates $x=r\cos\varphi$, $y=r\sin\varphi$ in (6.1), we obtain the equation

$$\left(\partial_{rr}+r^{-1}\partial_r+r^{-2}\partial_{\varphi\varphi}+\omega^2\right)\Psi=0. \tag{7.1}$$

Clearly, this equation admits separable solutions of the form

$$J_{\pm m}(\omega r)s^m, \qquad s=e^{i\varphi}, \qquad m\in\mathbb{C}, \tag{7.2}$$

where $J_\nu(\omega r)$ is a Bessel function. Furthermore, a function $\Psi=f(r)s^p$ is a solution of (7.1) if and only if f satisfies Bessel's equation

$$\left(\frac{d^2}{dr^2}+r^{-1}\frac{d}{dr}-r^{-2}p^2+\omega^2\right)f=0. \tag{7.3}$$

Now suppose Ψ is a solution of (7.1) such that for some fixed $m\in\mathbb{C}$, $(rs)^{-m}\Psi(r,s)$ is analytic in the complex variables (r,s) in a neighborhood of $(0,0)$. Expanding in a power series in s,

$$\Psi(r,s)=\sum_{n=0}^{\infty} f_n(r)s^{m+n},$$

we find that inside the domain of convergence of the series, $f_n(r)$ is a solution of Bessel's equation (7.3) with $p=m+n$. Moreover, since $r^{-m}f_n(r)$ is analytic at $r=0$, we see from Appendix B, Section 5, that

$$f_n(r)=c_n J_{m+n}(\omega r), \qquad c_n\in\mathbb{C}.$$

ISBN-0-201-13503-5

Thus,

$$\Psi(r,s) = \sum_{n=0}^{\infty} c_n J_{m+n}(\omega r) s^{m+n} \qquad (7.4)$$

where the power series converges in a neighborhood of $(r,s)=(0,0)$; that is, Ψ is a *generating function* for Bessel functions. If Ψ is known, one can ordinarily compute the constants c_n by evaluating the right-hand side of (7.4) for special choices of r or by using operator identities satisfied by Ψ to obtain recurrence relations for the c_n.

Our observation has an interesting converse. Suppose we are given a convergent power series in the form of the right-hand side of (7.4). Then the sum of this series must be a solution of the complex Helmholtz equation (7.1). Thus the solutions of (7.1) (with appropriate analyticity properties) are just the possible generating functions (7.4) for Bessel functions. (These observations can easily be extended to generating functions that are of the more general form of Laurent expansions.)

In order to make this procedure effective we need an efficient method for obtaining explicit analytic solutions of (7.1). From our point of view an obvious answer is to take the separated solutions corresponding to each of the six systems listed in Table 3. Indeed, if Φ is a separated solution corresponding to the symmetry operator S, $S\Phi=\lambda\Phi$, and $g \in E(2)^c$, then the separated solution $\Psi=\mathbf{T}(g)\Phi$ corresponds to the operator $S' = \mathbf{T}(g)S\mathbf{T}(g^{-1})$ on the same orbit: $S'\Psi=\lambda\Psi$. Now Ψ can be substituted into (7.4) to yield a generating function for the $J_\nu(\omega r)$. Although Weisner nowhere states explicitly the relationship between symmetry operators and separation of variables, in his paper [134] he gives operator characterizations of separated solutions from all orbits except 3 and 6 to construct generating functions for Bessel functions.

Before plunging into this theory it is useful to compute the action of the operators $\mathbf{T}(g)$ on functions expressed in terms of polar coordinates. We switch to the complex coordinates r,s where

$$x = r\cos\varphi = \left(\frac{r}{2}\right)(s+s^{-1}),$$

$$y = r\sin\varphi = \left(\frac{r}{2i}\right)(s-s^{-1}), \qquad s=e^{i\varphi}. \qquad (7.5)$$

Furthermore, we choose a new basis $\{P^\pm, P^0\}$ for $\mathscr{E}(2)^c$ defined in terms of the old basis by

$$P^+ = P_1 + iP_2, \qquad P^- = -P_1 + iP_2, \qquad P^0 = iM. \qquad (7.6)$$

The basis commutation relations are

$$[P^+, P^-] = 0, \qquad [P^0, P^\pm] = \pm P^\pm. \qquad (7.7)$$

ISBN-0-201-13503-5

From (6.3), (7.5), and (7.6) we obtain

$$P^{\pm} = s^{\pm 1}\left(\pm \partial_r - \frac{s}{r}\partial_s\right), \qquad P^0 = s\partial_s = -i\partial_\varphi. \qquad (7.8)$$

Along with our new basis for $\mathcal{E}(2)^c$ we choose a new parametrization of the group $E(2)^c$:

$$g = \exp(\theta M)\exp(aP_1 + bP_2) = \exp(\tau P^0)\exp(\alpha P^+ + \beta P^-), \qquad (7.9)$$

$$\theta = i\tau, a = \alpha - \beta, b = i(\alpha + \beta).$$

In terms of the coordinates $[\tau, \alpha, \beta]$, group multiplication becomes

$$g[\tau, \alpha, \beta]\, g[\tau', \alpha', \beta'] = g[\tau + \tau', \alpha e^{-\tau'} + \alpha', \beta e^{\tau'} + \beta'], \qquad (7.10)$$

and the operators $\mathbf{T}(g)$ take the form

$$\mathbf{T}(g)\Psi(r,s) = \Psi\Big(r[1 + 2\alpha e^\tau s/r]^{1/2}[1 - 2\beta e^{-\tau}/rs]^{1/2},$$

$$se^\tau[1 - 2\beta e^{-\tau}/rs]^{1/2}[1 + 2\alpha e^\tau s/r]^{-1/2}\Big), \qquad (7.11)$$

well defined for $|2\beta e^{-\tau}/rs| < 1$, $|2\alpha e^\tau s/r| < 1$.

The value of the constant ω is of little importance in the following discussion, so we will henceforth set $\omega = 1$.

As follows from (7.2), the eigenfunctions of P^0 for a fixed $m \in \mathbb{C}$, which are also solutions of (7.1), take the form $J_{\pm m}(r)s^m$. To be definite, we choose the eigenfunction

$$\Psi_m(r,s) = J_m(r)s^m, \qquad P^0\Psi_m = m\Psi_m. \qquad (7.12)$$

The commutation relation $[P^0, P^+] = P^+$ implies $[P^0, P^+]\Psi_m = P^+\Psi_m$ or

$$P^0(P^+\Psi_m) = (m+1)P^+\Psi_m. \qquad (7.13)$$

Since P^+ is a symmetry operator, we see that $P^+\Psi_m$ is either zero or a solution of (7.1) that is an eigenfunction of P^0 with eigenvalue $m+1$. Thus $P^+\Psi_m$ must be a linear combination of the functions (7.2) with m replaced by $m+1$. From (B.17) we find explicitly

$$P^+\Psi_m(r,s) = -\Psi_{m+1}(r,s) = -J_{m+1}(r)s^{m+1}.$$

(This result can easily be checked by differentiating the power series expression (B.14) for $J_m(r)$ term by term.) Similarly, the commutation relation $[P^0, P^-] = -P^-$ implies

$$P^0(P^-\Psi_m) = (m-1)P^-\Psi_m, \qquad (7.14)$$

ISBN-0-201-13503-5

so that P^- lowers the eigenvalue m by one. Here, (B.17) yields

$$P^-\Psi_m(r,s) = -\Psi_{m-1}(r,s) = -J_{m-1}(r)s^{m-1}.$$

Now let m_0 be a complex number such that $0 \leqslant \mathrm{Re}\, m_0 < 1$ and consider the set $\{\Psi_m\}$ of all eigenvectors (7.12) such that $m = m_0 + n$, $n = 0,\ \pm 1,\ \pm 2,\ldots$. The action of $\mathcal{E}(2)^c$ on this set is

$$P^0\Psi_m = m\Psi_m, \qquad P^\pm\Psi_m = -\Psi_{m\pm 1}, \qquad m = m_0 + n. \qquad (7.15)$$

These relations define a representation of $\mathcal{E}(2)^c$ in the sense defined in Chapter 2 of [82]. Moreover, they show the intimate relationship between the recurrence relations (B.17) for Bessel functions and the representation theory of $\mathcal{E}(2)^c$. Note also that the order m of the Bessel function $J_m(r)$ is now an arbitrary complex number rather than just an integer as in Section 1.3.

As shown in [82] the Lie algebra representation (7.15) can be extended to a *local* group representation of $E(2)^c$, local in the sense that the group operators $\mathbf{T}(g)$ and the representation property $\mathbf{T}(gg') = \mathbf{T}(g)\mathbf{T}(g')$ are well defined and valid only for group elements $g,\ g'$ in a sufficiently small neighborhood of the identity element.

On one hand, the action of the operators $\mathbf{T}(g)$ on the basis functions Ψ_m can be calculated from expression (7.11) for α, β, τ sufficiently close to 0. On the other hand, we can write

$$\mathbf{T}(g) = \exp(\tau P^0)\exp(\alpha P^+ + \beta P^-) \qquad (7.16)$$

where $P^0, P^\pm$ are given by (7.8), and use expressions (1.17), (7.15) to expand $\mathbf{T}(g)\Psi_m$ as an infinite series in $\{\Psi_{m+n}\}$. The coefficients in the expansion are completely determined by the Lie algebra relations (7.15). For example,

$$\mathbf{T}(g(\tau,0,0))\Psi_m(r,s) = \Psi_m(r,e^\tau s) = \exp(\tau P^0)\Psi_m(r,s) = e^{\tau m}\Psi_m(r,s).$$

Although we could use relations (7.15) directly to expand $\mathbf{T}(g)\Psi_m$ in terms of our eigenbasis, it is more convenient to find a simpler model of the relations (7.15). Such a model was constructed in [82, Chapter 3], analogous to our $L_2(S_1)$ model for the solution space of the Helmholtz equation. The Lie algebra generators are Lie derivatives in one complex variable z,

$$P^+ = -z, \qquad P^- = -z^{-1}, \qquad P^0 = z(d/dz), \qquad (7.17)$$

and the basis functions $f_m(z) = z^m$, $m = m_0 + n$, satisfy

$$P^0 f_m = mf_m, \qquad P^\pm f_m = -f_{m\pm 1}. \qquad (7.18)$$

ISBN-0-201-13503-5

Clearly the operators (7.17) satisfy the commutation relations (7.7). The operators $\mathbf{T}(g)$, (7.16) with $P^0, P^{\pm}$ replaced by (7.17), act on the space $\mathcal{F}_0$ of all functions $f(z)$ analytic in a deleted neighborhood of $z=0$, that is, not necessarily analytic at the point $z=0$ and not necessarily single valued as one loops the origin. The operators are easily computed:

$$\mathbf{T}(g)f(z)=\exp(-\alpha e^{\tau}z-\beta e^{-\tau}/z)f(e^{\tau}z), \qquad f\in\mathcal{F}_0. \tag{7.19}$$

We define the *matrix elements* $T_{lj}(g)$ of $\mathbf{T}(g)$ with respect to the basis $\{f_m, m=m_0+n\}$ by

$$\mathbf{T}(g)f_{m_0+j}(z)=\sum_{l=-\infty}^{\infty}T_{lj}(g)f_{m_0+l}(z) \tag{7.20}$$

or (factoring out z^{m_0} from both sides of this expression)

$$\exp[-\alpha e^{\tau}z-\beta e^{-\tau}/z+(m_0+j)\tau]z^j=\sum_{l=-\infty}^{\infty}T_{lj}(g)z^l, \tag{7.21}$$

$$g=g[\tau,\alpha,\beta], j=0, \pm 1, \pm 2,\dots.$$

Note that we have derived a generating function for the matrix elements. Explicitly computing the coefficient of z^l in the Laurent series expansion for the left-hand side of (7.21), we obtain

$$T_{lj}(g)=\frac{\exp[(m_0+l)\tau]}{|j-l|!}(-1)^{j-l}\alpha^{(l-j+|j-l|)/2}\beta^{(j-l+|j-l|)/2}$$
$$\times {}_0F_1(|j-l|+1;\alpha\beta), \qquad l,j=0, \pm 1, \pm 2,\dots. \tag{7.22}$$

For $\alpha\beta=0$ the ${}_0F_1$ becomes $+1$ and the matrix elements become elementary functions. However, for $\alpha\beta\neq 0$ the matrix elements are closely related to Bessel functions of integral order. Indeed, introducing new parameters ρ,ω defined by

$$\rho=2|\alpha\beta|^{1/2}\exp[i(\arg\alpha+\arg\beta+\pi)/2],$$

$$\omega=|\alpha/\beta|^{1/2}\exp[i(\arg\alpha-\arg\beta-\pi)/2], \tag{7.23}$$

$$-\pi<\arg\alpha, \quad \arg\beta\leqslant\pi, \quad \alpha=\rho\omega/2, \quad \beta=-\rho/2\omega,$$

we find

$$T_{lj}(g)=\exp[(m_0+j)\tau](-\omega)^{l-j}J_{l-j}(\rho). \tag{7.24}$$

(Note that (7.21) is a generalization of the expansion (3.20). From the

ISBN-0-201-13503-5

group representation property we immediately obtain the identities

$$T_{lj}(gg') = \sum_{s=-\infty}^{\infty} T_{ls}(g)T_{sj}(g'), \qquad (7.25)$$

which are guaranteed to hold for g, g' sufficiently close to the identity element. Moreover, it is shown in [82, Chapter 3] that these identities actually hold for *all* complex values of the six parameters, where group multiplication is defined by (7.10). (Here it is necessary to make the proviso that if $m_0 \neq 0$, we can no longer identify group elements that differ only by an integral multiple of $2\pi i$ in the τ parameter, as was the case with $E(2)^c$. Relations (7.25) are associated with global representations of the universal covering group of $E(2)^c$ rather than with $E(2)^c$ itself.) Expressions (7.25) constitute a generalization of the identities (3.44), (3.54) with $j = 2$. A number of examples of these identities can be found in [82].

The matrix elements (7.22) are uniquely determined by the Lie algebra relations (7.18) for g sufficiently close to the identity element, so they must also be valid for our Helmholtz equation model (7.8), (7.11), (7.12), (7.15). Thus,

$$\mathbf{T}(g)\Psi_{m_0+j}(r,s) = \sum_{l=-\infty}^{\infty} T_{lj}(g)\Psi_{m_0+l}(r,s) \qquad (7.26)$$

or

$$J_m(r(1+2\alpha s/r)^{1/2}(1-2\beta/sr)^{1/2}) \left(\frac{1-2\beta/sr}{1+2\alpha s/r} \right)^{m/2}$$

$$= \sum_{n=-\infty}^{\infty} \frac{(-1)^n}{|n|!} \beta^{(-n+|n|)/2} \alpha^{(n+|n|)/2} {}_0F_1(|n|+1; \alpha\beta)J_{m+n}(r)s^n, \qquad (7.27)$$

$$m \in \mathcal{C}, \quad |2\alpha s/r| < 1, \quad |2\beta/rs| < 1.$$

The domain of validity follows from the fact that the right-hand side is a Laurent expansion in s. Such an expansion converges in an annulus $\rho_1 < |s| < \rho_2$ where ρ_1 and ρ_2 are determined by the singularities of the function on the left-hand side of (7.27) [1].

Some special cases of this identity are of interest. If $\beta = 0$, $s = 1$, equation (7.27) becomes

$$J_m\left[r(1+2\alpha/r)^{1/2} \right](1+2\alpha/r)^{-m/2} = \sum_{n=0}^{\infty} \frac{(-\alpha)^n}{n!} J_{m+n}(r), \qquad (7.28)$$

$$|2\alpha/r| < 1,$$

while if $\alpha = 0$, $s = 1$, it becomes

$$J_m\left[r(1+2\beta/r)^{1/2} \right](1+2\beta/r)^{m/2} = \sum_{n=0}^{\infty} \frac{\beta^n}{n!} J_{m-n}(r), \qquad (7.29)$$

$$|2\beta/r| < 1.$$

ISBN-0-201-13503-5

Note that these formulas are expansions of $\exp(\gamma P^{\pm})\Psi_m$ in terms of the basis $\{\Psi_{m_0+n}\}$. The formulas do not follow directly from a study of the real group $E(2)$ because the $P^{\pm}$ do not belong to the real Lie algebra $\mathscr{E}(2)$.

For $\alpha\beta\neq0$, (7.27) reduces to Graf's addition theorem [37, p. 44],

$$J_m\left[r(1+\rho\omega/r)^{1/2}(1+\rho/\omega r)^{1/2}\right]\left(\frac{1+\rho/\omega r}{1+\rho\omega/r}\right)^{m/2}$$

$$= \sum_{n=-\infty}^{\infty}(-\omega)^n J_n(\rho)J_{m+n}(r), \qquad |\rho\omega/r|<1, \qquad |\rho/\omega r|<1. \quad (7.30)$$

Let us note that expressions (7.26) are special cases of Weisner's equation (7.4) where $\Psi(r,s)=\mathbf{T}(g)\Psi_m(r,s)$ and g is in a suitably small neighborhood of the identity element. In these special cases we were able to use the model (7.17) to directly compute the expansion coefficients c_n.

In general, however, we need the full power of Weisner's method to compute the expansion coefficients. For example, consider $\mathbf{T}(g)\Psi_m$ where Ψ_m is the eigenfunction (7.12) and $g=g[0,0,-1]$. Then we can write

$$\mathbf{T}(g)\Psi_m=\exp(-P^-)\Psi_m=(r^2+2r/s)^{-m/2}J_m\left[(r^2+2r/s)^{1/2}\right](2+rs)^m$$

$$=\Phi(r,s). \qquad (7.31)$$

Since $z^{-m}J_m(z)$ is an entire function of z, Φ has a Laurent expansion in s about $s=0$:

$$\Phi(r,s)= \sum_{n=-\infty}^{\infty}c_nJ_n(r)s^n, \qquad |rs|<2. \qquad (7.32)$$

To determine the expansion coefficients we first set $r=0$ in (7.32) to obtain $c_0=1/\Gamma(m+1)$. Furthermore, since $P^0\Psi_m=m\Psi_m$, we have $L\Phi=m\Phi$ where $L=\exp(-P^-)P^0\exp(P^-)=P^0-P^-$. Thus $(P^0-P^-)\Phi=m\Phi$, and applying L term by term to the right-hand side of (7.32) we obtain $c_{n+1}=(m-n)c_n$ for all n. We conclude that $c_n=1/\Gamma(m-n+1)$ and

$$(r^2+\frac{2r}{s})^{-m/2}J_m[(r^2+\frac{2r}{s})^{1/2}](2+rs)^m = \sum_{n=-\infty}^{\infty}\frac{J_n(r)s^n}{\Gamma(m-n+1)}, \qquad (7.33)$$

$$|rs|<2.$$

This identity does not follow from our computation of the matrix elements (7.22) because here we have fixed g and allowed r and s to become small. If we had chosen r sufficiently large, we would have obtained (7.29).

To derive some more complicated formulas we will now make use of Table 3 and characterize the solution Ψ, (7.4), of the complex Helmholtz

ISBN-0-201-13503-5

equation by requiring that it be an eigenfunction of an operator $S \in \mathcal{S}^{(2)}$ corresponding to one of the separable orbits listed on Table 3. We present two of the examples contained in Weisner's paper [134].

Consider the case where Ψ is a solution of the complex Helmholtz equation satisfying

$$\{ M, P_1 \}\Psi = i(4\lambda - 1)\Psi, \qquad \lambda \in \mathcal{C}. \tag{7.34}$$

(The eigenvalue is written as $i(4\lambda - 1)$ for convenience in the formulas to follow.) Since $\{M, P_1\}$ lies on orbit 4 (Table 3) and corresponds to 4 in Table 2, it follows that Ψ must be a separated solution in parabolic coordinates $\{u, v\}$:

$$x = uv, \qquad iy = \tfrac{1}{2}(u^2 + v^2).$$

Setting $\xi = -u^2$, $\eta = -v^2$, and relating the $\{\xi, \eta\}$ coordinates to the $\{r, s\}$ coordinates, we find

$$\xi + \eta = -r(s - s^{-1}), \qquad \xi - \eta = 2ir, \tag{7.35}$$

and Ψ separates in the $\{\xi, \eta\}$ system. The solutions of the separated equations are parabolic cylinder functions. However, it is more convenient to choose a basis for the separated solutions in the form

$$e^{-z/2} {}_1F_1 \left(\begin{matrix} \lambda \\ \tfrac{1}{2} \end{matrix} \middle| z \right), \qquad e^{-z/2} z^{1/2} {}_1F_1 \left(\begin{matrix} \lambda + 1/2 \\ \tfrac{3}{2} \end{matrix} \middle| z \right), \qquad z = \xi, \eta.$$

The relation between this basis and the parabolic cylinder function basis is given by (B.9iii). To be definite, we set

$$\Psi = \exp\left[-\tfrac{1}{2}(\xi + \eta) \right] {}_1F_1 \left(\begin{matrix} \lambda \\ \tfrac{1}{2} \end{matrix} \middle| \xi \right) {}_1F_1 \left(\begin{matrix} \lambda \\ \tfrac{1}{2} \end{matrix} \middle| \eta \right).$$

Expressing ξ, η in terms of r, s by (7.35) and expanding in a Laurent series about $s = 0$, we find

$$\exp\left[\frac{r}{2}(s - s^{-1}) \right] {}_1F_1 \left(\begin{matrix} \lambda \\ \tfrac{1}{2} \end{matrix} \middle| \frac{r}{2s} - \frac{1}{2}rs + ir \right) {}_1F_1 \left(\begin{matrix} \lambda \\ \tfrac{1}{2} \end{matrix} \middle| \frac{r}{2s} - \frac{1}{2}rs - ir \right)$$

$$= \sum_{n=-\infty}^{\infty} c_n J_n(r) s^n. \tag{7.36}$$

ISBN-0-201-13503-5

Setting $s = 2a/b$, $r = b$ in (7.36), letting $b \to 0$, and using (B.14), we obtain

$$e^a \left[{}_1F_1 \left(\begin{matrix} \lambda \\ \frac{1}{2} \end{matrix} \middle| -a \right) \right]^2 = {}_1F_1 \left(\begin{matrix} \lambda \\ \frac{1}{2} \end{matrix} \middle| -a \right) {}_1F_1 \left(\begin{matrix} \frac{1}{2} - \lambda \\ \frac{1}{2} \end{matrix} \middle| a \right) = \sum_{n=0}^{\infty} c_n a^n / n!, \quad (7.37)$$

where the first equality follows from the transformation formula for ${}_1F_1$. Also, since (7.36) is invariant under the replacement $s \leftrightarrow -s^{-1}$, we have $c_n = c_{-n}$. It follows that we can compute $c_{\pm n}$ by expanding the ${}_1F_1$ product, (7.37), in a power series in the variable a and determining the coefficient of a^n. The result is

$$c_{\pm n} = {}_3F_2 \left(\begin{matrix} \lambda, n, -n \\ \frac{1}{2}, \frac{1}{2} \end{matrix} \middle| 1 \right), \quad n = 0, 1, 2, \ldots . \quad (7.38)$$

Substitution of (7.38) into (7.36) leads to a nontrivial identity. Identities for the other possible choices of basis are derived in [134].

For our final example we consider the case where Ψ satisfies the Helmholtz equation and

$$\left((P^0)^2 - (P^+)^2 \right) \Psi = \nu^2 \Psi. \quad (7.39)$$

This operator corresponds to system 5 in Table 3 and Ψ is separable in the variables $\{u, v\}$ where

$$uv = x + iy = rs, \quad (u^2 + v^2)/uv = x - iy = rs^{-1},$$

that is,

$$u = \tfrac{1}{2} \left[(r^2 + 2rs)^{1/2} - (r^2 - 2rs)^{1/2} \right],$$

$$v = \tfrac{1}{2} \left[(r^2 + 2rs)^{1/2} + (r^2 - 2rs)^{1/2} \right]. \quad (7.40)$$

Here we choose the square roots such that $v = r$ and $u = 0$ when $s = 0$. The separated solution Ψ takes the form $J_{\pm\nu}(u)J_{\pm\nu}/v)$. More generally, the function $\exp(\alpha P^+)\Psi = \Psi'$ satisfies the equation

$$S'\Psi' = \nu^2 \Psi', \quad S' = (P^0)^2 - \alpha \{ P^0, P^+ \} + (\alpha^2 - 1)(P^+)^2, \quad (7.41)$$

and, as follows from (7.11), Ψ' takes the form $J_{\pm\nu}(u')J_{\pm\nu}(v')$ where

$$u' = \tfrac{1}{2} \left[(r^2 + 2(1+\alpha)rs)^{1/2} - (r^2 - 2(1-\alpha)rs)^{1/2} \right],$$

$$v' = \tfrac{1}{2} \left[(r^2 + 2(1+\alpha)rs)^{1/2} + (r^2 - 2(1-\alpha)rs)^{1/2} \right]. \quad (7.42)$$

ISBN-0-201-13503-5

Choosing the product of two $(+\nu)$ solutions to be definite, we see that Ψ' has a power series expansion in s,

$$J_\nu(u')J_\nu(v') = \sum_{n=0}^{\infty} c_n J_{\nu+n}(r)s^{\nu+n}, \qquad |\alpha| < 1, \qquad \nu \neq -1, -2, \dots . \quad (7.43)$$

Setting $r=a$, $s=b/a$, letting $a\to 0$, and comparing coefficients of like powers of b on both sides of the resulting equation, we find

$$c_n = \frac{2^{-\nu}\left[i(1-\alpha^2)^{1/2}\right]^n}{\Gamma(\nu+1)n!} \, {}_2F_1\left[\begin{array}{c} -n, n+2\nu+1 \\ \nu+1 \end{array} \middle| \frac{\alpha+i(1-\alpha^2)^{1/2}}{2i(1-\alpha^2)^{1/2}} \right]. \quad (7.44)$$

For $\alpha = -1$ we also find

$$J_\nu\left(\tfrac{1}{2}\left[r-(r^2-4rs)^{1/2}\right]\right)J_\nu\left(\tfrac{1}{2}\left[r+(r^2-4rs)^{1/2}\right]\right)$$

$$= \sum_{n=0}^{\infty} \frac{(-1)^n(-2\nu-2n)_n}{\Gamma(\nu+n+1)n!} J_{\nu+n}(r)\left(\frac{s}{2}\right)^{\nu+n} \quad (7.45)$$

where $(a)_n$ is Pochhammer's symbol (B.3).

Exercises

1. Substitute the matrix elements (3.54) into the identity (3.44) to obtain an addition theorem for Bessel functions, known as Graf's addition theorem. (For extensions and applications of this result see Chapter 11 of [130a] as well as identity (7.27).)

2. Work out the bilinear expansions of the Bessel function (3.55) in terms of the separable solutions $\Psi^{(j)}(\mathbf{x})$ of the Helmholtz equation for $j=2,3,4$. Verify that the expansion for $j=2$ is a special case of Graf's addition theorem. Show that the $j=4$ expansion leads to an integral identity of the form

$$2\int_{-\pi}^{\pi} J_0\left\{\omega\left[(x-x')^2+(y-y')^2\right]^{1/2}\right\} ce_n(\beta',q)\,d\beta'$$

$$= |C_n|^2 Ce_n(\alpha,q)\,\overline{Ce}_n(\alpha',q)ce_n(\beta,q), \qquad n=0,1,2,\dots,$$

with a similar result involving se_n.

3. Verify that the full symmetry algebra of the Klein–Gordon equation (4.1) is $\mathcal{E}(1,1)\oplus\{1\}$ where a basis for $\mathcal{E}(1,1)$ is given by (4.2).

4. Show that the symmetry algebra $\mathcal{E}(2)^c$ of the complex Helmholtz equation decomposes into exactly three orbits under the adjoint action of $\tilde{E}(2)^c$.

ISBN-0-201-13503-5

5. One advantage of local Lie theory methods over global group methods is that the local theory applies to so-called special functions of the second kind. The Bessel functions of the second kind $Y_m(r)$ are defined by

$$Y_m(r) = \left[J_m(r)\cos(m\pi) - J_{-m}(r) \right] / \sin(m\pi)$$

or the limit of this expression for m an integer [130a, Chapter 3]. Show that $Y_m(r)$ satisfies the same differential equation and recurrence relations (B.16), (B.17) as does $J_m(r)$. (Indeed for each complex m, $\{J_m(r), Y_m(r)\}$ is a basis of solutions for Bessel's equation (7.3).) Conclude from this that the functions $\Psi_m(r,s) = Y_m(r)s^m$ satisfy the relations (7.15), hence the identities (7.26) where the matrix elements $T_{ij}(g)$ are given by (7.22). Obtain the identities for $Y_m(r)$ analogous to (7.28)–(7.30). (See [130a, Chapter 11].)

6. The simple Bessel polynomials $f_m(r)$ are defined by

$$f_m(r) = {}_2F_0\left(\begin{array}{c} -m, m+1 \\ \underline{\quad} \end{array} \middle| -\frac{r}{2} \right), \qquad m = 1, 2,, \ldots,$$

with $f_{-m}(r) = f_{m-1}(r)$ and $f_{-1}(r) = f_0(r) = 1$; see (B.18). These polynomials satisfy the recurrence relations

$$r^2 f_m'(r) + (1 - mr) f_m(r) = f_{m-1}(r),$$
$$r^2 f_m'(r) + \left[1 + (m+1)r \right] f_m(r) = f_{m+1}(r).$$

Verify that the functions $\Psi_m(r,s) = f_m(r)s^m$ satisfy relations (7.15) where

$$P^0 = s\partial_s, \quad P^- = -s^{-1}(r^2\partial_r + 1 - rs\,\partial_s), \quad P^+ = -s(r^2\partial_r + r + 1 + rs\,\partial_s).$$

Conclude from this that

$$\mathbf{T}(g)\Psi_j = \sum_{l=-\infty}^{\infty} T_{lj}(g)\Psi_l$$

where the matrix elements T_{lj} are given by (7.22) with $m_0 = 0$, Compute the operators $\mathbf{T}(g)$ and obtain the identities for Bessel polynomials that are analogous to (7.27)–(7.30). (For more details, see [78, Chapter 3].)

ISBN-0-201-13503-5

CHAPTER 2 _____

The Schrödinger and Heat Equations

2.1 Separation of Variables for the Schrödinger Equation $(i\,\partial_t + \partial_{xx})\Psi(t,x)=0$

In the quantum-mechanical study of a nonrelativistic system in two-dimensional space-time, consisting of a particle (mass m) subject to a potential $V(x)$, it is postulated that the state of the system at time t is completely determined by a state function $\Psi(t,x)$ which is a solution of the *time-dependent Schrödinger equation*

$$i\hbar\,\partial_t\Psi = -\frac{\hbar^2}{2m}\,\partial_{xx}\Psi + V(x)\Psi, \tag{1.1}$$

where $\hbar = h/2\pi$ and h is Planck's constant [70]. (The constants $\hbar$ and $\hbar^2/2m$, although very important in physics, are simply a nuisance in this book, so we will henceforth choose units such that $\hbar = \hbar^2/2m = 1$.) Among the most important Schrödinger equations are those for which the potential function $V(x)$ takes one of the forms in Table 5. For systems (1)–(4) the variable x ranges over the real line, while for (5)–(7) we assume x is nonnegative. (These latter equations arise from Schrödinger equations in higher-dimensional space-time which separate in polar or spherical coordinates. In (5)–(7) $x=r$, the radial coordinate [70].) Anderson *et al.* [3] and Boyer [18] have classified all Schrödinger equations (1.1) that admit nontrivial symmetry algebras. (Clearly all Schrödinger equations admit the two-dimensional complex symmetry algebra with basis ∂_t and $E=1$. By "nontrivial" we mean that the symmetry algebra is at least three dimensional.) They have shown that the only such equations are those with potentials (1)–(7). These potentials can be characterized in terms of symmetry groups.

ENCYCLOPEDIA OF MATHEMATICS and Its Applications, Gian-Carlo Rota (ed.).
Vol. 4: Willard Miller, Jr., Symmetry and Separation of Variables

Table 5 Potentials $V(x)$ with Nontrivial Symmetries

$V(x)$	Name of system
(1) 0	Free particle
(2) kx^2, $k>0$	Harmonic oscillator
(3) $-kx^2$, $k>0$	Repulsive oscillator
(4) ax, $a\neq0$	Free fall (linear potential)
(5) a/x^2, $a\neq0$	Radial free particle
(6) a/x^2+kx^2, $a\neq0, k>0$	Radial harmonic oscillator
(7) a/x^2-kx^2, $a\neq0, k>0$	Radial repulsive oscillator

In the next three sections we shall study these seven equations and uncover the surprising relations between them and the connection with separation of variables.

We write the free-particle Schrödinger equation in the form

$$Q\Psi=0, \qquad Q=i\partial_t+\partial_{xx}. \tag{1.2}$$

To compute the symmetry algebra of this equation, we follow the method described in Section 1.1. That is, we find all linear differential operators

$$L=a(t,x)\partial_x+b(t,x)\partial_t+c(t,x),$$

a, b, c, analytic in a suitable region $\mathcal{D}$ of the x–t plane, such that $L\Psi$ satisfies (1.2) whenever Ψ, analytic in $\mathcal{D}$, satisfies (1.2). A necessary and sufficient condition for L to belong to the symmetry algebra is

$$[L,Q]=R_L(t,x)Q \tag{1.3}$$

for some function R_L analytic in $\mathcal{D}$. By equating coefficients of ∂_{xx}, ∂_t, ∂_x, and 1 on both sides of (1.3), we obtain a system of differential equations for a, b, c, and R. Details of the straightforward computation can be found in [3, 15, 18]. The final result is that the symmetry operators L form a six-dimensional complex Lie algebra $\mathcal{G}_2^c$ with basis

$$K_2=-t^2\partial_t-tx\partial_x-t/2+ix^2/4, \qquad K_1=-t\partial_x+ix/2,$$
$$K_0=i, \qquad K_{-1}=\partial_x, \qquad K_{-2}=\partial_t, \qquad K^0=x\partial_x+2t\partial_t+1/2 \tag{1.4}$$

and commutation relations

$$[K^0,K_j]=jK_j \quad (j=\pm2,\pm1,0), \qquad [K_{-1},K_1]=\tfrac12 K_0,$$
$$[K_{-1},K_2]=K_1, \qquad [K_{-2},K_1]=-K_{-1}, \qquad [K_{-2},K_2]=-K^0. \tag{1.5}$$

The reader should now be able to appreciate expression (1.3), since it has enabled us to compute symmetry operators for (1.2) that are not im-

ISBN-0-201-13503-5

mediately obvious. Furthermore, some of the operators are not purely differential but involve multipliers. The geometrical significance of K_0, K_{-1}, and K_{-2} is obvious and K^0 is the generator for a dilatation symmetry $\Psi(t,x) \rightarrow \Psi(\alpha^2 t, \alpha x)$. However, K_1 is the generator of a Galilean transformation, not an obvious symmetry, and the geometrical interpretation of K_2 is unknown to the writer. Furthermore, K^0 and K_2 do not commute with Q even though they map solutions to solutions, so they correspond to operators L in (1.3) with $R_L \not\equiv 0$.

Since x and t are real variables and since we wish to exponentiate the symmetry operators (1.4) to obtain group symmetries, we restrict ourselves to the *real* six-dimensional Lie algebra $\mathcal{G}_2$ with basis (1.4). (Note that we cannot throw out the identity operator K_0, since K_0 occurs as the commutator $2[K_{-1}, K_1]$.) A second useful basis for $\mathcal{G}_2$ is $\{C_j, L_k, E\}$ where

$$C_1 = K_{-1}, \quad C_2 = K_1, \quad L_3 = K_{-2} - K_2,$$
$$L_1 = K^0, \quad L_2 = K_{-2} + K_2, \quad E = K_0. \tag{1.6}$$

The commutation relations become

$$[L_1, L_2] = -2L_3, \quad [L_3, L_1] = 2L_2, \quad [L_2, L_3] = 2L_1,$$
$$[C_1, C_2] = \tfrac{1}{2}E, \quad [L_3, C_1] = C_2, \quad [L_3, C_2] = -C_1, \tag{1.7}$$
$$[L_2, C_1] = [C_2, L_1] = -C_2, \quad [L_1, C_1] = [L_2, C_2] = -C_1$$

where E generates the center of $\mathcal{G}_2$.

To explain the structure of $\mathcal{G}_2$ we recall some facts about the group $SL(2, R)$ of all real 2×2 matrices A with determinant $+1$.

$$A = \begin{pmatrix} \alpha & \beta \\ \gamma & \delta \end{pmatrix}, \quad \alpha\delta - \gamma\beta = 1, \quad \alpha, \beta, \gamma, \delta \in R. \tag{1.8}$$

As is well known [46, 82], the Lie algebra $sl(2, R)$ of $SL(2, R)$ consists of all 2×2 real matrices $\mathcal{C}$ with trace zero,

$$A = \begin{pmatrix} a & b \\ c & -a \end{pmatrix}, \quad a, b, c \in R. \tag{1.9}$$

This Lie algebra is three dimensional and the matrices

$$\mathcal{C}_1 = \begin{pmatrix} 1 & 0 \\ 0 & -1 \end{pmatrix}, \quad \mathcal{C}_2 = \begin{pmatrix} 0 & 1 \\ 1 & 0 \end{pmatrix}, \quad \mathcal{C}_3 = \begin{pmatrix} 0 & -1 \\ 1 & 0 \end{pmatrix} \tag{1.10}$$

form a basis with commutation relations

$$[\mathcal{C}_1, \mathcal{C}_2] = -2\mathcal{C}_3, \quad [\mathcal{C}_3, \mathcal{C}_1] = 2\mathcal{C}_2, \quad [\mathcal{C}_2, \mathcal{C}_3] = 2\mathcal{C}_1. \tag{1.11}$$

It follows immediately that the symmetry operators L_k form a basis for a subalgebra of $\mathcal{G}_2$ isomorphic to $sl(2,R)$.

Furthermore, the operators C_1, C_2, E form a basis for a subalgebra of $\mathcal{G}_2$ isomorphic to the Weyl algebra $\mathcal{W}_1$. The Weyl group W_1 consists of all real 3×3 matrices

$$B(u,v,\rho) = \begin{bmatrix} 1 & v & 2\rho + uv/2 \\ 0 & 1 & u \\ 0 & 0 & 1 \end{bmatrix}, \qquad u,v,\rho \in R, \qquad (1.12)$$

with group multiplication

$$B(u,v,\rho)B(u',v',\rho') = B(u+u', v+v', \rho+\rho'+(vu'-uv')/4). \quad (1.13)$$

The Lie algebra $\mathcal{W}_1$ has basis

$$C_1 = \begin{bmatrix} 0 & 1 & 0 \\ 0 & 0 & 0 \\ 0 & 0 & 0 \end{bmatrix}, \qquad C_2 = \begin{bmatrix} 0 & 0 & 0 \\ 0 & 0 & 1 \\ 0 & 0 & 0 \end{bmatrix}, \qquad E = \begin{bmatrix} 0 & 0 & 2 \\ 0 & 0 & 0 \\ 0 & 0 & 0 \end{bmatrix}$$

with commutation relations

$$[C_1, C_2] = \tfrac{1}{2} E, \qquad [C_k, E] = 0. \qquad (1.14)$$

Using standard results from Lie theory (Theorem A.3), we can exponentiate the differential operators of $\mathcal{G}_2$ to obtain a local Lie group G_2 of symmetry operators. The action of the Weyl group W_1 is given by operators

$$\mathbf{T}(u,v,\rho) = \exp([\rho + uv/4]E)\exp(uC_2)\exp(vC_1)$$

where

$$\mathbf{T}(u,v,\rho)\Phi(t,x) = \exp[i\rho + i(uv + 2ux - u^2 t)/4]\Phi(t, x+v-ut) \quad (1.15)$$

and Φ belongs to the space $\mathcal{F}$ of analytic functions with domain $\mathcal{D}$. The group multiplication property is given by (1.13). The action of $SL(2,R)$ is given by

$$\mathbf{T}(A)\Phi(t,x) = \exp\left[i\left(\frac{x^2\beta/4}{\delta+t\beta}\right)\right](\delta+t\beta)^{-1/2}\Phi\left(\frac{\gamma+t\alpha}{\delta+t\beta}, \frac{x}{\delta+t\beta}\right) \quad (1.16)$$

ISBN-0-201-13503-5

where $A \in SL(2,R)$ is represented in the form (1.8). Here,

$$\mathbf{T}\begin{pmatrix} 1 & \beta \\ 0 & 1 \end{pmatrix} = \exp(\beta K_2), \qquad \mathbf{T}\begin{pmatrix} 1 & 0 \\ \gamma & 1 \end{pmatrix} = \exp(\gamma K_{-2}),$$

$$\mathbf{T}\begin{pmatrix} e^\alpha & 0 \\ 0 & e^{-\alpha} \end{pmatrix} = \exp(\alpha K^0), \qquad \mathbf{T}\begin{pmatrix} \cos\theta & -\sin\theta \\ \sin\theta & \cos\theta \end{pmatrix} = \exp(\theta L_3), \quad (1.17)$$

$$\mathbf{T}\begin{pmatrix} \cosh\varphi & \sinh\varphi \\ \sinh\varphi & \cosh\varphi \end{pmatrix} = \exp(\varphi L_2).$$

Now $SL(2,R)$ acts on W_1 via the adjoint representation

$$\mathbf{T}(A^{-1})\mathbf{T}(u,v,\rho)\mathbf{T}(A) = \mathbf{T}(u\delta + v\beta, u\gamma + v\alpha, \rho), \qquad (1.18)$$

so the full symmetry group G_2, the *Schrödinger group* in two-dimensional space-time, is obtained as a semidirect product of $SL(2,R)$ and W_1 [18, 59]:

$$g = (A, \mathbf{w}) \in G_2, \quad A \in SL(2,R),$$

$$\mathbf{w} = (u,v,\rho) \in W_1, \qquad \mathbf{T}(g) = \mathbf{T}(A)\mathbf{T}(\mathbf{w}), \qquad (1.19)$$

$$\mathbf{T}(g)\mathbf{T}(g') = \mathbf{T}(AA')\{\mathbf{T}(A'^{-1})\mathbf{T}(\mathbf{w})\mathbf{T}(A')\}\mathbf{T}(\mathbf{w}') = \mathbf{T}(gg').$$

It follows from our general theory that $\mathbf{T}(g)$ maps solutions Ψ of (1.2) into solutions $\mathbf{T}(g)\Psi$. However, G_2 is only a local symmetry group, for not only do we have the domain problem in defining $\mathbf{T}(g)\Phi$, as discussed in Section 1.1, but also expression (1.16) makes no sense when $\delta + t\beta = 0$. Expression (1.16) follows from the exponentiation of Lie derivatives only if $|t\beta/\delta| < 1$. For $|t\beta/\delta| > 1$ this expression still defines a symmetry, but one that is not directly obtainable from the symmetry algebra.

The Schrödinger group G_2 acts on the Lie algebra $\mathcal{G}_2$ of symmetry operators K via the adjoint representation

$$K \to K^g = \mathbf{T}(g)K\mathbf{T}(g^{-1})$$

and this action splits $\mathcal{G}_2$ into G_2 orbits. For our purposes the operator $K_0 = i$, which generates the center $\{K_0\}$ of $\mathcal{G}_2$, is trivial, so we merely determine the orbit structure of the factor space $\mathcal{G}_2' = \mathcal{G}_2/\{K_0\}$. The results are as follows. Let

$$K = a_2 K_2 + a_1 K_1 + a_0 K^0 + a_{-1} K_{-1} + a_{-2} K_{-2}$$

be a nonzero element of $\mathcal{G}_2'$ and set $\alpha = a_2 a_{-2} + a_0^2$. It is straightforward to show that α is invariant under the adjoint representation. In Table 6 we give a complete set of orbit representatives. That is, K lies on the same G_2

ISBN-0-201-13503-5

orbit as a real multiple of exactly one of the five operators in the list.

$$\text{Case 1} \quad (\alpha < 0) \quad K_{-2} - K_2 = L_3;$$

$$\text{Case 2} \quad (\alpha > 0) \quad K^0; \quad\quad\quad\quad\quad (1.20)$$

$$\text{Case 3} \quad (\alpha = 0) \quad K_2 + K_{-1}, K_{-2}, K_{-1}.$$

Note that there are five orbits.

Since K_{-2} and K_{-1} commute, they can be simultaneously diagonalized. Furthermore, $K_{-2}\Psi = iK^2_{-1}\Psi$ for all solutions Ψ of $Q\Psi = 0$. Thus we associate the same coordinate system $\{t, x\}$ with both of these orbits and end up with only four separable coordinate systems.

One can also compute the second-order symmetries of $Q\Psi = 0$ and show that the free-particle Schrödinger equation is class I. However, all separable coordinate systems for the equation turn out to be associated with orbits of first-order symmetries. This is related to the fact that the Schrödinger equation is only first order in the variable t.

For this equation it is useful (and necessary) to consider R-separable solutions. To explain this concept we choose a nonzero analytic function $R(t, x) = \exp(i\Re(t, x))$ and write $\Psi = R\Phi$ where Ψ satisfies the Schrödinger equation $Q\Psi = 0$. Writing the differential equation in terms of Φ, we find $Q'\Phi = 0$ where $Q' = R^{-1}QR$ is the transformed differential operator. Now suppose the new equation $Q'\Phi = 0$ admits separable solutions $\Phi_\lambda = U_\lambda(u)V_\lambda(v)$ in terms of a $\{u, v\}$ coordinate system. If $R = a(u)b(v)$—that is, if R factors in the $\{u, v\}$ coordinates—then $\Psi_\lambda = a(u)U_\lambda(u)b(v)V_\lambda(v)$ is a separable solution of $Q\Psi_\lambda = 0$ and we have obtained nothing new. However, if $R(u, v)$ does not factor, then we have obtained a new family of R-separable solutions $\Psi_\lambda = \exp(iR(u, v))U_\lambda(u)V_\lambda(v)$. Thus R-separability is a generalization of ordinary separability. R-separable solutions of one equation $Q\Psi = 0$ correspond to ordinary separable solutions of an equivalent equation $Q'\Phi = 0$, $Q' = R^{-1}QR$.

We have not introduced the notion of R-separability earlier because the equations studied in Chapter 1 admit no R-separable solutions that are not already separable in the ordinary sense. However, the situation changes for the Schrödinger operators. The existence of R-separable solutions is clearly related to the existence of symmetry operators K that do not commute with Q, even though they map solutions into solutions.

In [59], Kalnins and the author have computed all coordinate systems that permit R-separation of variables for equation (1.2) and have shown that the associated R-separated solutions $\Psi_\lambda = \exp(iR(u, v))U_\lambda(u)V_\lambda(v)$ can be characterized as eigenfunctions of some $K \in \mathcal{G}_2$, $K\Psi_\lambda = i\lambda\Psi_\lambda$, $Q\Phi_\lambda = 0$. The association between orbits in $\mathcal{G}'_2$ and separable coordinates is given in Table 6.

ISBN-0-201-13503-5

Table 6 R-Separable Coordinates for the Equation $(i\partial_t + \partial_{xx})\Psi(t,x) = 0$

Operator	Coordinates $\{u,v\}$	Multiplier $R = e^{i\mathfrak{R}}$	Separated Solutions
1 K_{-1}, K_{-2}	$x = u$	$\mathfrak{R} = 0$	Product of exponentials
2a $K_{-2} - K_1$	$x = u + v^2/2$	$\mathfrak{R} = uv/2$	Airy times exponential function
2b $K_2 + K_{-1}$	$x = uv + 1/2v$	$\mathfrak{R} = (u^2 v - u/v)/4$	Airy times exponential functions
3a K^0	$x = u\sqrt{v}$	$\mathfrak{R} = 0$	Parabolic cylinder times exponential function
3b $K_2 + K_{-2}$	$x = u\lvert 1 - v^2 \rvert^{1/2}$	$\mathfrak{R} = \pm u^2 v/4$ ($+$ if $\lvert v \rvert > 1$, $-$ if $\lvert v \rvert < 1$)	Parabolic cylinder times exponential function
4 $K_2 - K_{-2}$	$x = u(1 + v^2)^{1/2}$	$\mathfrak{R} = u^2 v/4$	Hermite times exponential function

For all coordinate systems $\{u,v\}$ in Table 6, $v = t$. As stated earlier, there are only four types of separable coordinates and these are associated with four nontrivial G_2 orbits in $\mathcal{G}_2'$. (Here we are identifying the two orbits with commuting representatives K_{-1}, K_{-2}.) However, the table contains six entries, and each of the six separable systems appears to be distinct from the rest. The explanation for this relates to our definition of equivalent coordinate systems. We regard two systems as equivalent if one system can be mapped into the other by a G_2 transformation $\mathbf{T}(g)$. However, such transformations, particularly (1.16), can sometimes have a rather complicated form, so that two equivalent systems will appear very different. Since the operator K_2 has a rather obscure physical significance, it is difficult to interpret the physical or geometrical relationship between two systems that are related by the exponential of this operator.

However, there is a five-parameter subgroup of G_2 whose physical significance is well understood [73]. This is the Galilei group plus dilations with Lie algebra basis $\{K_{-2}, K_{+1}, K_0, K^0\}$. If we regard systems from the point of view of equivalence under the Galilei group plus dilatations, we find that G_2 orbits 2 and 3 each split into two Galilean–dilatation orbits. This accounts for the six systems listed in Table 6. (However, the classification is based more on significance for separation of variables than accuracy for Galilean–dilatation orbits. Indeed, $2a$ splits into two Galilean–dilatation orbits $K_{-2} \pm K_1$ and 2b splits into $K_2 \pm K_{-1}$. These subcases yield coordinates that differ only in the sign of a parameter, and we choose not to distinguish between them.)

We can describe the equivalences on orbits 2 and 3 in terms of the operator $J = \exp[(\pi/4)(K_2 - K_{-2})] = \exp(-(\pi/4)L_3)$.

$$J\Phi(t,x) = \frac{2^{1/4}}{(1+t)^{1/2}} \exp\left(\frac{ix^2/4}{1+t}\right) \Phi\left(\frac{t-1}{t+1}, \frac{x\sqrt{2}}{t+1}\right) \qquad (1.21)$$

ISBN-0-201-13503-5

Note that $J^2 = \exp[(\pi/2)(K_2 - K_{-2})] = \exp((-\pi/2)L_3)$, and

$$J^2\Phi(t,x) = \frac{\exp(ix^2/4t)}{\sqrt{t}}\Phi\left(\frac{-1}{t},\frac{x}{t}\right), \qquad J^8\Phi = -\Phi, \qquad J^{16}\Phi = \Phi. \quad (1.22)$$

A direct computation yields

$$J(K_{-2}+K_2)J^{-1} = K^0, \qquad J^2(K_1 - K_{-2})J^{-2} = K_{-1}+K_2, \quad (1.23)$$

which proves the G_2 equivalence of systems 2a, 2b and 3a, 3b.

We now show that the operators (1.4) can be interpreted as a Lie algebra of skew-Hermitian operators on the Hilbert space $L_2(R)$ of complex-valued Lebesgue square-integrable functions on the real line (Chapter 1, (5.2)). To do this we consider the operators (1.4) formally restricted to the solution space of (1.2). Then we can replace ∂_t by $i\partial_{xx}$ in these expressions and consider t as a fixed parameter. It is easy to show that the resulting operators restricted to the domain $\mathcal{D} \subset L_2(R)$ of infinitely differentiable functions with compact support are skew-symmetric. Moreover, each of these operators, when multiplied by i, has a unique self-adjoint extension. Indeed, the operators (1.4) are real linear combinations of

$$\begin{array}{lll} \mathcal{K}_2 = ix^2/4, & \mathcal{K}_1 = ix/2, & \\ \mathcal{K}_{-1} = \partial_x, & \mathcal{K}_{-2} = i\partial_{xx}, & \mathcal{K}_0 = i, \quad \mathcal{K}^0 = x\partial_x + \tfrac{1}{2}, \end{array} \quad (1.24)$$

and $i\mathcal{K}_j, i\mathcal{K}^0$ have unique self-adjoint extensions. When the parameter t is set equal to zero, K_j becomes $\mathcal{K}_j$ and K^0 becomes $\mathcal{K}^0$. It follows that the script operators also satisfy commutation relations (1.5).

From spectral theory [111, Chapter VIII], we know that to each skew-Hermitian $\mathcal{K} \in \mathcal{G}_2$ there corresponds a one-parameter group $U(\alpha) = \exp(\alpha\mathcal{K})$ of unitary operators on $L_2(R)$. This group in turn acts on $\mathcal{G}_2$ via $\mathcal{K} \to U(\alpha)\mathcal{K}U(-\alpha)$. In particular, the following result is of importance in quantum mechanics:

$$\exp(a\mathcal{K}_{-2})f(x) = \text{l.i.m.}(4\pi ia)^{-1/2}\int_{-\infty}^{\infty}\exp\left[-(x-y)^2/4ia\right]f(y)\,dy, \quad (1.25)$$

$$f \in L_2(R), a \neq 0.$$

(Here $(ia)^{1/2} = e^{i\pi/4}|a|^{1/2}$ for $a>0$ and $e^{-i\pi/4}|a|^{1/2}$ for $a<0$. See [67, p. 493] for a proof of (1.25).) We can verify that

$$\exp(t\mathcal{K}_{-2})\mathcal{K}_j\exp(-t\mathcal{K}_{-2}) = K_j,$$
$$\exp(t\mathcal{K}_{-2})\mathcal{K}^0\exp(-t\mathcal{K}_{-2}) = K^0. \quad (1.26)$$

ISBN-0-201-13503-5

(A formal proof is easily obtained from the commutation relations (1.5), although a rigorous proof specifying domains is somewhat more difficult.)

Now if $f \in L_2(R)$ and f belongs to the domain of the self-adjoint operator $\mathcal{K}_{-2}$, then $\Psi(t,x) = \exp(t\mathcal{K}_{-2})f(x)$ satisfies $\partial_t\Psi = \mathcal{K}_{-2}\Psi$ or $i\partial_t\Psi = -\partial_{xx}\Psi$ for almost every t, and $\Psi(0,x)=f(x)$. We see that $\exp(t\mathcal{K}_{-2})$ is the operator of time translation in quantum mechanics [70, 109]. Moreover, this operator is unitary, so

$$\int_{-\infty}^{\infty} \Psi_1(t,x)\overline{\Psi}_2(t,x)\,dx = \langle\exp(t\mathcal{K}_{-2})f_1, \exp(t\mathcal{K}_{-2})f_2\rangle$$

$$= \langle f_1,f_2\rangle = \int_{-\infty}^{\infty} f_1(x)\bar{f}_2(x)\,dx, \qquad (1.27)$$

independent of t. We have introduced a Hilbert space structure on the solutions of (1.2) that agrees exactly with the usual Hilbert space of states corresponding to a free-particle system. Moreover, the mappings (1.26) relating the time-zero (script) operators to the time-t (italic) operators are the usual transformations relating the Heisenberg and Schrödinger pictures in quantum theory [109, 118]. It is also easy to show that the unitary operators $\exp(\alpha K) = \exp(t\mathcal{K}_{-2})\exp(\alpha\mathcal{K})\exp(-t\mathcal{K}_{-2})$ map a Hilbert space solution Ψ of (1.2) into $\Phi = \exp(\alpha K)\Psi$, which also satisfies (1.2). Thus the unitary operators $\exp(\alpha K)$ are symmetries of (1.2).

Later we will see that the $\mathcal{K}$ operators generate a global unitary representation of a covering group $\tilde{G}_2$ of G_2, although not of G_2 itself. Assuming this for the moment, let $U(g), g \in \tilde{G}_2$, be the corresponding unitary operators, and set $T(g) = \exp(t\mathcal{K}_{-2})U(g)\exp(-t\mathcal{K}_{-2})$. Again it is easy to demonstrate that the $T(g)$ are unitary symmetries of (1.2) and that the associated infinitesimal operators are $K = \exp(t\mathcal{K}_{-2})\mathcal{K}\exp(-t\mathcal{K}_{-2})$.

Now consider the operator $\mathcal{L}_3 = \mathcal{K}_{-2} - \mathcal{K}_2 = i\partial_{xx} - ix^2/4 \in \mathcal{G}_2$. If $f \in L_2(R)$, then $\Psi(t,x)=\exp(t\mathcal{L}_3)f(x)$ satisfies $\partial_t\Psi = \mathcal{L}_3\Psi$ or $i\partial_t\Psi = -\partial_{xx}\Psi + x^2\Psi/4$ and $\Psi(0,x)=f(x)$. Similarly, the unitary operators $V(g) = \exp(t\mathcal{L}_3)U(g)\exp(-t\mathcal{L}_3)$ are symmetries of this equation, the Schrödinger equation for the harmonic oscillator, (2) in Table 5. (Here we have normalized k to the value $\frac{1}{4}$.) One can verify that the associated infinitesimal operators $\exp(t\mathcal{L}_3)\mathcal{K}\exp(-t\mathcal{L}_3)$ can be expressed as first-order differential operators in x and t. (In particular these operators will be real linear combinations of the basis operators (1.24) with coefficients that depend on t. Considering these operators as acting on the solution space of the harmonic oscillator Schrödinger equation, we can replace $i\partial_{xx}$ by $\partial_t + ix^2/4$ wherever it occurs.) Conversely, if K' is a first-order symmetry operator for the time-dependent harmonic oscillator Schrödinger equation, we can show that at time $t=0$, K' reduces to a real linear combination of the operators (1.24). It follows that the symmetry algebras of the equations with potentials (1) and (2) in Table 5 are both isomorphic to $\mathcal{G}_2$ with basis

ISBN-0-201-13503-5

(1.24). For the free-particle equation the symmetries are $K = \exp(t\mathcal{K}_{-2})\mathcal{K}\exp(-t\mathcal{K}_{-2})$, while for the harmonic oscillator equation the symmetries are $K' = \exp(t\mathcal{L}_3)\mathcal{K}\exp(-t\mathcal{L}_3)$. In each case the $\mathcal{K}$ operators are identical. Moreover, for fixed $\mathcal{K}$, the operators K and K' are unitary equivalent, $K' = A(t)KA(t)^{-1}$, although the unitary operator $A(t) = \exp(t\mathcal{L}_3)\exp(-t\mathcal{K}_{-2})$ depends on t.

Continuing in this manner, we consider the operator $\mathcal{L}_2 = \mathcal{K}_{-2} + \mathcal{K}_2 = i\partial_{xx} - ix^2/4 \in \mathcal{G}_2$. If $f \in L_2(R)$, then $\Psi(t,x) = \exp(t\mathcal{L}_2)f(x)$ satisfies $\partial_t\Psi = \mathcal{L}_2\Psi$ or $i\partial_t\Psi = -\partial_{xx}\Psi - x^2\Psi/4$ and $\Psi(0,x) = f(x)$. The operators $\mathbf{W}(g) = \exp(t\mathcal{L}_2)\mathbf{U}(g)\exp(-t\mathcal{L}_2)$ form the unitary symmetry group of this equation, repulsive harmonic oscillator potential ((3) in Table 5), and the associated infinitesimal operators $\exp(t\mathcal{L}_2)\mathcal{K}\exp(-t\mathcal{L}_2)$ are first order in x and t. Finally, we consider the operator $\mathcal{U} = \mathcal{K}_{-2} - \mathcal{K}_1 = i\partial_{xx} - ix/2 \in \mathcal{G}_2$. If $f \in L_2(R)$, then $\Psi(t,x) = \exp(t\mathcal{U})f(x)$ satisfies $\partial_t\Psi = \mathcal{U}\Psi$ or $i\partial_t\Psi = -\partial_{xx}\Psi + x\Psi/2$ and $\Psi(0,x) = f(x)$. The unitary operators $\mathbf{X}(g) = \exp(t\mathcal{U})\mathbf{U}(g)\exp(-t\mathcal{U})$ are symmetries of this Schrödinger equation corresponding to a linear potential and the infinitesimal operators $\exp(t\mathcal{U})\mathcal{K}\exp(-t\mathcal{U})$ are first order in x and t.

Note further from (1.20) that the operators $\mathcal{K}_{-2}$, $\mathcal{L}_3$, $\mathcal{L}_2$, $\mathcal{K}_{-2} - \mathcal{K}_1$ corresponding to the free-particle, attractive and repulsive harmonic oscillator, and linear potential Hamiltonians lie on the same G_2 orbits as the four representatives $\mathcal{K}_{-2}$, $\mathcal{L}_3$, $\mathcal{K}^0$, and $\mathcal{K}_2 + \mathcal{K}_{-1}$, respectively. Thus, these four Hamiltonians correspond exactly to the four systems of coordinates in which equation (1.2) separates. We see that these Hamiltonians form a complete set of orbit representatives in $\mathcal{G}_2$.

It now follows that the Schrödinger equations (1)–(4) in Table 5 have isomorphic symmetry algebras. In each case if we compute the symmetry operators at time $t = 0$, we obtain the Lie algebra $\mathcal{G}_2$ with basis (1.24). Although we first obtained this symmetry algebra through a study of the Schrödinger equation (1), we could equally have obtained it by studying (2), (3), or (4). Moreover, we see from the preceding paragraphs how to construct the (time-dependent) unitary operators on $L_2(R)$ that map a solution of any one of these equations to a solution of another equation. The four equations can and should be studied as a unit.

The connection between orbits and separation of variables can now be made clear. Suppose $\Psi(t,x)$ is a solution of the free-particle equation

$$i\partial_t\Psi = -\partial_{xx}\Psi. \tag{1.28}$$

This equation clearly separates in the variables $\{t,x\}$ and these variables are "naturally" associated with (1.28). Now we have seen that the operator $A(t) = \exp(t\mathcal{L}_3)\exp(-t\mathcal{K}_{-2}) = \exp(-t\mathcal{K}_{-2})\exp(tL_3)$ maps Ψ to a solution $\Phi(t,x) = A(t)\Psi(t,x)$ of the harmonic oscillator equation

$$i\partial_t\Phi = -\partial_{xx}\Phi - x^2\Phi/4. \tag{1.29}$$

ISBN-0-201-13503-5

Explicitly,

$$\Phi(t,x) = (\cos t)^{-1/2} \exp(-ix^2 \tan(t)/4)\Psi(\tan t, x/\cos t).$$

Now equation (1.29) "naturally" separates in the variables $\{t,x\}$, so we can find solutions Ψ of (1.28) in the form

$$\Psi(t,x) = (1+v^2)^{-1/4} \exp(iu^2v/4)\Phi(\tan^{-1}v, u), \qquad x = u(1+v^2)^{1/2}, t = v. \tag{1.30}$$

Since (1.29) separates in $\{\tan^{-1}v, u\}$, hence $\{v,u\}$, it follows that equation (1.28) R-separates in the coordinates $\{v,u\}$ where the multiplier $R = e^{i\Re}$ is given by $\Re = iu^2v/4$. (The factor $(1+v^2)^{-1/4}$ can be absorbed in the separated solution.) Thus we have explained the existence of the coordinates 4 in Table 6, associated with the operator $\mathcal{K}_{-2} - \mathcal{K}_2$. In a similar manner we can associate a "natural" coordinate system with each of our four Hamiltonians, thus exhausting the possible R-separable coordinate systems inequivalent with respect to G_2.

Note that if two operators lie on the same G_2 orbit, then the first operator is unitary equivalent to a real constant times the second operator. Thus two suitable normalized operators on the same orbit have the same spectrum. In particular, if $\mathcal{K}, \mathcal{K}' \in \mathcal{G}_2$ with $\mathcal{K}' = U(g)\mathcal{K}U(g^{-1})$ and the self-adjoint operator $i\mathcal{K}$ has a complete set of (possibly generalized) eigenvectors $f_\lambda(x)$ with

$$i\mathcal{K}f_\lambda = \lambda f_\lambda, \qquad \langle f_\lambda, f_\mu \rangle = \delta_{\lambda,\mu} \tag{1.31}$$

where

$$\langle h_1, h_2 \rangle = \int_{-\infty}^{\infty} h_1(x)\bar{h}_2(x)\,dx, \qquad h_j \in L_2(R), \tag{1.32}$$

then for $f'_\lambda = U(g)f_\lambda$ we have

$$i\mathcal{K}'f'_\lambda = \lambda f'_\lambda, \qquad \langle f'_\lambda, f'_\mu \rangle = \delta_{\lambda,\mu} \tag{1.33}$$

and the f'_λ form a complete set of eigenvectors for $i\mathcal{K}'$ [77]. These remarks imply that if we wish to compute the spectrum corresponding to each operator $\mathcal{K} \in \mathcal{G}_2$, it is enough to determine the spectra of the four Hamiltonians listed earlier. Moreover, we may be able to choose another operator $\mathcal{K}$ on the same G_2 orbit as a given Hamiltonian such that the spectral decomposition of $\mathcal{K}$ is especially easy. The spectral decomposition of the Hamiltonian and the corresponding eigenfunction expansions then follow from those of $\mathcal{K}$ by application of a group operator $U(g)$.

ISBN-0-201-13503-5

As a special case of these remarks consider the operator $\mathcal{K}_{-2} = i\partial_{xx}$. If $\{f_\lambda\}$ is the basis of generalized eigenvectors for some operator $\mathcal{K} \in \mathcal{G}_2$, then $\{\Psi_\lambda = \exp(t\mathcal{K}_{-2})f_\lambda\}$ is the basis of generalized eigenvectors for $K = \exp(t\mathcal{K}_{-2})\mathcal{K}\exp(-t\mathcal{K}_{-2})$ and the Ψ_λ satisfy the free-particle Schrödinger equation (1.28). Similar remarks hold for the other Hamiltonians.

We begin our explicit computations by determining the spectral resolution of the operator $\mathcal{L}_3 = \mathcal{K}_{-2} - \mathcal{K}_2$. The results are well known [141]. The eigenfunction equation is

$$i\mathcal{L}_3 f = \lambda f, \qquad (-\partial_{xx} + x^2/4)f = \lambda f,$$

and the normalized eigenfunctions are

$$f_n^{(4)}(x) = [n!(2\pi)^{1/2}2^n]^{-1/2}\exp(-x^2/4)H_n(x/\sqrt{2}), \qquad \lambda_n = n + \tfrac{1}{2}, \quad (1.34)$$

$$n = 0,1,2,\ldots, \langle f_n^{(4)}, f_m^{(4)}\rangle = \delta_{nm},$$

where $H_n(x)$ is a Hermite polynomial, (B.12). The $\{f_n^{(4)}\}$ form an ON basis for $L_2(R)$.

From (1.34) we see that

$$\exp(2\pi\mathcal{L}_3)f_n^{(4)} = \exp\left[-2\pi i\left(n + \tfrac{1}{2}\right)\right]f_n^{(4)} = -f_n^{(4)}$$

so $\exp(2\pi\mathcal{L}_3) = -E$ where E is the identity operator on $L_2(R)$. However, from (1.17) if the operators $\exp(\alpha\mathcal{K})$ generate a global unitary representation of G_2 on $L_2(R)$, we should have $\exp(2\pi\mathcal{L}_3) = E$. In fact, it can be shown that the $\mathcal{K}$ operators exponentiate to a global irreducible representation of the simply connected covering group $\tilde{G}_2$ of G_2.

To describe this covering group we first consider the topology of the group manifold $SL(2,R)$, (1.8).

$$A = \begin{pmatrix} \alpha & \beta \\ \gamma & \delta \end{pmatrix} \in SL(2,R), \qquad \alpha\delta - \beta\gamma = 1.$$

Setting

$$2a = (\alpha + \delta) + i(\gamma - \beta), \qquad 2b = (-\alpha + \delta) + i(\gamma + \beta), \quad (1.35)$$

we see that the complex numbers a,b satisfy the identity

$$|a|^2 - |b|^2 = 1. \tag{1.36}$$

Conversely, if $a = a_1 + ia_2$, $b = b_1 + ib_2$, and a,b satisfy (1.36), then relations (1.35) can be uniquely inverted to yield an $A \in SL(2,R)$ with parameters $\alpha = a_1 - b_1$, $\beta = -a_2 + b_2$, $\gamma = a_2 + b_2$, $\delta = a_1 + b_1$. It follows from (1.36) that

ISBN-0-201-13503-5

topologically $SL(2,R)$ can be identified with the hyperboloid

$$a_1^2 + a_2^2 - b_1^2 - b_2^2 = 1.$$

Another parametrization of $SL(2,R)$ is due to Bargmann [10]. He sets

$$\mu = b/a, \qquad \omega = \arg a, \qquad -\pi < \omega \leqslant \pi \,(\text{mod}\, 2\pi). \qquad (1.37)$$

It follows from (1.36) that $|\mu| < 1$. Furthermore,

$$a = e^{i\omega}\left(1 - |\mu|^2\right)^{-1/2}, \qquad b = e^{i\omega}\mu\left(1 - |\mu|^2\right)^{-1/2}. \qquad (1.38)$$

We can now write $A \equiv (\mu, \omega)$, $|\mu| < 1$, $-\pi < \omega \leqslant \pi$, and parametrize $SL(2,R)$ in terms of μ and ω. The group product can be expressed as follows. If $A = (\mu, \omega)$, $A' = (\mu', \omega')$, then $AA' = (\mu'', \omega'')$ where

$$\mu'' = \left(\mu + \mu' e^{-2i\omega}\right)\left(1 + \bar{\mu}\mu' e^{-2i\omega}\right)^{-1},$$
$$\omega'' = \omega + \omega' + (1/2i)\ln\left[\left(1 + \bar{\mu}\mu' e^{-2i\omega}\right)\left(1 + \mu\bar{\mu}' e^{2i\omega}\right)^{-1}\right], \qquad (1.39)$$

$\ln z$ is defined by its principal value ($\ln re^{i\theta} = \ln r + i\theta$, $r > 0$, $-\pi < \theta \leqslant \pi$), and ω' is defined mod 2π. It is easy to check that μ, ω are appropriate Lie group parameters [115]. We now have a topological characterization of $SL(2,R)$ as the product of the open unit disk $|\mu| < 1$ and the circle $-\pi < \omega \leqslant \pi$, mod 2π.

The universal covering group $\widetilde{SL}(2,R)$ of $SL(2,R)$ is the Lie group with elements

$$\widetilde{SL}(2,R) = \{\{\mu, \omega\}: |\mu| < 1, \ -\infty < \omega < \infty\}.$$

Here, distinct values of ω correspond to distinct group elements. Group multiplication is defined by (1.39) except that ω'' is no longer defined mod 2π. There is a homomorphism of $\widetilde{SL}(2,R)$ onto $SL(2,R)$ given by $\{\mu, \omega\} \to (\mu, \omega)$ and the elements $\{0, 2\pi n\}$, $n = 0, \pm 1, \pm 2, \ldots$, of $\widetilde{SL}(2,R)$ are exactly those which map onto the identity element $(0,0)$ of $SL(2,R)$.

Finally, it is easy to verify that any element of $\widetilde{SL}(2,R)$ can be factored in the form

$$\{\mu, \omega\} = \{0, -\theta/2\}\{r, 0\}\{0, \omega + \theta/2\}, \qquad \mu = re^{i\theta}, \qquad (1.40)$$

and if $r > 0$, $-\pi < \theta \leqslant \pi$, this factorization is unique.

ISBN-0-201-13503-5

It is now straightforward to show that the $\mathcal{K}$ operators exponentiate to a global unitary irreducible representation of the simply connected covering group $\tilde{G}_2$ of G_2. Indeed, from the known recurrence formulas for the Hermite polynomials one can check that the operators $\mathcal{L}_1$, $\mathcal{L}_2$, $\mathcal{L}_3$ acting on the $f^{(4)}$ basis define a reducible representation of $sl(2,R)$ belonging to the discrete series. We will work out these recurrence formulas in Section 2.2.) The value of the Casimir operator is $\frac{1}{4}(\mathcal{L}_1{}^2 + \mathcal{L}_2{}^2 - \mathcal{L}_3{}^2) = -\frac{3}{16}$. As first shown by Bargmann ([10]; see also [115]), this Lie algebra representation extends to a global unitary reducible representation of $\widetilde{SL}(2,R)$. Similarly, the operators $\mathcal{C}_1, \mathcal{C}_2, \mathcal{L}_3$ acting on the $f^{(4)}$ basis define the irreducible representation $(\lambda, l) = (-\frac{1}{2}, 1)$ of the Lie algebra of the harmonic oscillator group S [82]. Again this Lie algebra representation is known to generate a global unitary irreducible representation of S [80, 86]. Since from (1.40) we see that every operator from $\widetilde{SL}(2,R)$ can be written in the form $\exp(-(\theta/2)\mathcal{L}_3)\exp(-\tau\mathcal{L}_1)\exp\{[(\theta/2)+\omega]\mathcal{L}_3\}$ with $2\tau = \ln[(1+r)/(1-r)]$ where $\exp(\theta\mathcal{L}_3)$ also belongs to S, and since $\mathcal{L}$ is a first-order operator whose exponential is easily determined, we can check that the identity (1.18) holds in general. (That is, we replace $\mathbf{T}(A)$ by $\exp(-(\theta/2)\mathcal{L}_3)\exp(-\tau\mathcal{L}_1)\exp\{[(\theta/2)+\omega]\mathcal{L}_3\}$ and use (1.35), (1.37), (1.38), (1.40) to express $\alpha,\beta,\gamma,\delta$ in terms of θ,τ,ω on the right-hand side of (1.18).) Then expressions (1.19) define $\tilde{G}_2$ as a semidirect product of $\widetilde{SL}(2,R)$ and W_1. Thus our representation of $\mathcal{G}_2$ extends to a global unitary representation $\mathbf{U}$ of $\tilde{G}_2$ that is irreducible since $\mathbf{U}|S$ is already irreducible.

The unitary operators $\mathbf{U}(g)$ on $L_2(R)$ are easily computed. The operators

$$\mathbf{U}(u,v,\rho) = \exp([\rho + uv/4]\mathcal{E})\exp(u\mathcal{C}_2)\exp(v\mathcal{C}_1)$$

defining a representation of W_1 take the form

$$\mathbf{U}(u,v,\rho)h(x) = \exp[i(\rho + uv/4 + ux/2)]h(x+v), \qquad h \in L_2(R). \quad (1.41)$$

The operators $\mathbf{U}\{\mu,\omega\}$, $\{\mu,\omega\} \in \widetilde{SL}(2,R)$, are more complicated. Here $\exp(a\mathcal{K}_{-2})$ is given by (1.25) and it is elementary to show

$$\exp(b\mathcal{K}^0)h(x) = \exp(b/2)h(e^b x), \qquad \exp(c\mathcal{K}_2)h(x) = \exp(icx^2/4)h(x). \quad (1.42)$$

Relations (1.17), (1.39) imply

$$\exp(\varphi\mathcal{L}_2) = \exp(\tanh(\varphi)\mathcal{K}_2)\exp(\sinh(\varphi)\cosh(\varphi)\mathcal{K}_{-2})\exp(-\ln\cosh(\varphi)\mathcal{K}^0),$$

ISBN-0-201-13503-5

so (1.25) and (1.42) yield

$$\exp(\varphi \mathcal{L}_2)h(x) = \frac{\exp\left[(ix^2/4)\tanh\varphi\right]}{(4\pi i \sinh\varphi)^{1/2}}\text{l.i.m.}$$

$$\int_{-\infty}^{\infty} \exp\left[\frac{-(x-y\cosh\varphi)^2}{4i\sinh\varphi\cosh\varphi}\right]h(y)\,dy, \qquad (1.43)$$

$$\varphi \neq 0.$$

A similar computation for $\exp(\theta \mathcal{L}_3)$ gives

$$\exp(\theta \mathcal{L}_3)h(x) = \frac{\exp\left[(ix^2/4)\cot\theta\right]}{(4\pi i \sin\theta)^{1/2}}\text{l.i.m.}$$

$$\int_{-\infty}^{\infty} \exp\left[\frac{-(y^2\cos\theta-2xy)}{4i\sin\theta}\right]h(y)\,dy, \qquad (1.44a)$$

$$0<|\theta|<\pi,$$

$$\exp(2\pi \mathcal{L}_3)h(x) = -h(x). \qquad (1.44b)$$

The general group operator $U(g)$ can be obtained from these results.

From (1.25) we see that the ON basis functions $f_n^{(4)}(x)$ map to the ON basis functions $F_n^{(4)}(t,x)=\exp(t\mathcal{K}_{-2})f_n^{(4)}(x)$ or

$$F_n^{(4)}(t,x) = \left\{n!2^n\left[2\pi(1+t^2)\right]^{1/2}\right\}^{-1/2}\exp\left(\frac{i}{4}\frac{x^2t}{1+t^2} - \frac{x^2}{4(1+t^2)}\right.$$

$$\left. -i\left(n+\frac{1}{2}\right)\tan^{-1}t\right)H_n\left\{x/\left[2(1+t^2)\right]^{1/2}\right\}, \quad n=0,1,2,\ldots, \qquad (1.45)$$

which are solutions of (1.28). This result can be derived from (1.30) or 4, Table 6. Indeed we know that variables $\{u,v\}$ R-separate in the integral expression for (1.45) where $u=x/(1+t^2)^{1/2}$, $v=t$, and $\mathcal{R}=iu^2v/4$. Applying the standard methods discussed in Chapter 1, we obtain expressions (1.45).

Next we study the spectral theory for the orbit containing the operators $\mathcal{K}_{-2}+\mathcal{K}_2$ (repulsive oscillator) and $\mathcal{K}^0$. Since the spectral analysis for $\mathcal{K}^0$ is more elementary, we study it first. (The corresponding results for

ISBN-0-201-13503-5

$\mathcal{K}_{-2}+\mathcal{K}_2$ will then follow by application of the unitary operator $\mathcal{G}=\exp((-\pi/4)\mathcal{L}_3)$, (1.21), (1.23).) The eigenfunction equation is

$$i\mathcal{K}^0 f=\lambda f,\qquad \mathcal{K}^0=x\,\partial_x+\tfrac{1}{2}.$$

The spectral resolution for this operator is well known [128]. It is obtained by considering $L_2(R)$ as the direct sum $L_2(R-)\oplus L_2(R+)$ of square-integrable functions on the negative and positive reals, respectively, and taking the Mellin transform of each component. Then $i\mathcal{K}^0$ transforms into multiplication by the transform variable. The spectrum is continuous and covers the real axis with multiplicity two. The generalized eigenfunctions are

$$f_\lambda^{(3)\pm}(x)=(2\pi)^{-1/2}x_\pm^{-i\lambda-\frac{1}{2}},\qquad -\infty<\lambda<\infty,$$
$$\langle f_\lambda^{(3)\pm},f_\mu^{(3)\pm}\rangle=\delta(\lambda-\mu),\qquad \langle f_\lambda^{(3)\pm},f_\mu^{(3)\mp}\rangle=0,\tag{1.46}$$

where

$$x_+^\alpha=\begin{cases}x^\alpha & \text{if } x>0,\\ 0 & \text{if } x<0;\end{cases}\qquad x_-^\alpha=\begin{cases}0 & \text{if } x>0,\\ (-x)^\alpha & \text{if } x<0.\end{cases}$$

From (1.25) we find $F_\lambda^{(3)\pm}(t,x)=\exp(t\mathcal{K}_{-2})f_\lambda^{(3)\pm}(x)$ where

$$F_\lambda^{(3)\pm}(t,x)=\exp\left(-\frac{x^2}{8it}+\frac{\pi\lambda}{4}+\frac{i\pi}{8}\right)\frac{(2t)^{-i\lambda/2+\frac{1}{4}}}{(8\pi^2 it)^{1/2}}\Gamma\left(\frac{1}{2}-i\lambda\right)$$
$$\times D_{i\lambda-\frac{1}{2}}\left(-\frac{xe^{-i\pi/4}}{(2t)^{1/2}}\right),\qquad t>0,\tag{1.47}$$

$\Gamma(z)$ is a gamma function (Appendix B, Section 1) and $D_\nu(z)$ is a parabolic cylinder function (Appendix B, Section 4). (This follows from (1.25) by displacement of the integration contour from the positive real axis to a ray making an angle of $\pi/4$ with the real axis. We also use the fact that, from 3a in Table 6, we have pure separation of variables in the coordinates $u=x/\sqrt{t}$, $v=t$.) Also we have

$$F_\lambda^{(3)+}(t,x)=F_{-\lambda}^{(3)+}(-t,x),\qquad F_\lambda^{(3)-}(t,x)=F_\lambda^{(3)+}(t,-x).\tag{1.48}$$

It follows immediately from (1.46) that the $\{F_\lambda^{(3)\pm}\}$ form a basis for $L_2(R)$ with orthogonality relations

$$\langle F_\lambda^{(3)\pm},F_\mu^{(3)\pm}\rangle=\delta(\lambda-\mu),\qquad \langle F_\lambda^{(3)\pm},F_\mu^{(3)\mp}\rangle=0\tag{1.49}$$

for each fixed t. Application of these orthogonality and completeness

ISBN-0-201-13503-5

relations to expand an arbitrary $h \in L_2(R)$ yields the Hilbert space version of Cherry's theorem [28, 37], which is an expansion in terms of parabolic cylinder functions. Note that our expansion is simply related to the spectral resolution of the operator

$$K^0 = 2t\,\partial_t + x\,\partial_x + \tfrac{1}{2} = 2it\,\partial_{xx} + x\,\partial_x + \tfrac{1}{2}.$$

The next orbit we consider contains the operators $\mathcal{K}_{-2} - \mathcal{K}_1$ (linear potential) and $\mathcal{K}_2 + \mathcal{K}_{-1}$. Since the spectral analysis for the second operator is simpler, we study it. (The corresponding results for $\mathcal{K}_{-2} - \mathcal{K}_1$ will follow upon application of the unitary operator $\mathcal{J}^2 = \exp[(-\pi/2)\mathcal{L}_3]$, (1.21)–(1.23).) The eigenfunction equation is

$$i(\mathcal{K}_2 + \mathcal{K}_{-1})f = \lambda f, \qquad \mathcal{K}_2 + \mathcal{K}_{-1} = ix^2/4 + \partial_x.$$

The spectral resolution is easily obtained from the Fourier integral theorem. The spectrum is continuous and covers the real axis, and a basis of generalized eigenfunctions is

$$f_\lambda^{(2)}(x) = (2\pi)^{-1/2}\exp\left[-i(\lambda x + x^3/12)\right], \qquad -\infty < \lambda < \infty,$$
$$\langle f_\lambda^{(2)}, f_\mu^{(2)}\rangle = \delta(\lambda - \mu). \tag{1.50}$$

We find that

$$F_\lambda^{(2)}(t,x) = \exp\left[-\frac{i}{4}\left(\pi + \frac{1}{8v^2} - u^2 v + \frac{u}{v} + \frac{4\lambda}{v}\right)\right]2^{1/6}\mathrm{Ai}[2^{2/3}(\tfrac{u}{2} + \lambda)], \tag{1.51}$$

$$x = uv + (2v)^{-1}, \ t = v,$$

where $\mathrm{Ai}(z)$ is an Airy function

$$\mathrm{Ai}(z) = \pi^{-1}(z/3)^{1/2}K_{1/3}(2z^{3/2}/3), \qquad |\arg z| < 2\pi/3. \tag{1.52}$$

As usual, we derive (1.51) by R-separation of variables in (1.25). The $\{F_\lambda^{(2)}\}$ are basis functions for the operator $K_2 + K_{-1} = -it^2\partial_{xx} + (1 - tx)\partial_x - t/2 + ix^2/4$.

Finally, for the operator $\mathcal{K}_{-1} = \partial_x$ a complete set of generalized eigenfunctions is

$$f_\lambda^{(1)}(x) = (2\pi)^{-1/2}e^{-i\lambda x}, \qquad -\infty < \lambda < \infty,$$
$$i\mathcal{K}_{-1}f_\lambda^{(1)} = \lambda f_\lambda^{(1)}, \qquad \langle f_\lambda^{(1)}, f_\mu^{(1)}\rangle = \delta(\lambda - \mu). \tag{1.53}$$

Furthermore,

$$F_\lambda^{(1)}(t,x) = \exp(t\mathcal{K}_{-2})f_\lambda^{(1)}(x) = (2\pi)^{-1/2}\exp\left[i(\lambda^2 t - \lambda x)\right]. \tag{1.54}$$

ISBN-0-201-13503-5

If $\{f_\lambda(x)\}$ is a basis of (generalized) eigenfunctions of some $\mathcal{K} \in \mathcal{G}_2$ and $F_\lambda(t,x) = \exp(t\mathcal{K}_{-2})f_\lambda(x)$, then $F_\lambda(\tau,x) = \exp([\tau - t]\mathcal{K}_{-2})F_\lambda(t,x)$ and we have the Hilbert space expansions

$$k(t,x-y) = \int F_\lambda(t,x)\bar{f}_\lambda(y)\,d\lambda,$$

$$k(\tau - t, x - y) = \int F_\lambda(\tau,x)\bar{F}_\lambda(t,y)\,d\lambda \qquad (1.55)$$

where the integration domain is the spectrum of $i\mathcal{K}$ and

$$k(t,x) = (4\pi i t)^{-\frac{1}{2}}\exp(-x^2/4it)$$

is the kernel of the integral operator $\exp(t\mathcal{K}_{-2})$. These expansions are known as continuous generating functions [35, 136].

Now we compute the overlap functions $\langle f_\lambda^{(i)}, f_\mu^{(j)} \rangle$ which allow us to expand eigenfunctions $f_\lambda^{(i)}$ in terms of eigenfunctions $f_\mu^{(j)}$. Since $\langle U(g)f_\lambda^{(i)}, U(g)f_\mu^{(j)} \rangle = \langle f_\lambda^{(i)}, f_\mu^{(j)} \rangle$, the same expressions allow us to expand eigenfunctions $U(g)f_\lambda^{(i)}$ in terms of eigenfunctions $U(g)f_\mu^{(j)}$. In particular, for $U(g) = \exp(t\mathcal{K}_{-2})$ we have $\langle F_\lambda^{(i)}, F_\mu^{(j)} \rangle = \langle f_\lambda^{(i)}, f_\mu^{(j)} \rangle$ for fixed t, and this permits us to expand one basis of solutions for the free-particle Schrödinger equation in terms of another basis.

We give here some overlaps of special interest.

$$\langle f_n^{(4)}, f_\lambda^{(3)\pm} \rangle = (2)^{(3n/2) + i\lambda - \frac{1}{2}}\frac{\Gamma\left(i\lambda/2 + \frac{1}{4} + n/2\right)}{2\pi(2^n n!)^{1/2}}$$

$$\times \left\{\begin{array}{c} +1 \\ (-1)^n \end{array}\right\}\, {}_2F_1\left[\begin{array}{c} -n/2, \frac{1}{2} - n/2 \\ \frac{3}{4} - i\lambda/2 - n/2 \end{array}\bigg|\begin{array}{c} 1 \\ 2 \end{array}\right]. \qquad (1.56)$$

For the calculation of the overlaps $\langle f_n^{(4)}, f_\lambda^{(2)} \rangle$ it is convenient to give a generating function rather than an explicit expression. The result is

$$2^{2/3}\exp\left\{-i\left[\tfrac{1}{6} + \lambda + (2y)^{1/2}\right]\right\}\mathrm{Ai}\left\{2^{2/3}\left[\tfrac{1}{4} - i\lambda - i(2y)^{1/2}\right]\right\}$$

$$= \sum_{n=0}^{\infty}\frac{\left(\sqrt{2}\,iy\right)^n}{(n!)^{1/2}}\langle f_n^{(4)}, f_\lambda^{(2)} \rangle. \qquad (1.57)$$

This expression follows from the form of the generating function for Hermite polynomials that we will derive in Section 2.2.

$$\langle f_n^{(4)}, f_\lambda^{(1)} \rangle = \left[n!(-2)^n\pi\right]^{-1/2}\exp(-\lambda^2)H_n\left[(2\lambda)^{1/2}\right], \qquad (1.58)$$

$$\langle f_\lambda^{(2)}, f_\mu^{(1)} \rangle = 2^{2/3}\mathrm{Ai}\left[2^{2/3}(\mu - \lambda)\right]. \qquad (1.59)$$

Additional overlaps can be found in [59].

ISBN-0-201-13503-5

The computation of mixed-basis matrix elements $\langle U(g)f_\lambda^{(i)}, f_\mu^{(j)} \rangle$ permits the derivation of many more expansions relating solutions of the Schrödinger equations. For example, we have

$$\langle \exp(t\mathcal{K}_{-2})f_n^{(4)}, f_\mu^{(3)+} \rangle = \langle f_n^{(4)}, \exp(-t\mathcal{K}_{-2})f_\mu^{(3)+} \rangle$$

$$= \frac{(2)^{(3n/2)+i\mu-\frac{1}{2}}(1+it)^{i\mu/2}}{(2\pi)^{3/4}(2^n n!)^{1/2}(1-it)^{n/2+\frac{1}{4}}} \exp\left[-i\left(n+\tfrac{1}{2}\right)\tan^{-1}t\right]$$

$$\cdot \Gamma\left(\frac{i\mu}{2}+\frac{1}{4}+\frac{n}{2}\right){}_2F_1\left[\begin{array}{c} -n/2, \tfrac{1}{2}-n/2 \\ \tfrac{3}{4}-i\mu/2-n/2 \end{array}\middle| \frac{1-it}{2}\right], \qquad (1.60)$$

with a similar result for $f_\mu^{(3)-}$. This expression allows us to expand Hermite polynomials as an integral over parabolic cylinder functions and parabolic cylinder functions in series of Hermite polynomials.

The matrix elements $\langle U(g)f_n^{(4)}, f_m^{(4)} \rangle = \langle T(g)F_n^{(4)}, F_m^{(4)} \rangle$ are easily computed and of great interest. However, in Section 2.2 we will apply Weisner's method to the complex heat equation and derive expansions for Hermite polynomials that yield these matrix elements as special cases.

It is also of great interest to compute the matrix elements with respect to the basis $\{f_\lambda^{(3)\pm}\}$ of generalized eigenvectors for $\mathcal{K}^0$. In this case the addition theorem for the matrix elements becomes an integral. In [128] Vilenkin has computed these matrix elements for the subgroup of $\tilde{G}_2$ whose Lie algebra has basis $\{\mathcal{K}_1, \mathcal{K}_{-1}, \mathcal{K}_0, \mathcal{K}^0\}$. The group operators are $U(a,b,c,\tau)$ where

$$U(a,b,c,\tau)h(x) = \exp(a\mathcal{K}_1)\exp(c\mathcal{K}_0)\exp(\tau\mathcal{K}^0)\exp(b\mathcal{K}_{-1})h(x)$$

$$= \exp(\tau/2 + iax/2 + ic)h(e^\tau x + b), \qquad (1.61)$$

$$a,b,c,\tau \in R, \ h \in L_2(R).$$

(The rule for group multiplication can easily be determined from (1.61).) Vilenkin shows that the matrix elements of the operator $U(a,b,c,\tau)$ in the $\{f_\lambda^{(3)\pm}\}$ basis can be expressed in terms of confluent hypergeometric functions ${}_1F_1$ and that the resulting addition theorems yield many interesting integral identities for these functions. Furthermore, just as in the analogous case for the group $E(1,1)$ (Section 1.5), we can allow the parameter a in (1.61) to become complex and derive more general integral identities. For these results see Vilenkin ([128]; also [81]).

2.2 The Heat Equation $(\partial_t - \partial_{xx})\Phi = 0$

The heat equation in two-dimensional space-time (with units appropriately normalized) is

$$Q\Phi = 0, \qquad Q = \partial_t - \partial_{xx}, \tag{2.1}$$

where t, x are the real time and space variables, respectively [107]. Clearly, this equation can be obtained from the Schrödinger equation by replacing t in (1.2) by $-it$, so the symmetry algebras of these equations are closely related. Indeed a simple computation shows that the symmetry algebra of (2.1) is six dimensional, with basis

$$
\begin{aligned}
H_2 &= t^2\partial_t + tx\,\partial_x + t/2 + x^2/4, \qquad H_1 = t\partial_x + x/2, \\
H_0 &= 1, \qquad H_{-1} = \partial_x, \qquad H_{-2} = \partial_t, \qquad H^0 = x\,\partial_x + 2t\,\partial_t + \tfrac{1}{2}
\end{aligned}
\tag{2.2}
$$

and commutation relations (H_0 commutes with everything),

$$\left[H^0, H_j\right] = jH_j, \quad j = \pm 2, \pm 1, 0, \qquad \left[H_1, H_2\right] = \left[H_{-1}, H_{-2}\right] = 0,$$

$$\left[H_{-1}, H_2\right] = H_1, \qquad \left[H_{-1}, H_1\right] = \tfrac{1}{2}H_0, \tag{2.3}$$

$$\left[H_{-2}, H_1\right] = H_{-1}, \qquad \left[H_{-2}, H_2\right] = H^0$$

We denote by $\mathcal{G}_2'$ the real Lie algebra with basis (2.2).

As usual we can exponentiate the elements of $\mathcal{G}_2'$ to obtain a local Lie group G_2' of operators acting on the space $\mathcal{F}$ of functions $\Psi(t, x)$ analytic in some given domain $\mathcal{D}$ of the x–t plane. The operators H_{-1}, H_1, H_0 form a basis for the Weyl algebra $\mathcal{W}_1$ and the corresponding action of the Weyl group W_1 is given by operators

$$\mathbf{T}(u, v, \rho) = \exp([\rho + uv/4]H_0)\exp(uH_1)\exp(vH_{-1}) \tag{2.4}$$

with multiplication

$$\mathbf{T}(u, v, \rho)\mathbf{T}(u', v', \rho') = \mathbf{T}(u + u', v + v', \rho + \rho' + (vu' - uv')/4) \tag{2.5}$$

where

$$\mathbf{T}(u, v, \rho)\Psi(t, x) = \exp\left[\rho + (uv + 2ux + u^2t)/4\right]\Psi(t, x + v + ut), \qquad \Psi \in \mathcal{F}.$$

The operators H_2, H_{-2}, H^0 form a basis for a subalgebra isomorphic to

ISBN-0-201-13503-5

$sl(2, R)$ and the corresponding action of $SL(2, R)$ is given by operators

$$\mathbf{T}(A)\Psi(t,x) = \exp\left(-\frac{x^2\beta/4}{\delta + t\beta}\right)(\delta + t\beta)^{-1/2}\Psi\left(\frac{\gamma + t\alpha}{\delta + t\beta}, \frac{x}{\delta + t\beta}\right) \quad (2.6)$$

where

$$A = \begin{pmatrix} \alpha & \beta \\ \gamma & \delta \end{pmatrix} \in SL(2, R).$$

Here,

$$\mathbf{T}\begin{pmatrix} 1 & \beta \\ 0 & 1 \end{pmatrix} = \exp(-\beta H_2), \qquad \mathbf{T}\begin{pmatrix} 1 & 0 \\ \gamma & 1 \end{pmatrix} = \exp(\gamma H_{-2}),$$

$$\mathbf{T}\begin{pmatrix} e^\alpha & 0 \\ 0 & -\alpha \end{pmatrix} = \exp(\alpha H^0). \quad (2.7)$$

The group $SL(2, R)$ acts on W_1 via the adjoint representation

$$\mathbf{T}^{-1}(A)\mathbf{T}(u, v, \rho)\mathbf{T}(A) = \mathbf{T}(u\delta - v\beta, v\alpha - u\gamma, \rho). \quad (2.8)$$

We can now define the symmetry group G_2' as a semidirect product of $SL(2, R)$ and W_1:

$$g = (A, \mathbf{w}) \in G_2', \quad A \in SL(2, R), \quad \mathbf{w} = (u, v, \rho) \in W_1,$$

$$\mathbf{T}(g) = \mathbf{T}(A)\mathbf{T}(\mathbf{w}), \quad (2.9)$$

$$\mathbf{T}(g)\mathbf{T}(g') = \mathbf{T}(AA')\left[\mathbf{T}^{-1}(A')\mathbf{T}(\mathbf{w})\mathbf{T}(A')\right]\mathbf{T}(\mathbf{w}') = \mathbf{T}(gg').$$

Clearly, the operators $\mathbf{T}(g)$ map solutions of (2.1) into solutions. Furthermore, G_2' acts on the Lie algebra $\mathcal{G}_2'$ of differential operators H via the adjoint representation

$$H \rightarrow H^g = \mathbf{T}(g)H\mathbf{T}^{-1}(g)$$

and this action splits $\mathcal{G}_2'$ into G_2' orbits.

It is straightforward to show that there are five orbits in $\mathcal{G}_2'/\{H_0\}$ under the adjoint representation (just as in Section 2.1 we ignore the center of $\mathcal{G}_2'$) with corresponding orbit representatives $H^0, H_2 + H_{-2}, H_{-2} + H_1, H_{-2}, H_{-1}$. Since $H_{-2} = (H_{-1})^2$ for solutions of the heat equation, only four R-separable coordinate systems are associated with the five orbits. The results are presented in Table 7. (See [59] for more details.) For each system $\{u, v\}$ we have $t = v$.

Table 7 R-separable Coordinates for the Equation $(\partial_t - \partial_{xx})\Phi(t,x)=0$

Operator H	Coordinates $\{u,v\}$	Multiplier $R=e^{\mathcal{R}}$	Separated solutions
1 H_{-1}, H_{-2}	$x=u$	$\mathcal{R}=0$	Product of exponentials
2 $H_{-2}+H_1$	$x=u+v^2/2$	$\mathcal{R}=-uv/2$	Airy times exponential function
3 H^0	$x=u\sqrt{v}$	$\mathcal{R}=0$	Hermite times exponential function
4 H_2+H_{-2}	$x=u(1+v^2)^{1/2}$	$\mathcal{R}=u^2v/4$	Parabolic cylinder times exponential function

The eigenfunctions of H^0 are of special interest. From Table 7, the eigenfunctions separate in the variables $u=x/\sqrt{t}$, $v=t$. Moreover, the solutions $\Phi_n(t,x)$ of the heat equation that satisfy $H^0\Phi_n=(n+\frac{1}{2})\Phi_n$, $n=0,1,2,\ldots$, are the *heat polynomials*

$$\Phi_n(t,x)=\left(i\sqrt{t}/2\right)^n H_n\left(ix/2\sqrt{t}\right). \tag{2.10}$$

(These functions are easily seen to be polynomials in t and x.) Rosenbloom and Widder [113] have presented a complete theory of the expansion of solutions of the heat equation in terms of heat polynomials.

Although the symmetries (2.6) are not very well known, there is a special case that has been of great importance in the theory of the heat equation. If in (2.6) we set

$$A_0=2^{-1/2}\begin{pmatrix} 1 & 1 \\ -1 & 1 \end{pmatrix}, \qquad A_0^8=\begin{pmatrix} 1 & 0 \\ 0 & 1 \end{pmatrix},$$

we find the symmetries

$$\mathbf{T}(A_0)\Psi(t,x)=\exp\left(\frac{-x^2/4}{1+t}\right)\left(\frac{\sqrt{2}}{1+t}\right)^{1/2}\Psi\left(\frac{t-1}{t+1}, \frac{(2x)^{1/2}}{t+1}\right),$$

$$\mathbf{T}(A_0^2)\Psi(t,x)=\exp\left(\frac{-x^2}{4t}\right)t^{-1/2}\Psi\left(\frac{-1}{t}, \frac{x}{t}\right). \tag{2.11}$$

The symmetry $\mathbf{T}(A_0^2)$ is called the *Appell transform* [4, 13]. We have embedded this transform in a Lie symmetry group.

It is well known that if $f(x)$ is a bounded continuous function defined on the real line, then there is exactly one solution $\Psi(t,x)$ of the heat equation (2.1), bounded and continuous in (t,x) for all $x\in R$ and $t\geq 0$ and continuously differentiable in t, twice continuously differentiable in x for all $x\in R$ and $t>0$, such that $\Psi(0,x)=f(x)$ [107]. This solution is given by the expression

$$\Psi(t,x)=(4\pi t)^{-1/2}\int_{-\infty}^{\infty}\exp\left[-(x-y)^2/4t\right]f(y)\,dy=I'(f). \tag{2.12}$$

ISBN-0-201-13503-5

Moreover,

$$\Psi(t,x) = \left[4\pi(t-\tau)\right]^{-1/2} \int_{-\infty}^{\infty} \exp\left[-(x-y)^2/4(t-\tau)\right]\Psi(\tau,y)\,dy, \qquad (2.13)$$

$$t > \tau,$$

which shows how to obtain the solution Ψ at time t given a knowledge of Ψ at an *earlier* time $\tau < t$.

Although some expansion theorems for solutions of (2.1) can be obtained through use of the time-independent form

$$(\Psi, \Phi) = \int_{-\infty}^{\infty} \Psi(t,x)\overline{\Phi}(-t,x)\,dx$$

where Ψ, Φ are solutions of the heat equation (see [113]), the operators (2.4)–(2.6) are not all unitary. There appears to be no convenient Hilbert space structure for this problem. Nonetheless, in analogy with our work for the Schrödinger equations, we can find another model of the group action that is very useful. To obtain this model we consider the operators (2.2) restricted to the solution space of the heat equation. Then we can replace ∂_t by ∂_{xx} in expressions (2.2) and consider $t \geqslant 0$ as a fixed parameter. We now think of the H operators as the symmetry operators at a fixed time t. At time $t = 0$ these operators become

$$\mathcal{K}_2 = x^2/4, \qquad \mathcal{K}_1 = x/2, \qquad \mathcal{K}_0 = 1,$$
$$\mathcal{K}_{-1} = \partial_x, \qquad \mathcal{K}_{-2} = \partial_{xx}, \qquad \mathcal{K}^0 = x\partial_x + \tfrac{1}{2} \qquad (2.14)$$

and when restricted, say, to the space $\mathcal{F}_0$ of infinitely differentiable functions $f(x)$ on R with compact support, the $\mathcal{K}$ operators satisfy the usual commutation relations (2.3).

A deeper understanding of this procedure follows from the observation that (2.12) has the interpretation

$$\Psi(t,x) = I'(f) = \exp(t\partial_{xx})f(x) = \exp(t\mathcal{K}_{-2})f(x), \qquad (2.15)$$

$$f \in \mathcal{F}_0, t > 0,$$

in analogy with (1.25). Then with integration by parts, we can check that

$$H\exp(t\mathcal{K}_{-2}) = \exp(t\mathcal{K}_{-2})\mathcal{K} \qquad (2.16)$$

where $H \in \mathcal{G}_2'$ and $\mathcal{K}$ is obtained from H by setting $t = 0$. (Precisely, if $\Psi(t,x) = I'(f)$, then $H\Psi(t,x) = I'(\mathcal{K}f)$.) Note that (2.16) is the counterpart of (1.26), except that here we avoid use of the unbounded operator $\exp(-t\mathcal{K}_{-2})$, $t > 0$. The theory leading to (2.16) appears to have been first studied by Hida [49].

ISBN-0-201-13503-5

Similarly we can derive results of the form

$$\exp(aH)\exp(t\mathcal{K}_{-2})=\exp(t\mathcal{K}_{-2})\exp(a\mathcal{K}). \qquad (2.17)$$

Just as in the preceding section, we can show that the equations

$$\partial_t\Psi(t,x)=\left(\partial_{xx}+ax\,\partial_x+b\,\partial_x+cx^2+dx+e\right)\Psi(t,x), \qquad a,\ldots,e\in R \quad (2.18)$$

all have symmetry algebras isomorphic to $\mathcal{G}_2'$ and that in fact these equations are all equivalent.

We present an example, due to Rosencrans [114], which exhibits this equivalence and shows how to make use of it to solve the Cauchy problem for each of the equations (2.18). (We shall present a formal argument. The rigorous validity of the result can easily be checked.)

We wish to determine the bounded solution $\Phi(t,x)$ of the heat equation with linear drift

$$\partial_t\Phi=\partial_{xx}\Phi-kx\,\partial_x\Phi, \qquad k>0, \qquad (2.19)$$

for all $t>0$ such that $\Phi(0,x)=f(x)$ where $f(x)$ is bounded and continuous on the real line. Now (2.19) reads $\partial_t\Phi=(\mathcal{K}_{-2}-k\mathcal{K}^0+(k/2)\mathcal{K}_0)\Phi$, $\Phi(0,x)=f(x)$, or

$$\Phi(t,x)=\exp\left[i\left(\mathcal{K}_{-2}-k\mathcal{K}^0+(k/2)\mathcal{K}_0\right)\right]f(x).$$

Since the $\mathcal{K}$ operators satisfy the same commutation relations as the H operators, we can use expressions (2.7) and group multiplication in $SL(2,R)$ to evaluate products of exponentials of the operators $\mathcal{K}_{-2}$, $\mathcal{K}^0$, $\mathcal{K}_0$. We find

$$\exp\left\{t\left[\mathcal{K}_{-2}-k\mathcal{K}^0+(k/2)\mathcal{K}_0\right]\right\}$$
$$=\exp\left[(tk/2)\mathcal{K}_0\right]\exp(-tk\mathcal{K}^0)\exp\left\{\left[(1-e^{-2kt})/2k\right]\mathcal{K}_{-2}\right\}$$
$$=\exp\left[(tk/2)\mathcal{K}_0\right]\exp\left[(e^{2kt}-1)/2k\,\mathcal{K}_{-2}\right]\exp(-tk\mathcal{K}^0). \qquad (2.20)$$

From (2.12), (2.15), and the easily verified relation

$$\exp(-tk\mathcal{K}^0)h(x)=\exp(-tk/2)h\left(\exp(-tk)x\right)$$

we obtain

$$\Phi(t,x)=\left\{\frac{2\pi}{k}\left[1-\exp(-2kt)\right]\right\}^{-1/2}\int_{-\infty}^{\infty}\exp\left\{\frac{k\left[\exp(-tk)x-y\right]^2}{2\left[1-\exp(-2kt)\right]}\right\}f(y)\,dy$$

$$(2.21)$$

as the solution to the Cauchy problem for (2.19).

ISBN-0-201-13503-5

Next we study the complex heat equation, that is, equation (2.1) with the variables t, x complex. It is easy to show that the symmetry algebra $\mathcal{G}_2^c$ of this equation is six dimensional with basis (2.2) and commutation relations (2.3). However, now the Lie algebra consists of all *complex* linear combinations of the basis elements. We can exponentiate the elements of $\mathcal{G}_2^c$ to obtain a local Lie group G_2^c of operators acting on the space $\mathcal{F}$ of functions $\Psi(t,x)$ analytic in some given domain $\mathcal{D}$ in the complex x–t plane. The group action is given by (2.4)–(2.9) where the parameters u, v, ρ are allowed to take arbitrary complex values and the matrices $A = \begin{pmatrix} \alpha & \beta \\ \gamma & \delta \end{pmatrix}$ now range over the group $SL(2, \mathcal{C})$ of all complex matrices with determinant $+1$: $\alpha, \beta, \gamma, \delta \in \mathcal{C}$, $\alpha\delta - \beta\gamma = 1$. As usual, the operators $\mathbf{T}(g), g \in G_2$, map solutions of the complex heat equation into solutions. Furthermore, G_2^c acts on the Lie algebra $\mathcal{G}_2^c$ of Lie derivatives H via the adjoint representation

$$H \to H^g = \mathbf{T}(g) H \mathbf{T}^{-1}(g)$$

and splits $\mathcal{G}_2^c$ (as well as $\mathcal{G}_2^c / \{H_0\}$) into G_2^c orbits. It is straightforward to show that there are exactly four orbits in $\mathcal{G}_2^c / \{H_0\}$ with orbit representatives $H^0, H_{-2} + H_1, H_{-2}, H_{-1}$. (The distinct orbits in $\mathcal{G}_2^c / \{H_0\}$ with representatives H^0 and $H_2 + H_{-2}$ become equivalent when the group G_2' is extended to G_2^c by complexification.) Since $H_{-2} = (H_{-1})^2$ when acting on solutions of the complex heat equation, there are only three R-separable coordinate systems associated with the four orbits. (It can be shown that these are the only R-separable systems admitted by the complex heat equation. Here an admissible coordinate system $\{u, v\}$ must be such that $u(t,x)$, $v(t,x)$ are complex analytic functions of (t,x) with nonzero Jacobian. Two separable systems are equivalent if one system can be mapped into the other by an element of G_2^c.) The results appear in Table 8, where $t = v$ for each separable system $\{u, v\}$.

Table 8 R-Separable Coordinates for the Complex Heat Equation

	Operator H	Coordinates $\{u,v\}$	Multiplier $R = e^{\mathcal{R}}$	Separated solutions
1	H_{-1}, H_{-2}	$x = u$	$\mathcal{R} = 0$	Product of exponentials
2	$H_{-2} + H_1$	$x = u + v^2/2$	$\mathcal{R} = -uv/2$	Airy times exponential function
3	H^0	$x = u\sqrt{v}$	$\mathcal{R} = 0$	Hermite times exponential function

Note that the complex heat equation is the complexification of both the real heat equation and the free-particle Schrödinger equation. The effect of the complexification in terms of separation of variables is that orbits 1 and 2 in Tables 6 and 7 correspond to orbits 1 and 2 in Table 8, while orbits 3 and 4 in Tables 6 and 7 collapse to the single orbit 3 in Table 8.

ISBN-0-201-13503-5

To derive identities relating separated solutions of the complex heat equation we can use Weisner's method and expand arbitrary analytic solutions in terms of the Hermite functions (orbit 3, Table 8). This was carried out in detail by Weisner [135] and the Lie algebraic aspects are covered in [82], so here we discuss only a few of the features of these expansions.

As suggested by (2.10) and 3, Table 8, for Hermite polynomial solutions of (2.1) the coordinates $\{s,z\}$ are appropriate, where

$$s = -i\sqrt{t}\,/2, \qquad z = ix/2\sqrt{t}\,. \tag{2.22}$$

In terms of these coordinates the operators (2.2) become

$$H_2 = -2s^2(z\,\partial_z + s\,\partial_s + 1 - 2z^2), \qquad H_1 = s(-\partial_z + 2z), \qquad H_0 = 1,$$

$$H_{-1} = \frac{s^{-1}}{4}\,\partial_z, \qquad H_{-2} = \frac{s^{-2}}{8}(z\,\partial_z - s\,\partial_s), \qquad H^0 = s\,\partial_s + \frac{1}{2}, \tag{2.23}$$

and the heat equation reads

$$(\partial_{zz} - 2z\,\partial_z + 2s\,\partial_s)\Phi(z,s) = 0. \tag{2.24}$$

Consider the solutions Φ of (2.24) which are eigenfunctions of H^0:

$$H^0\Phi = \left(n + \tfrac{1}{2}\right)\Phi \Rightarrow \Phi = f_n(z)s^n.$$

Substituting these solutions into (2.24) and comparing the resulting ordinary differential equation in z with (B.10), we find that the functions

$$\Phi_n(z,s) = H_n(z)s^n, \qquad \tilde{\Phi}_n(z,s) = e^{z^2}H_{-n-1}(iz)s^n \tag{2.25}$$

form a basis of simultaneous solutions where the *Hermite functions* $H_n(z)$ are defined by

$$H_n(z) = 2^{n/2}\exp(z^2/2)D_n(\sqrt{2}\,z), \qquad n \in \mathcal{C}, \tag{2.26}$$

and $D_n(z)$ is a parabolic cylinder function. If $n = 0,1,2,\ldots$, then $H_n(z)$ is the *Hermite polynomial* (B.12).

To understand the significance of the polynomial solutions, let us consider the system of equations

$$H_{-1}\Phi = 0, \qquad Q\Phi = 0,$$

which has the solution $\Phi \equiv 1$, unique to within a multiplicative constant. We will use this elementary solution and our knowledge of the symmetry

ISBN-0-201-13503-5

algebra $\mathscr{G}_2^c$ to construct other solutions. If $\Phi(z,s)$ is in analytic function of (z,s), we know from standard Lie theory that

$$\exp(\alpha H_1)\Phi(z,s) = \exp(2\alpha zs - \alpha^2 s^2)\Phi(z - \alpha s, s) = \sum_{n=0}^{\infty} \frac{\alpha^n}{n!}(H_1)^n\Phi(z,s).$$

Furthermore, if Φ is a solution of the complex heat equation, then $\exp(\alpha H_1)\Phi$ is a solution (provided it is well defined). Putting our solution $\Phi \equiv 1$ in this expression, we find

$$\exp(2\alpha zs - \alpha^2 s^2) = \sum_{n=0}^{\infty} \frac{\alpha^n}{n!}\Phi_n(z,s),$$

$$\Phi_0 = 1, \quad \Phi_n = (H_1)^n\Phi_0, \quad n = 1,2,\ldots. \tag{2.27}$$

Now consider the action of the symmetry operators H_j on the Φ_n. An elementary induction argument based on $[H_{-1}, H_1] = \frac{1}{2}H_0$ shows $[H_{-1},(H_1)^n] = (n/2)(H_1)^{n-1}$, $n = 1,2,\ldots$. Applying both sides of this identity to Φ_0, we obtain

$$H_{-1}\Phi_n = (n/2)\Phi_{n-1}, \qquad n = 1,2,\ldots. \tag{2.28}$$

(This expression makes sense for $n=0$ if we define $\Phi_n \equiv 0$ for $n < 0$.) By definition of Φ_n we also have

$$H_1\Phi_n = \Phi_{n+1}, \qquad n = 0,1,\ldots, \tag{2.29}$$

and, from $[H^0, H_1] = H_1$,

$$H^0\Phi_n = \left(n + \tfrac{1}{2}\right)\Phi_n, \qquad \Phi_n = f_n(z)s^n. \tag{2.30}$$

It follows from (2.30) that the $f_n(z)$ are expressible in terms of Hermite functions. Indeed, comparing (2.28), (2.29) with the recurrence relations (B.13), we find

$$\Phi_n(z,s) = H_n(z)s^n, \qquad n = 0,1,2,\ldots, \tag{2.31}$$

the Hermite polynomial solutions. Substituting (2.31) into (2.27) and setting $s = 1$, we obtain the fundamental generating function for Hermite polynomials

$$\exp(2\alpha z - \alpha^2) = \sum_{n=0}^{\infty} \frac{\alpha^n}{n!}H_n(z). \tag{2.32}$$

ISBN-0-201-13503-5

In addition to the recurrence formulas (2.28), (2.29) for the Hermite polynomials we can use the commutation relations to derive

$$H_{-2}\Phi_n = (n/4)(n-1)\Phi_{n-2}, \qquad H_2\Phi_n = \Phi_{n+2}, \qquad H_0\Phi_n = \Phi_n, \qquad (2.33)$$

$$n = 0, 1, 2, \ldots .$$

We can obtain many other identities obeyed by Hermite polynomials if we apply the group operators $\mathbf{T}(g), g \in G_2^c$ (see (2.9) to a basis element Φ_m and expand the result in terms of the $\{\Phi_n\}$ basis:

$$\mathbf{T}(g)\Phi_m(z,s) = \sum_{n=0}^{\infty} T_{nm}(g)\Phi_n(z,s), \qquad m = 0, 1, 2, \ldots . \qquad (2.34)$$

This procedure is practical provided we can compute the matrix elements $T_{nm}(g)$. However, for g close to the group identity element, these matrix elements can be computed directly from the Lie algebra relations (2.28)–(2.30) and (2.33).

To perform the computation it is convenient to construct a simpler model of the Lie algebra representation. We choose $f_n(w) = w^n$ and

$$H_{-1} = \frac{1}{2}\frac{d}{dw}, \qquad H_1 = w, \qquad H_2 = w^2,$$

$$H^0 = w\frac{d}{dw} + \frac{1}{2}, \qquad H_0 = 1, \qquad H_{-2} = \frac{1}{4}\frac{d^2}{dw^2}. \qquad (2.35)$$

These operators satisfy the commutation relations of $\mathcal{G}_2^c$ and their action on the basis functions $f_n(w)$ agrees with the action (2.28)–(2.30), (2.33) on the Φ_n basis. In terms of this model we define the matrix elements $T_{nm}(\alpha,\beta)$, $R_{nm}(\alpha,\beta)$:

$$\exp(\alpha H_1)\exp(\beta H_2)f_m(w) = \sum_{n=0}^{\infty} T_{nm}(\alpha,\beta)f_n(w), \qquad (2.36a)$$

$$\exp(\alpha H_1)\exp(\beta H_{-1})f_m(w) = \sum_{n=0}^{\infty} R_{nm}(\alpha,\beta)f_n(w); \qquad (2.36b)$$

that is, we apply the group operators $\exp(\alpha H)\exp(\beta H')$ to a basis function w^m and expand the resulting analytic functions in power series about $w=0$. The matrix elements are model independent. We will compute them using the simple model (2.35), then apply the results to the heat equation. Elementary Lie theory yields

$$\exp(\alpha H_1)f(w) = \exp(\alpha w)f(w), \qquad \exp(\beta H_2)f(w) = \exp(\beta w^2)f(w),$$

$$\exp(\beta H_{-1})f(w) = f(w + \beta/2).$$

ISBN-0-201-13503-5

Thus (2.36) becomes

$$\exp(\alpha w + \beta w^2) = \sum_{n=0}^{\infty} T_{nm}(\alpha, \beta) w^{n-m}, \qquad (2.37a)$$

$$\exp(\alpha w)(w + \beta/2)^m = \sum_{n=0}^{\infty} R_{nm}(\alpha, \beta) w^n. \qquad (2.37b)$$

These are well-known generating functions (2.32), (7.30), which yield

$$T_{nm}(\alpha, \beta) = \begin{cases} \dfrac{(-i\sqrt{\beta})^{n-m}}{(n-m)!} H_{n-m}\left(\dfrac{i\alpha}{2\sqrt{\beta}}\right), & n \geqslant m; \\ 0, & n < m; \end{cases} \qquad (2.38)$$

$$R_{nm}(\alpha, \beta) = \left(\frac{\beta}{2}\right)^{m-n} L_n^{(m-n)}\left(\frac{-\alpha\beta}{2}\right),$$

where $L_n^{(\alpha)}(z)$ is a Laguerre polynomial (see (B.9i)).

Now we exponentiate the operators (2.23). In addition to (2.27) we obtain

$$\exp(\alpha H_2)\Phi(z,s) = (1 + 4\alpha s^2)^{-1/2} \exp\left(\frac{4\alpha z^2 s^2}{1 + 4\alpha s^2}\right) \Phi\left[\frac{z}{(1 + 4\alpha s^2)^{1/2}},\right.$$

$$\left. \times \frac{s}{(1 + 4\alpha s^2)^{1/2}}\right], \qquad |4\alpha s^2| < 1,$$

$$\exp(\beta H_{-2})\Phi(z,s) = \Phi\left[\frac{z}{(1 - \beta/4s^2)^{1/2}}, s\left(1 - \frac{\beta}{4s^2}\right)^{1/2}\right], \qquad |\beta/4s^2| < 1,$$

$$\exp(\gamma H_{-1})\Phi(z,s) = \Phi(z + \gamma/4s, s),$$

$$\exp(\delta H^0)\Phi(z,s) = \exp(\delta/2)\Phi(z, e^{\delta}s),$$

$$\exp(\varphi H_0)\Phi(z,s) = e^{\varphi}\Phi(z,s). \qquad (2.39)$$

Substituting (2.38) and (2.39) into (2.36), we find (after simplification)

$$(1 - s^2)^{-(m+1)/2} \exp\left[\frac{2zs\alpha - (z^2 + \alpha^2)s^2}{1 - s^2}\right] H_m\left(\frac{z - s\alpha}{(1 - s^2)^{1/2}}\right)$$

$$= \sum_{n=0}^{\infty} \frac{(s/2)^n}{n!} H_n(\alpha) H_{m+n}(z), \qquad |s| < 1, \qquad (2.40a)$$

ISBN-0-201-13503-5

$$\exp(-s^2\alpha^2 - 2zs\alpha)H_m(z + s\alpha - \beta/s)s^m$$

$$= \sum_{n=0}^{\infty} (-\beta)^{m-n}L_n^{(m-n)}(-\alpha\beta)H_n(z)s^n. \qquad (2.40b)$$

Expression (2.40a) is a generalization of Mehler's theorem [37, p. 194], to which it reduces in the case in which $m=0$ ($H_0(z)=1$),

$$(1-s^2)^{-1/2}\exp\left[\frac{2zs\alpha - (z^2 + \alpha^2)s^2}{1-s^2}\right] = \sum_{n=0}^{\infty}\frac{(s/2)^n}{n!}H_n(\alpha)H_n(z), \qquad (2.41)$$

$$|s| < 1.$$

For $\beta = 0$, $s = 1$, expression (2.40b) simplifies to

$$\exp(-\alpha^2 - 2z\alpha)H_m(z + \alpha) = \sum_{n=0}^{\infty}\frac{(-\alpha)^n}{n!}H_{m+n}(z) \qquad (2.42)$$

and for $\alpha = 0$, $s = 1$ it becomes

$$H_m(z - \beta) = \sum_{n=0}^{m}\binom{m}{n}\beta^{m-n}H_n(z) \qquad (2.43)$$

where $\binom{m}{n}$ is a binomial coefficient (see (B.1)). By computing additional matrix elements $T_{mn}(g)$ of G_2^c, we could derive further generating functions for the Hermite polynomials [135, 82].

We now discuss the Hermite function, nonpolynomial solutions of the complex heat equation, that is, the eigenfunctions $\Phi_n(z,s)$, (2.25), with $n \in \not\!\!C$, $n \neq 0, 1, 2, \ldots$. In particular, we will study the eigenfunctions

$$\Phi_\lambda(z,s) = H_\lambda(z)s^\lambda, \qquad \lambda \in \not\!\!C, \qquad (2.44)$$

where λ is not an integer. The Φ_λ satisfy

$$H^0\Phi_\lambda = (\lambda + \tfrac{1}{2})\Phi_\lambda, \qquad Q\Phi_\lambda = 0. \qquad (2.45)$$

The commutation relations $[H^0, H_j] = jH_j, j = 0, \pm 1, \pm 2$ imply that the operators H_j map a solution of (2.45) corresponding to eigenvalue λ into a solution corresponding to eigenvalue $\lambda + j$. Indeed, using the fundamental recurrence formulas (B.13) it is straightforward to show

$$H_{-1}\Phi_\lambda = \tfrac{\lambda}{2}\Phi_{\lambda-1}, \qquad H_{-2}\Phi_\lambda = \tfrac{\lambda}{4}(\lambda - 1)\Phi_{\lambda-2},$$

$$H_1\Phi_\lambda = \Phi_{\lambda+1}, \qquad H_2\Phi_\lambda = \Phi_{\lambda+2}, \qquad H_0\Phi_\lambda = \Phi_\lambda. \qquad (2.46)$$

ISBN-0-201-13503-5

These relations are the same as (2.28), (2.29), (2.33) except here λ is not an integer. By applying the operators H_j to a given Φ_λ we can thus obtain an infinite ladder of solutions $\Phi_{\lambda+n}$ where n runs over all integers.

For a study of the transformation properties of these solutions under G_2^c it is convenient to consider the operators $\exp(\alpha H_1)\exp(\beta H_2), \exp(\alpha H_1)$ $\exp(\beta H_{-1})$, and $\exp(\alpha H_2)\exp(\beta H_{-2})$. (We can obtain the general group action as a product of three such operators and the trivial $\exp(\gamma H^0)$ $\exp(\delta H_0)$.) The matrix elements are defined by

$$\exp(\alpha H_1)\exp(\beta H_2)\Phi_{\lambda+m} = \sum_{n=-\infty}^{\infty} \hat{T}_{nm}(\alpha,\beta)\Phi_{\lambda+n}, \qquad (2.47a)$$

$$\exp(\alpha H_1)\exp(\beta H_{-1})\Phi_{\lambda+m} = \sum_{n=-\infty}^{\infty} \hat{R}_{nm}(\alpha,\beta)\Phi_{\lambda+n}, \qquad (2.47b)$$

$$\exp(\alpha H_2)\exp(\beta H_{-2})\Phi_{\lambda+m} = \sum_{n=-\infty}^{\infty} \hat{S}_{nm}(\alpha,\beta)\Phi_{\lambda+n}. \qquad (2.47c)$$

From (2.46) it is easy to see that the $T_{nm}(\alpha,\beta)$ are identical with the matrix elements $T_{nm}(\alpha,\beta)$, (2.38), except that now m and n may take negative integral values. Thus, (2.47a) becomes (2.40a) and H_m replaced by $H_{\lambda+m}$, H_{m+n} replaced by $H_{\lambda+m+n}$, where m can take on all integer values. This is a further generalization of Mehler's theorem.

To compute the matrix elements $\hat{R}_{nm}(\alpha,\beta)$ we choose a simpler model of some of the relations (2.46). Indeed, the choices $h_{\lambda+m}|(w) = w^m$, $m = 0, \pm 1, \pm 2, \ldots$,

$$H_1 = w, \qquad H_{-1} = \frac{1}{2}\frac{d}{dw} + \frac{\lambda}{2w}, \qquad H_0 = 1$$

satisfy $[H_1, H_{-1}] = -\frac{1}{2}H_0$ and

$$H_1 h_{\lambda+m} = h_{\lambda+m+1}, \qquad H_{-1}h_{\lambda+m} = \left(\frac{\lambda+m}{2}\right)h_{\lambda+m-1},$$

in agreement with (2.46). In this model

$$\exp(\alpha H_1)\exp(\beta H_{-1})h_{\lambda+m}(w) = \exp(\alpha w)(1+\beta/2w)^{\lambda+m}w^m$$

$$= \sum_{n=-\infty}^{\infty} \hat{R}_{nm}(\alpha,\beta)w^n. \qquad (2.48)$$

Computing the coefficient of w^n, we find

$$\hat{R}_{nm}(\alpha,\beta) = \left(\frac{\beta}{2}\right)^{m-n} L_{\lambda+n}^{(m-n)}\left(\frac{-\alpha\beta}{2}\right) \qquad (2.49)$$

ISBN-0-201-13503-5

where $L_\lambda^{(\nu)}(z)$ is a generalized Laguerre function,

$$L_\lambda^{(\nu)}(z) = \frac{\Gamma(\nu+\lambda+1)}{\Gamma(\nu+1)\Gamma(\lambda+1)} {}_1F_1\left(\begin{array}{c|c} -\lambda \\ \nu+1 \end{array} z\right), \qquad (2.50)$$

proportional to a general ${}_1F_1$. Thus, (2.47b) becomes

$$\exp(-s^2\alpha^2 - 2zs\alpha)H_{\lambda+m}(z+s\alpha-\beta/s)s^m$$

$$= \sum_{n=-\infty}^{\infty} (-\beta)^{m-n}L_{\lambda+n}^{(m-n)}(-\alpha\beta)H_{\lambda+n}(z)s^n, \qquad (2.51)$$

$$m = 0, \pm 1, \pm 2, \ldots .$$

To compute the matrix elements $\hat{S}_{nm}(\alpha,\beta)$ we choose another model:

$$h_{\lambda+m}(w) = \Gamma\left(\frac{\lambda+m+2}{2}\right)w^m, \qquad H_2 = \frac{w^3}{2}\frac{d}{dw} + \frac{(\lambda+2)}{2}w^2,$$

$$H_{-2} = \frac{1}{2w}\frac{d}{dw} + \frac{\lambda-1}{2w^2}, \qquad H^0 = w\frac{d}{dw} + \lambda + \frac{1}{2}.$$

With these operators

$$\exp(\alpha H_2)\exp(\beta H_{-2})h_{\lambda+m}(w)$$

$$= \Gamma\left(\frac{\lambda+m+2}{2}\right)w^m(1-\alpha w^2)^{-(\lambda+m+2)/2}$$

$$\times \left(1+\frac{\beta}{(1-\alpha\beta)w^2}\right)^{(\lambda+m-1)/2}(1-\alpha\beta)^{(\lambda+m-1)/2}$$

$$= \sum_{n=-\infty}^{\infty}\hat{S}_{nm}(\alpha,\beta)\Gamma\left(\frac{\lambda+n+2}{2}\right)w^n,$$

$$|\alpha w^2| < 1, \quad |\beta| < |(1-\alpha\beta)w^2|, \qquad (2.52)$$

so

$$\hat{S}_{nm}(\alpha,\beta) = \frac{(1-\alpha\beta)^{(\lambda+m-1)/2}\alpha^{(n-m)/2}}{\Gamma((n-m+2)/2)}$$

$$\times {}_2F_1\left(\begin{array}{c|c} \dfrac{\lambda+n+2}{2}, \dfrac{1-\lambda-m}{2} \\ \dfrac{n-m+2}{2} \end{array} \dfrac{-\alpha\beta}{1-\alpha\beta}\right) \qquad \text{if} \quad n-m \text{ even,}$$

$$\hat{S}_{nm}(\alpha,\beta) = 0 \qquad \text{if} \quad n-m \text{ odd.} \qquad (2.53)$$

ISBN-0-201-13503-5

(We use the fact that $_2F_1(a,b;\,c;\,z)/\Gamma(c)$ is an entire function of c to make sense of these expressions for $m > n$.) Thus (2.47c) becomes

$$(1+4\alpha s^2)^{-(\lambda+m+1)/2}\left(1-\alpha\beta-\frac{\beta}{4s^2}\right)^{(\lambda+m)/2}\exp\left(\frac{4\alpha z^2 s^2}{1+4\alpha s^2}\right)$$

$$\times H_{\lambda+m}\left[z\left[(1+4\alpha s^2)\left(1-\alpha\beta-\frac{\beta}{4s^2}\right)\right]^{-1/2}\right]s^m$$

$$= \sum_{n=-\infty}^{\infty}\hat{S}_{nm}(\alpha,\beta)H_{\lambda+n}(z)s^n, \qquad \left|\frac{\beta}{(1-\alpha\beta)}\right| < |4s^2| < |\alpha|^{-1}. \quad (2.54)$$

Next we present a simple application of the general Weisner method to a case where the expansion coefficients in a Hermite polynomial basis cannot be computed from the Lie algebra alone. Consider the function $\Psi(z,s) = \exp(-4H_{-2})\Phi_\lambda(z,s)$ where Φ_λ is given by (2.25), (2.26) with $\lambda = n \in \mathbb{C}$ and $|s| < 1$. Then

$$\Psi(z,s) = H_\lambda\left[\frac{sz}{(1+s^2)^{1/2}}\right](1+s^2)^{\lambda/2} = \sum_{j=0}^{\infty} f_j(z)s^j, \qquad |s| < 1. \quad (2.55)$$

(Note that this expansion is not a special case of (2.54). Since $Q\Psi = 0$, it follows that $Q(f_j(z)s^j) = 0$ for each j, hence $f_j(z)$ is a linear combination of the H^0 basis functions Φ_j and $\tilde{\Phi}_j$; see (2.25). However, since $H_\lambda(w)$ is an entire function of w, it follows easily from (2.55) that the highest power of z occurring in $f_j(z)$ is z^j. Thus $f_j(z) = c_j H_j(z)$. Setting $z = w^{-1}, s = wv$ in (2.55) and letting w go to zero, we obtain

$$H_\lambda(v) = \sum_{j=0}^{\infty} c_j(2v)^j.$$

However, the special case of (2.51) with $\beta = -v, s = 1, m = 0$ yields

$$H_\lambda(z+v) = \sum_{j=0}^{\infty}\binom{\lambda}{j}H_{\lambda-j}(z)(2v)^j. \quad (2.56)$$

Thus $c_j = \binom{\lambda}{j}H_{\lambda-j}(0)$. This result suggests the existence of a more general generating function. Indeed, consideration of $\exp(4wH_{-1} - 4H_{-2})\Phi_\lambda$ leads to the generating function

$$(1+s^2)^{\lambda/2}H_\lambda\left[\frac{w+zs}{(1+s^2)^{1/2}}\right] = \sum_{j=0}^{\infty}\binom{\lambda}{j}H_{\lambda-j}(w)H_j(z)s^n, \qquad |s| < 1. \quad (2.57)$$

ISBN-0-201-13503-5

For derivations of this and many other generating functions for Hermite polynomials, including some from the Airy basis, see [135].

2.3 Separation of Variables for the Schrödinger Equation $(i \partial_t + \partial_{xx} - a/x^2)\Psi = 0$

Next we apply the methods discussed in Section 2.1 to the radial Schrödinger equation for a free particle

$$i \partial_t \Psi = -\partial_{xx}\Psi + \frac{a}{x^2}\Psi. \qquad (3.1)$$

Here a is a nonzero real constant, t is real, and $x > 0$. As mentioned in the discussion following Table 5, this equation arises for certain values of $a > 0$ from free-particle Schrödinger equations for higher-dimensional spaces in which spherical coordinates have been introduced and all of the angular variables separated out (e.g., [70, p. 108]). Here $x = r$, the radial coordinate. We shall show that a group-theoretic study of (3.1) leads naturally to the Schrödinger equations for the radial harmonic oscillator and radial repulsive oscillator. Thus, our analysis of equations (1.2) and (3.1) will incorporate all seven potentials listed in Table 5.

A straightforward computation shows that the complex symmetry algebra of (3.1) is three dimensional, with basis

$$K_{-2} = \partial_t, \qquad K_2 = -t^2 \partial_t - tx \partial_x - \frac{t}{2} + \frac{ix^2}{4}, \qquad K^0 = 2t\partial_t + x\partial_x + \tfrac{1}{2} \quad (3.2)$$

and commutation relations

$$\left[K^0, K_{\pm 2}\right] = \pm 2K_{\pm 2}, \qquad \left[K_2, K_{-2}\right] = K^0.$$

For the alternate basis $\{L_j\}$ where

$$L_1 = K^0, \qquad L_2 = K_{-2} + K_2, \qquad L_3 = K_{-2} - K_2$$

we have the relations

$$\left[L_1, L_2\right] = -2L_3, \qquad \left[L_3, L_1\right] = 2L_2, \qquad \left[L_3, L_2\right] = -2L_1. \quad (3.3)$$

Comparing these results with (1.4), (1.5), (1.7), we see that the real Lie algebra generated by the basis elements is $sl(2, R)$ and that the corresponding local group action of $SL(2, R)$ on functions $\Phi(t, x)$ is given by the operators $T(A)$, (1.16). The explicit relations between the group and Lie algebra operators follow from (1.17). (Note, however, that in (1.16) we must require $x > 0$.)

ISBN-0-201-13503-5

The group $SL(2,R)$ acts on $sl(2,R)$ via the adjoint representation and splits the Lie algebra into orbits. Let

$$K = a_2 K_2 + a_{-2} K_{-2} + a_0 K^0 \in sl(2,R)$$

and set $\alpha = a_2 a_{-2} + a_0^2$. It is straightforward to check that α is invariant under the adjoint action and that K lies on the same $SL(2,R)$ orbit as exactly one of these three operators:

$$
\begin{array}{llll}
\text{Case 1} & (\alpha < 0) & K_{-2} - K_2 = L_3; & \\
\text{Case 2} & (\alpha > 0) & K^0; & (3.4) \\
\text{Case 3} & (\alpha = 0) & K_2. &
\end{array}
$$

Thus, there are three orbits.

The computation of all R-separable coordinate systems for (3.1) is easily carried out, due to the fact that an R-separable system must also be R-separable for the free-particle equation (set $a=0$ in (3.1)). Thus the possible systems are those listed in Table 6, subject to the additional requirement that they are still R-separable when the potential a/x^2 is added to the free-particle Hamiltonian. We find that only orbit 2 in Table 6 is lost. The results are presented in Table 9, where as usual we have $t = v$.

Table 9 R-Separable Coordinates for the Equation $(i\partial_t + \partial_{xx} - a/x^2)\Psi(t,x) = 0$

	Operator K	Coordinates $\{u,v\}$	Multiplier $R = e^{i\Re}$	Separated solutions
1a	K_{-2}	$x = u$	$\Re = 0$	Bessel times exponential function
1b	K_2	$x = uv$	$\Re = u^2 v/4$	Bessel times exponential function
2a	K^0	$x = u\sqrt{v}$	$\Re = 0$	Whittaker times exponential function
2b	$K_2 + K_{-2}$	$x = u[\pm(1-v^2)]^{1/2}$	$\Re = \pm u^2 v/4$	Whittaker times exponential function
3	$K_2 - K_{-2}$	$x = u(1+v^2)^{1/2}$	$\Re = u^2 v/4$	Laguerre times exponential function

Note that we have listed two coordinate systems on each of the orbits 1 and 2 even though the systems are $SL(2,R)$ equivalent. These systems are inequivalent with respect to the subgroup of "obvious symmetries" generated by time translation and dilatations. The exact relationship is

$$J(K_2 + K_{-2})J^{-1} = K^0, \qquad J^2 K_{-2} J^{-2} = -K_2, \qquad (3.5)$$

where J and J^2 are given by (1.21), (1.22).

In analogy with our argument in Section 2.1 we can interpret the operators (3.2) as a Lie algebra of skew-symmetric operators on the Hilbert

ISBN-0-201-13503-5

space $L_2(R+)$ of complex-valued Lebesgue square-integrable functions on the positive real line, $0 < x < \infty$. This is accomplished by considering t as a fixed parameter and replacing ∂_t with $i\partial_{xx} - ia/x^2$ in expressions (3.2). The resulting operators, when multiplied by i and restricted to the domain of infinitely differentiable functions with compact support in $R+$, are, via Weyl's theorem [122, p. 297], seen to be essentially self-adjoint provided $a \geqslant 3/4$. In the remainder of this section we assume that the constant a satisfies this inequality. We see then that the operators are real linear combinations of the skew-symmetric operators

$$\mathcal{K}_{-2} = i\partial_{xx} - ia/x^2, \qquad \mathcal{K}_2 = ix^2/4, \qquad \mathcal{K}^0 = x\partial_x + \tfrac{1}{2} \qquad (3.6)$$

to which they reduce when $t = 0$. Similarly, the skew-symmetric operators

$$\mathcal{L}_1 = \mathcal{K}^0 = x\partial_x + \tfrac{1}{2}, \qquad \mathcal{L}_2 = \mathcal{K}_{-2} + \mathcal{K}_2 = i\partial_{xx} - ia/x^2 + ix^2/4,$$
$$\mathcal{L}_3 = \mathcal{K}_{-2} - \mathcal{K}_2 = i\partial_{xx} - ia/x^2 - ix^2/4, \qquad (3.7)$$

satisfy relations (3.3) and the L_j reduce to $\mathcal{L}_j$ when $t = 0$.

In analogy with (1.26) we find

$$\exp(t\mathcal{K}_{-2})\mathcal{K}_j \exp(-t\mathcal{K}_{-2}) = K_j,$$
$$\exp(t\mathcal{K}_{-2})\mathcal{L}_j \exp(-t\mathcal{K}_{-2}) = L_j, \qquad (3.8)$$

where $\exp(t\mathcal{K}_{-2})$ is a unitary operator on $L_2(R+)$. Thus for any $f \in L_2(R+)$ the function $\Psi(t,x) = \exp(t\mathcal{K}_{-2})f(x)$ satisfies $\partial_t\Psi = \mathcal{K}_{-2}\Psi$ or $i\partial_t\Psi = -\partial_{xx}\Psi + a\Psi/x^2$ and $\Psi(0,x) = f(x)$. Also the unitary operators $\exp(K) = \exp(t\mathcal{K}_{-2})\exp(\mathcal{K})\exp(-t\mathcal{K}_{-2})$, $\mathcal{K} \in sl(2,R)$ map solutions of the equation $\partial_t\Psi = \mathcal{K}_{-2}\Psi$ into other solutions.

We will soon demonstrate that the operators $\mathcal{K}_{\pm 2}, \mathcal{K}^0$ generate a global unitary irreducible representation of the universal covering group $\widetilde{SL}$ $(2,R)$ of $SL(2,R)$, (1.37)–(1.40), by operators $U(g), g \in \widetilde{SL}(2,R)$, on $L_2(R+)$. Assuming this, we see that the operators $T(g) = \exp(t\mathcal{K}_{-2})U(g)\exp(-t\mathcal{K}_{-2})$ define a group of unitary symmetries for equation (3.1), with associated infinitesimal operators $K = \exp(t\mathcal{K}_{-2})\mathcal{K}\exp(-t\mathcal{K}_{-2})$. This discussion shows the relationship between our Lie algebra of $\mathcal{K}$ operators and the Schrödinger equation for the radial free particle.

Next consider the operator $\mathcal{L}_3 \in sl(2,R)$. If $f \in L_2(R+)$, then $\Psi(t,x) = \exp(t\mathcal{L}_3)f(x)$ satisfies $\partial_t\Psi = \mathcal{L}_3\Psi$ or $i\partial_t\Psi = -\partial_{xx}\Psi + a\Psi/x^2 + x^2\Psi/4$, the Schrödinger equation for the radial harmonic oscillator. The unitary operators $V(g) = \exp(t\mathcal{L}_3)U(g)\exp(-t\mathcal{L}_3)$ are symmetries of this equation and the associated infinitesimal operators $\exp(t\mathcal{L}_3)\mathcal{K}\exp(-t\mathcal{L}_3)$ can be writ-

ISBN-0-201-13503-5

ten as first-order linear differential operators in x and t. Similarly, if $f \in L_2(R+)$, then $\Psi(t,x) = \exp(t\mathcal{L}_2)f(x)$ satisfies $\partial_t\Psi = \mathcal{L}_2\Psi$ or $i\partial_t\Psi = -\partial_{xx}\Psi + a\Psi/x^2 - x^2\Psi/4$, the Schrödinger equation for the repulsive radial oscillator. The operators $\mathbf{W}(g) = \exp(t\mathcal{L}_2)\mathbf{U}(g)\exp(-t\mathcal{L}_2)$ determine the symmetry group of this equation and the associated infinitesimal operators $\exp(t\mathcal{L}_2)\mathcal{K}\exp(-t\mathcal{L}_2)$ can be written as first order in x and t.

It follows from these remarks that the Schrödinger equations (5)–(7) in Table 5 have isomorphic symmetry algebras. For each of these equations the algebra of symmetry operators at time $t = 0$ is $sl(2,R)$ with basis (3.6). Although we first derived this symmetry algebra through a study of the Schrödinger equation (5), we could also have obtained it from (6) or (7) in Table 5. Moreover, we have shown how to map a solution of any of these equations to a solution of another equation.

From (3.4) we see that the operators $\mathcal{K}_{-2}$, $\mathcal{L}_3$, $\mathcal{L}_2$, corresponding to the radial free particle, attractive, and repulsive harmonic oscillator Hamiltonians, lie on the same $SL(2,R)$ orbits as the three orbit representatives $\mathcal{K}_2, \mathcal{L}_3$, and $\mathcal{K}^0$, respectively. Our three Hamiltonians correspond to the three orbits of $sl(2,R)$. The remarks concerning expressions (1.31)–(1.33) and the invariance of spectra for operators on an orbit carry over without change to this case, except that the inner product is now

$$\langle h_1, h_2 \rangle = \int_0^\infty h_1(x)\bar{h}_2(x)\,dx, \qquad h_j \in L_2(R+). \tag{3.9}$$

Note that if $\{f_\lambda\}$ is the basis of generalized eigenvectors for some $\mathcal{K} \in sl(2,R)$, then $\{\Psi_\lambda(t,x) = \exp(t\mathcal{K}_{-2})f_\lambda(x)\}$ is the basis of eigenvectors for $K = \exp(t\mathcal{K}_{-2})\mathcal{K}\exp(-t\mathcal{K}_{-2})$ and the Ψ_λ satisfy the Schrödinger equation for the radial free particle. Similar remarks hold for the other Hamiltonians.

We first present the well-known results for the spectrum of $\mathcal{L}_3$. The eigenfunction equation is

$$i\mathcal{L}_3 f = \lambda f, \qquad (-\partial_{xx} + a/x^2 + x^2/4)f = \lambda f,$$

and the normalized eigenfunctions are

$$f_n^{(3)}(x) = \left(\frac{n!2^{-\mu/2}}{\Gamma(n+1+\mu/2)}\right)^{1/2} \exp\left(-\frac{x^2}{4}\right)x^{(\mu+1)/2}L_n^{(\mu/2)}\left(\frac{x^2}{2}\right),$$

$$\lambda = \lambda_n = -2n - \frac{\mu}{2} - 1, \qquad a = \frac{(\mu^2 - 1)}{4}, \qquad \mu \geq 2, n = 0,1,2,\ldots, \tag{3.10}$$

where $L_n^{(\alpha)}(z)$ is a Laguerre polynomial (see (B.9i)). The $\{f_n^{(3)}\}$ form an ON basis for $L_2(R+)$ [123, p. 108].

ISBN-0-201-13503-5

Using known recurrence relations for the Laguerre polynomials, (4.9), we can check that the operators $\pounds_j$ acting on the $f^{(3)}$ basis define an irreducible representation of $sl(2,R)$ belonging to the discrete series. The Casimir operator is $\frac{1}{2}(\pounds_1{}^2+\pounds_2{}^2-\pounds_3{}^2)=-3/16+a/4$. As is well known [10, 115], this Lie algebra representation extends to a global unitary irreducible representation of $\widetilde{SL}(2,R)$. The matrix elements of the operators $U(g)$ in an $f^{(3)}$ basis can be found in [115] or [87].

We now compute the operators $U(g)$ directly. Clearly,

$$\exp(a\mathcal{K}^0)h(x)=\exp(a/2)h(e^a x)$$

$$\exp(a\mathcal{K}_2)h(x)=\exp(iax^2/4)h(x), \qquad h\in L_2(R+).$$

Furthermore,

$$\exp(\beta\pounds_3)h(x)=\frac{\exp[\mp i\pi(\mu+2)/4]}{2|\sin\beta|}\text{l.i.m.}\int_0^\infty (xy)^{1/2}$$

$$\cdot\exp\left(\pm\frac{i}{4}(x^2+y^2)|\cot\beta|\right)J_{\mu/2}\left(\frac{xy}{2|\sin\beta|}\right)h(y)\,dy, \qquad (3.11)$$

$$0<|\beta|<\pi,$$

where we take the upper sign for $\beta>0$ and the lower for $\beta<0$. The additional relation $\exp(\pi\pounds_3)=\exp[-i\pi(1+\mu/2)]$ allows us to determine $\exp(\beta\pounds_3)$ for any β. To prove these results we apply the integral operator (3.11) to an $f^{(3)}$ basis element, using the Hille–Hardy formula, (4.27), and the fact that $\exp(\beta\pounds_3)f_n^{(3)}=\exp[-i\beta(2n+\mu/2+1)]f_n^{(3)}$ to check its validity. Since (3.11) is valid on an ON basis and $\exp(\beta\pounds_3)$ is unitary, the expression must be true for all $h\in L_2(R+)$.

The group multiplication formula

$$\exp(\gamma\mathcal{K}_{-2})=\exp(-\sin(\theta)\cos(\theta)\mathcal{K}_2)\exp(\ln\cos(\theta)\mathcal{K}^0)\exp(\theta\pounds_3)$$

with $\gamma=\tan\theta$ and expressions (3.10), (3.11) easily yield

$$\exp(\gamma\mathcal{K}_{-2})h(x)=\frac{\exp[\mp i\pi(\mu+2)/4]}{2|\gamma|}\text{l.i.m.}\int_0^\infty (xy)^{1/2}$$

$$\cdot\exp\left(i\frac{(x^2+y^2)}{4\gamma}\right)J_{\mu/2}\left(\frac{xy}{2|\gamma|}\right)h(y)\,dy, \qquad (3.12)$$

where we take the upper sign for $\gamma>0$ and the lower for $\gamma<0$. A similar

ISBN-0-201-13503-5

group-theoretic calculation gives

$$\exp(\varphi \mathcal{E}_2)h(x) = \frac{\exp\left[\mp i\pi(\mu+2)/4\right]}{2|\sinh\varphi|}\,\mathrm{l.i.m.}\int_0^\infty (xy)^{1/2}$$

$$\cdot \exp\left(\frac{i}{4}(x^2+y^2)\coth\varphi\right)J_{\mu/2}\left(\frac{xy}{2|\sinh\varphi|}\right)h(y)\,dy. \quad (3.13)$$

From (3.12) we find that the basis functions $f_n^{(3)}(x)$ map to the ON basis functions $\Psi_n^{(3)}(t,x) = \exp(t\mathcal{K}_{-2})f_n^{(3)}(x)$:

$$\Psi_n^{(3)}(t,x) = (-2)^n \exp\left[\pm i\pi\frac{(\mu+2)}{4}\right]\left(\frac{x^2}{1+t^2}\right)^{(\mu+1)/4}(t-i)^{-\mu/4-3/4-n}$$

$$\times (t+i)^{\mu/4+1/4+n}\exp\left[\frac{x^2(it-1)}{4(1+t^2)}\right]L_n^{(\mu/2)}\left(\frac{1}{2}\frac{x^2}{1+t^2}\right), \quad t\neq 0, \quad (3.14)$$

which are R-separable solutions of (3.1). (Indeed we can derive (3.14) from our knowledge that the $\Psi_n^{(3)}$ are R-separable solutions of the form 3 in Table 9.)

The $SL(2,R)$ orbit containing the operator $\mathcal{E}_2$ (repulsive radial oscillator) also contains $\mathcal{K}^0$, so we merely study the spectral theory for $\mathcal{K}^0$. The results are well known [128]. The eigenfunction equation is

$$i\mathcal{K}^0 f = \lambda f, \qquad \mathcal{K}^0 = x\partial_x + \tfrac{1}{2}.$$

The spectrum is continuous and covers the real axis with multiplicity one. The generalized eigenfunctions are

$$f_\lambda^{(2)}(x) = (2\pi)^{-1/2}x^{i\lambda-\frac{1}{2}}, \quad -\infty < \lambda < \infty, \qquad \langle f_\lambda^{(2)}, f_\zeta^{(2)}\rangle = \delta(\lambda-\zeta). \quad (3.15)$$

Using (3.12) and separation of variables, we find $\Psi_\lambda^{(2)}(t,x) = \exp(t\mathcal{K}_{-2})f_\lambda^{(2)}(x)$ where

$$\Psi_\lambda^{(2)}(t,x) = (2\pi)^{-1/2}\Gamma\left(\frac{i\lambda}{2}-\frac{\mu}{4}+\frac{1}{2}\right)\exp\left[-\frac{\pi}{4}\left(\frac{i\mu}{2}+i-\lambda\right)\right]2^{i\lambda-\mu/2-1}t^{i\lambda/2-\frac{1}{4}}$$

$$\times\left(\frac{x^2}{t}\right)^{(\mu+1)/4}L_{i\lambda/2-\mu/2-1/2}^{(\mu/2)}\left(\frac{ix^2}{4t}\right), \qquad t>0,$$

$$\Psi_\lambda^{(2)}(-t,x) = \overline{\Psi}_{-\lambda}^{(2)}(t,x). \quad (3.16)$$

It follows from our procedure that the basis functions satisfy

$$\langle\Psi_\lambda^{(2)}(t,\cdot), \quad \Psi_\zeta^{(2)}(t,\cdot)\rangle = \delta(\lambda-\zeta)$$

and can be used to expand any $h\in L_2(R+)$.

ISBN-0-201 503-5

Finally, the orbit containing $\mathcal{K}_{-2}$, corresponding to the radial free particle, also contains $\mathcal{K}_2$. The spectral theory for $\mathcal{K}_2$ is elementary, since this operator is already diagonalized in our realization. The generalized eigenfunctions are (symbolically)

$$f_\lambda^{(1)}(x) = \delta(x-\lambda), \qquad i\mathcal{K}_2 f_\lambda^{(1)} = (\lambda^2/4) f_\lambda^{(1)}, \qquad \lambda \geqslant 0. \qquad (3.17)$$

The spectrum is continuous and covers the positive real axis with multiplicity one. We have $\Psi_\lambda^{(1)}(t,x) = \exp(t\mathcal{K}_{-2}) f_\lambda^{(1)}(x)$ or

$$\Psi_\lambda^{(1)}(t,x) = \frac{\exp\left[\mp i\pi(\mu+2)/4\right]}{2|t|} (x\lambda)^{1/2} \exp\left(\frac{i(x^2+\lambda^2)}{4t}\right) J_{\mu/2}\left(\frac{x\lambda}{2|t|}\right)$$

$$(3.18)$$

with $\langle \Psi_\lambda^{(1)}, \Psi_\zeta^{(1)} \rangle = \delta(\lambda - \zeta)$. Expansions in the basis $\{\Psi_\lambda^{(1)}\}$ are equivalent to the inversion theorem for the Hankel transform [141, p. 199]. The $\Psi_\lambda^{(1)}$ are basis functions for the operator K_2.

Each of our bases has a continuous generating function of the form (1.55) where now

$$k(t,x,y) = \frac{\exp\left[\pm i\pi(\mu+2)/4\right]}{2|t|} (xy)^{1/2} \exp\left[\frac{i(x^2+y^2)}{4t}\right] J_{\mu/2}\left(\frac{xy}{2|t|}\right)$$

$$(3.19)$$

(see [59]).

The overlap functions $\langle f_\lambda^{(i)}, f_\zeta^{(j)} \rangle$ have the same significance as described in Section 2.1. Because of the simplicity of the basis $f_\lambda^{(1)}$, the only nontrivial overlap is

$$\langle f_n^{(3)}, f_\lambda^{(2)} \rangle = \frac{1}{2}\left[\frac{\Gamma(n+1+\mu/2)}{\pi n!} 2^{\mu/2-2i\lambda-1}\right]^{1/2} \frac{\Gamma(-i\lambda/2+\mu/4+1/2)}{\Gamma(1+\mu/2)}$$

$$\times {}_2F_1\left(\begin{array}{c}-n, i\lambda/2+\mu/4+1/2\\ 1+\mu/2\end{array}\middle| 2\right). \qquad (3.20)$$

In particular, we note that the overlap functions are dependent on the representatives $f_\lambda^{(i)}, f_\zeta^{(j)}$ chosen on each orbit. The most general way to define an overlap function is as a mixed-basis matrix element $\langle f_\lambda^{(i)}, U(g) f_\zeta^{(j)} \rangle, g \in \widetilde{SL}(2,R)$. Some of these elements have been computed in [24].

ISBN-0-201-13503-5

2.4 The Complex Equation $(\partial_\tau - \partial_{xx} + a/x^2)\Phi(\tau, x) = 0$

Here we study the complexification of equation (3.1). The variables t, x are now allowed to take complex values and a is a given nonzero complex constant. Introducing the new variable $\tau = it$, we can write this equation in the form

$$\left(\partial_\tau - \partial_{xx} + a/x^2\right)\Phi(\tau, x) = 0. \tag{4.1}$$

The complex symmetry algebra for equation (4.1) is easily seen to be three dimensional with basis

$$\begin{aligned}
J^+ &= \tau^2\partial_\tau + \tau x\,\partial_x + \tau/2 + x^2/4, \\
J^- &= -\partial_\tau, \qquad J^0 = \tau\partial_\tau + \tfrac{1}{2}x\partial_x + \tfrac{1}{4}
\end{aligned} \tag{4.2}$$

and commutation relations

$$[J^0, J^\pm] = \pm J^\pm, \qquad [J^+, J^-] = 2J^0. \tag{4.3}$$

This Lie algebra is isomorphic to $sl(2, \mathbb{C})$, as we shall soon show. Thus the operators (4.2) can be exponentiated to yield a local representation of $SL(2, \mathbb{C})$ by operators $\mathbf{T}(A), A \in SL(2, \mathbb{C})$, acting on the solution space of (4.1). Furthermore, it is obvious that $SL(2, \mathbb{C})$ acts on the symmetry algebra $sl(2, \mathbb{C})$ via the adjoint representation and decomposes the algebra into $SL(2, \mathbb{C})$ orbits. A straightforward computation shows that there are exactly two orbits in $sl(2, \mathbb{C})$ with orbit representatives J^- and J^0. (The orbits in $sl(2, R)$, (3.4), with representatives $K_{-2} - K_2$ and K^0 both belong to the orbit of $sl(2, \mathbb{C})$ with representative J^0, while the orbit with representative K_2 belongs to the complex J^- (same as J^+) orbit.) Similarly, it can be shown that (4.1) R-separates in exactly two complex analytic coordinate systems. (As usual we consider two systems as equivalent if one can be mapped onto the other by one of the operators $\mathbf{T}(A), A \in SL(2, \mathbb{C})$. The separable systems are listed in Table 10.

Table 10 R-Separable Coordinates for the Complex Equation
$(\partial_\tau - \partial_{xx} + a/x^2)\Phi = 0$

	Operator J	Coordinates $\{u, v\}$	Multiplier R	Separated solutions
1	J^-	$x = u, \tau = v$	$R = 1$	Bessel times exponential function
2	J^0	$x = u\sqrt{v}, \tau = v$	$R = 1$	Laguerre times exponential function

To derive identities relating the various separated solutions of (4.1), we can apply Weisner's method and expand arbitrary analytic solutions in terms of the Laguerre functions (orbit 2). These results are worked out in detail in Chapter 5 of [82], so here we will be very brief.

ISBN-0-201-13503-5

As suggested by (3.16) and 2 (Table 10), the Laguerre function solutions of (4.1) correspond to the coordinates $\{s,z\}$ where

$$s = \tau, \qquad z = -x^2/4\tau. \tag{4.4}$$

Furthermore, it is convenient to transform (4.1), which we write $Q\Psi = 0$, to the equivalent equation $Q'\Psi' = 0$ where $\Psi' = R^{-1}\Psi = s^{1/4}z^\beta\Psi$ and $Q' = R^{-1}QR$. The symmetry algebra of the primed equation consists of the operators $J' = R^{-1}JR$ where J belongs to the symmetry algebra of (4.1). Explicitly, we have

$$J'^+ = s^2\partial_s + sz\,\partial_z - sz - ls, \qquad J'^- = -\partial_s + (z/s)\partial_z - l/s, \qquad J'^0 = s\,\partial_s,$$

$$\beta = l + \tfrac{1}{4}, \qquad a = 4(\tfrac{1}{16} - l^2), \qquad l \in \mathcal{C}, \tag{4.5}$$

and the differential equation $Q'\Psi' = 0$ reads

$$(z\,\partial_{zz} - (2l + z)\partial_z + s\,\partial_s + l)\Psi'(z,s) = 0. \tag{4.6}$$

(In the remainder of this section we will use only the operators (4.5) and equation (4.6), so we henceforth drop the primes.)

Now consider the solutions Ψ of (4.6) that are eigenfunctions of J^0:

$$J^0\Psi = m\Psi \Rightarrow \Psi = f_m(z)s^m.$$

Substituting this solution into (4.6), we see that the ordinary differential equation obtained by factoring out s^m is of the form (B.7). Thus the $f_m(z)$ are confluent hypergeometric functions. In particular, the functions

$$\Psi_m(z,s) = L_{m+l}^{(-2l-1)}(z)s^m \tag{4.7}$$

satisfy these equations.

Note that for $l \in \mathcal{C}$, $2l \neq 0, 1, 2,\ldots$, and $m = -l + n$, $n = 0, 1, 2,\ldots$, the solutions Ψ_m are well defined and reduce to polynomials in the variable z, the Laguerre polynomials $L_n^{(-2l-1)}(z)$, (B.9i). To understand the significance of these polynomial solutions it is helpful to consider the system of equations

$$J^-\Phi = 0, \qquad J^0\Phi = -l\Phi,$$

which has the solution $\Phi(z,s) = s^{-l}$, unique to within a multiplicative constant. It is easy to verify that Φ also satisfies the differential equation (4.6). Now we use our knowledge of the symmetry algebra of (4.6) to construct other solutions. If $\Phi(z,s)$ is any analytic function of (z,s), it is a

ISBN-0-201-13503-5

simple consequence of local Lie theory that

$$\exp(\alpha J^+)\Phi(z,s)=(1-\alpha s)^l \exp\left[-\frac{\alpha z s}{(1-\alpha s)}\right]\Phi\left(\frac{z}{1-\alpha s},\frac{s}{1-\alpha s}\right)$$

$$=\sum_{n=0}^{\infty}\frac{\alpha^n}{n!}(J^+)^n\Phi(z,s),\qquad |\alpha s|<1. \tag{4.8}$$

Furthermore, if Φ is a solution of (4.6), then so are $(J^+)^n\Phi$ and $\exp(\alpha J^+)\Phi$, provided they are well defined. Substituting our solution $\Phi = s^{-l}$ in this expression, we find

$$(1-\alpha s)^{2l}s^{-l}\exp\left[-\alpha z s/(1-\alpha s)\right]=\sum_{n=0}^{\infty}\alpha^n\Phi_n(z,s),$$

$$\Phi_n=(1/n!)(J^+)^n\Phi,\qquad \Phi_0=\Phi,\qquad |\alpha s|<1. \tag{4.8'}$$

Using the definition of Φ_n, the commutation relations (4.3), and a straightforward induction argument, we can derive the recurrence formulas

$$J^0\Phi_n=(n-l)\Phi_n,\quad J^+\Phi_n=(n+1)\Phi_{n+1},\quad J^-\Phi_n=(2l-n+1)\Phi_{n-1}, \tag{4.9}$$

$$\Phi_{-1}\equiv0, n=0,1,2,\dots.$$

Furthermore, comparison of these results with the recurrence formulas (B.8) and $\Phi_0=s^{-l}$ yields

$$\Phi_n(z,s)=\Psi_{n-l}(z,s)=L_n^{(-2l-1)}(z)s^{n-l},\qquad n=0,1,2,\dots. \tag{4.10}$$

Thus (4.8') becomes a generating function for Laguerre polynomials:

$$(1-\alpha)^{2l}\exp\left[-\alpha z/(1-\alpha)\right]=\sum_{n=0}^{\infty}\alpha^n L_n^{(-2l-1)}(z),\qquad |\alpha|<1. \tag{4.11}$$

To derive more identities for Laguerre polynomials we need to determine the operators $\mathbf{T}(A)$ that define the action of the local symmetry group $SL(2,\mathcal{C})$ on the solution space of (4.6). Recall that $SL(2,\mathcal{C})$ is the complex Lie group of complex 2×2 matrices A with determinant $+1$:

$$A=\begin{pmatrix} a & b \\ c & d \end{pmatrix},\qquad a,b,c,d\in\mathcal{C},\qquad ad-bc=1. \tag{4.12}$$

The Lie algebra $sl(2,\mathcal{C})$ of this group consists of all complex 2×2 matrices A with trace zero:

$$A=\begin{pmatrix} \alpha & \beta \\ \gamma & -\alpha \end{pmatrix},\qquad \alpha,\beta,\gamma\in\mathcal{C}.$$

ISBN-0-201-13503-5

As a basis for $sl(2,\mathbb{C})$ we choose the matrices

$$
\mathcal{J}^+ = \begin{pmatrix} 0 & -1 \\ 0 & 0 \end{pmatrix}, \qquad \mathcal{J}^- = \begin{pmatrix} 0 & 0 \\ -1 & 0 \end{pmatrix}, \qquad \mathcal{J}^0 = \begin{pmatrix} \frac{1}{2} & 0 \\ 0 & -\frac{1}{2} \end{pmatrix}, \qquad (4.13)
$$

with commutation relations

$$
[\mathcal{J}^0, \mathcal{J}^\pm] = \pm \mathcal{J}^\pm, \qquad [\mathcal{J}^+, \mathcal{J}^-] = 2\mathcal{J}^0.
$$

Since these relations coincide with (4.3) we see that the symmetry algebra of (4.1) is indeed isomorphic to $sl(2,\mathbb{C})$.

A straightforward computation (see [82, Section 1.4]) shows that if $A \in SL(2,\mathbb{C})$ is given by (4.12) with $d \neq 0$, then

$$
\begin{aligned}
A &= \exp(\beta \mathcal{J}^+)\exp(\gamma \mathcal{J}^-)\exp(\tau \mathcal{J}^0), \\
e^\tau &= d^{-2}, \beta = -b/d, \gamma = -cd.
\end{aligned} \qquad (4.14)
$$

This expression enables us to parametrize a neighborhood of the identity in $SL(2,\mathbb{C})$. Next we exponentiate the operators (4.5) to determine the corresponding local multiplier representation of $SL(2,\mathbb{C})$ by operators $\mathbf{T}(A)$ acting on analytic functions $\Phi(z,s)$. According to (4.14) we have

$$
\mathbf{T}(A) = \exp((-b/d)J^+)\exp(-cdJ^-)\exp(-2\ln d J^0) \qquad (4.15)
$$

for A in a suitably small neighborhood of the identity element. We have already computed $\exp(\alpha J^+)$ in (4.8). Similar computations yield

$$
\begin{aligned}
\exp(\gamma J^-)\Phi(z,s) &= (1-\gamma/s)^l \Phi\big(z(1-\gamma/s)^{-1}, s-\gamma\big), \\
\exp(\tau J^0)\Phi(z,s) &= \Phi(z, e^\tau s).
\end{aligned}
$$

Composing these operators, we find

$$
\mathbf{T}(A)\Phi(z,s) = (d+bs)^l \left(a+\frac{c}{s}\right)^l \exp\left[\frac{bzs}{(d+bs)}\right]
$$

$$
\times \Phi\left(\frac{zs}{(as+c)(d+bs)}, \frac{as+c}{d+bs}\right), \quad \left|\frac{c}{as}\right|<1, \left|\frac{bs}{d}\right|<1, \quad (4.16)
$$

defined for all analytic functions Φ such that the right-hand side makes sense. Note that $\mathbf{T}(A)$ maps an analytic solution of (4.6) to another solution.

ISBN-0-201-13503-5

We now apply the operators $\mathbf{T}(A)$ to a basis function Φ_m and expand the resulting function in terms of the $\{\Phi_n\}$ basis:

$$\mathbf{T}(A)\Phi_m(z,s)= \sum_{n=0}^{\infty} T_{nm}(A)\Phi_n(z,s). \tag{4.17}$$

For A close to the identity we can compute the matrix elements $T_{nm}(A)$ directly from the Lie algebra relations (4.9). The computation is simplified through the construction of another model of our Lie algebra representation (4.9). Following Section 5.2 of [82], we choose basis functions

$$f_n(w)=(n!)^{-1}\Gamma(n-2l)w^n, \qquad n=0,1,2,\ldots,$$

and operators

$$J^+ = w^2 \frac{d}{dw} - 2lw, \qquad J^- = -\frac{d}{dw}, \qquad J^0 = w\frac{d}{dw} - l. \tag{4.18}$$

It is easy to check that these operators and basis functions satisfy (4.3) and (4.9). Furthermore, by applying (4.15), we can show that the corresponding local group action of $SL(2,\mathcal{C})$ is determined by operators $\mathbf{T}(A)$ where

$$\mathbf{T}(A)f(w)=(bw+d)^{2l}f\left(\frac{aw+c}{bw+d}\right), \qquad |bw/d|<1. \tag{4.19}$$

The matrix elements are given by

$$\mathbf{T}(A)f_m(w)= \sum_{n=0}^{\infty} T_{nm}(A)f_n(w)$$

or

$$d^{2l-m}(1+bw/d)^{2l-m}(aw+c)^m\Gamma(m-2l)/m!$$

$$= \sum_{n=0}^{\infty} T_{nm}(A)\Gamma(n-2l)w^n/n!, \qquad |bw/d|<1. \tag{4.20}$$

Expanding the left-hand side of (4.20) in a power series in w and computing the coefficient of w^n, we obtain

$$T_{nm}(A)=\frac{a^n d^{2l-m}c^{m-n}\Gamma(m-2l)}{\Gamma(m-n+1)\Gamma(n-2l)} {}_2F_1\left(\begin{array}{c}-n,m-2l\\m-n+1\end{array}\middle|\frac{bc}{ad}\right). \tag{4.21}$$

ISBN-0-201-13503-5

Moreover, the local representation property $T(AA')=T(A)T(A')$ valid for A, A' in a suitably small neighborhood of the identity in $SL(2,\mathcal{C})$ implies

$$T_{nm}(AA')= \sum_{k=0}^{\infty} T_{nk}(A)T_{km}(A').$$

Substituting (4.10) and (4.21) into (4.17), we obtain the identities

$$(1+bs/d)^{2l-m}(1+c/as)^m \exp[bzs/(d+bs)]$$
$$\times L_m^{(-2l-1)}[z(1+bc)^{-1}(1+c/as)^{-1}(1+bs/d)^{-1}]$$
$$= \sum_{n=0}^{\infty} \frac{n!}{m!}(-sb/d)^{n-m}\Gamma(n-m+1)^{-1}{}_2F_1\left(\begin{array}{c}-m,n-2l\\n-m+1\end{array}\bigg|\frac{bc}{ad}\right)$$
$$\times L_n^{(-2l-1)}(z), \qquad |bs/d|<1, d=(1+bc)/a, \qquad (4.22)$$

valid for all integers $m\geqslant 0$ and for all $l\in\mathcal{C}$ such that $2l\neq 0,1,2,\ldots$. (In (4.21) and (4.22) we use the fact that ${}_2F_1\left(\begin{array}{c}\alpha,\beta\\\gamma\end{array}\bigg|z\right)/\Gamma(\gamma)$ is an entire function of α,β,γ to define this expression for γ a negative integer.)

If $a=d=s=1, c=0$, the identity simplifies to

$$(1-b)^{2l-m}\exp\left[-bz/(1-b)\right]L_m^{(-2l-1)}\left(z(1-b)^{-1}\right)$$

$$= \sum_{n=0}^{\infty}\binom{n+m}{n}b^n L_{m+n}^{(-2l-1)}(z), \qquad |b|<1.$$

For $m=0$ this last expression becomes (4.11) with $\alpha=b$. When $a=d=s=1, b=0$, then (4.22) simplifies to

$$(1+c)^m L_m^{(-2l-1)}\left(z(1+c)^{-1}\right)= \sum_{n=0}^{m}\binom{m-2l-1}{n}c^n L_{m-n}^{(-2l-1)}(z).$$

Similar identities can be derived for the basis functions (4.7) that are not polynomials in z, that is, $m+l\neq 0,1,2,\ldots$. For these results see Section 5.8 of [82].

To derive more general identities for Laguerre functions we use the full power of Weisner's method. If $\Psi(z,s)$ is an analytic solution of (4.6) with convergent Laurent series expansion

$$\Psi(z,s)= \sum_m f_m(z)s^m,$$

ISBN-0-201-13503-5

then the $f_m(z)$ are confluent hypergeometric (Laguerre) functions (linear combinations of $L_{m+l}^{(-2l-1)}(z)$ and $z^{2l+1}L_{m-l-1}^{(2l+1)}(z)$ for $2l$ not an integer). If in addition Ψ is analytic in a neighborhood of $z=0$, it follows that $f_m(z) = c_m L_{m+l}^{(-2l-1)}(z), c_n \in \mathcal{C}$. Thus,

$$\Psi(z,s) = \sum_m c_m L_{m+l}^{(-2l-1)}(z)s^m. \qquad (4.23)$$

The expansion (4.22) is an example of such an identity where we can use Lie algebraic techniques to explicitly compute the coefficients c_m. However, this is no longer true for the example $T(A)\Psi_p, p \in \mathcal{C}$, where

$$A = \begin{pmatrix} 0 & -1 \\ 1 & 1 \end{pmatrix}$$

and Ψ_p is given by (4.7). Then we have

$$s^{-l}(1-s)^{l-p}\exp\left[-\frac{zs}{1-s}\right]L_{p+l}^{(-2l-1)}\left(\frac{zs}{1-s}\right)$$

$$= \sum_{n=0}^{\infty} c_n L_n^{(-2l-1)}(z)s^{-l+n}, \qquad |s| < 1.$$

This expansion is not of the form (4.22) because A is bounded away from the identity and p is not necessarily an integer. We can easily evaluate the constants c_n by setting $z=0$. The result is

$$(1-s)^{l-p}\exp\left(-\frac{zs}{1-s}\right)L_{p+l}^{(-2l-1)}\left(\frac{zs}{1-s}\right)$$

$$= \sum_{n=0}^{\infty}(-s)^n \frac{\Gamma(l-p+1)\Gamma(p-l)}{\Gamma(n-2l)\Gamma(l-p-n+1)\Gamma(l+p+1)}L_n^{(-2l-1)}(z), \qquad (4.24)$$

$$|s| < 1,$$

a generating function for Laguerre polynomials.

For our next example we choose the generating function Ψ in (4.23) to be an eigenfunction of the operator J^-. Since J^- belongs to orbit 1 in Table 10, we see that in the coordinates x,τ (suitably transformed from (4.1) to (4.6)), we can choose Ψ as separable and expressible in terms of a Bessel function. Indeed, a simultaneous solution Ψ of (4.6) and $J^-\Psi = -\Psi$ is

$$\Psi(z,s) = s^{-l}e^s(zs)^{(2l+1)/2}J_{-2l-1}\left(2(zs)^{1/2}\right). \qquad (4.25)$$

ISBN-0-201-13503-5

Moreover the function

$$\Psi' = \mathbf{T}(A)\Psi(z,s) = s^{-l}(d+bs)^{-1}\exp\left[\frac{(a+bz)s+c}{d+bs}\right](zs)^{(2l+1)/2}$$

$$\times J_{-2l-1}\left(\frac{2(zs)^{1/2}}{d+bs}\right), \qquad \left|\frac{bs}{d}\right| < 1,$$

where A is given by (4.12), also satisfies (4.6) and the equation $(J^-)'\Psi' = -\Psi'$ where

$$(J^-)' = \mathbf{T}(A)J^-\mathbf{T}(A^{-1}) = -b^2J^+ + d^2J^- - 2bdJ^0.$$

Since $w^{-m}J_m(w)$ is an entire function of w for all $m \in \mathbb{C}$, $\Psi'(z,s)$ has a Laurent expansion in s of the form

$$\Psi'(z,s) = \sum_{n=0}^{\infty} c_n(A)L_n^{(-2l-1)}(z)s^n, \qquad |bs/d| < 1.$$

Setting $z = 0$, we find

$$(d+bs)^{2l}\exp\left(\frac{as+c}{d+bs}\right) = \sum_{n=0}^{\infty} c_n(A)\Gamma(n-2l)\frac{s^n}{n!}, \qquad \left|\frac{bs}{d}\right| < 1,$$

and comparing this expression with (4.22) in the case where $m=0$, we obtain

$$c_n(A) = \exp\left(\frac{ac}{1+bc}\right)\frac{n!}{\Gamma(n-2l)}a^{-2l}\left(-\frac{ab}{1+bc}\right)^n L_n^{(-2l-1)}\left(\frac{a}{b(1+bc)}\right),$$

$$n = 0, 1, 2, \dots, \quad 2l \neq 0, 1, 2, \dots .$$

For the group parameter $c=0$, the result of our computation is the identity

$$(1+abs)^{-1}\exp\left[\frac{(a+bz)s}{a^{-1}+bs}\right](a^2zs)^{(2l+1)/2}J_{-2l-1}\left(\frac{2(zs)^{1/2}}{a^{-1}+bs}\right)$$

$$= \sum_{n=0}^{\infty}\frac{n!}{\Gamma(n-2l)}(-abs)^n L_n^{(-2l-1)}\left(\frac{a}{b}\right)L_n^{(-2l-1)}(z), \quad |abs| < 1. \quad (4.26)$$

If $a=1, b=0$, (4.26) reduces to

$$e^s(zs)^{(2l+1)/2}J_{-2l-1}\left(2(zs)^{1/2}\right) = \sum_{n=0}^{\infty} L_n^{(-2l-1)}(z)s^n/\Gamma(n-2l),$$

ISBN-0-201-13503-5

while if $a = iy^{1/2}, b = iy^{-1/2}$ it reduces to the Hille–Hardy formula

$$(1-s)^{-1}(-yzs)^{(2l+1)/2}\exp\left[-\frac{s(y+z)}{1-s}\right]J_{-2l-1}\left(\frac{2i(yzs)^{1/2}}{1-s}\right)$$

$$= \sum_{n=0}^{\infty}\frac{n!}{\Gamma(n-2l)}L_n^{(-2l-1)}(y)L_n^{(-2l-1)}(z)s^n, \qquad |s| < 1, \quad (4.27)$$

see [37, p. 189].

2.5 Separation of Variables for the Schrödinger Equation $(i\partial_t + \partial_{xx} + \partial_{yy})\Psi = 0$

Now we apply the methods of Section 2.1 to time-dependent Schrödinger equations in two space variables:

$$i\partial_t\Psi = -\partial_{xx}\Psi - \partial_{yy}\Psi + V(x,y)\Psi \qquad (5.1)$$

where V is the potential function. Boyer [18] has classified all equations (5.1) that admit a nontrivial symmetry algebra of first-order differential operators. He has shown that (a) the maximal dimension for a symmetry algebra is nine, (b) this maximum occurs only for the four potentials V listed in Table 11, and (c) the algebras of maximal dimension are isomorphic. (There are actually four classes of such potentials corresponding to orbits in the symmetry algebra. We have simply listed a representative from each class in Table 11.) Niederer [102] has shown that the four equations with maximal symmetry are in fact equivalent. Here we will examine this equivalence explicitly and relate it to separation of variables.

As in Section 2.1 we begin with a study of the free-particle Schrödinger equation, which we write in the form

$$Q\Psi = 0, \qquad Q = i\partial_t + \partial_{x_1 x_1} + \partial_{x_2 x_2}, \qquad (x_1, x_2) = (x, y). \qquad (5.2)$$

The complex symmetry algebra $\mathcal{G}_3^c$ of this equation is nine dimensional

Table 11   Potentials $V(x,y)$ with Maximal Symmetry

V	Name of System
(1) 0	Free particle
(2) $k(x^2+y^2), k>0$	Harmonic oscillator
(3) $-k(x^2+y^2), k>0$	Repulsive oscillator
(4) $ax, a \neq 0$	Free fall (linear potential)

ISBN-0-201-13503-5

with basis

$$K_2 = -t^2 \partial_t - t(x_1 \partial_{x_1} + x_2 \partial_{x_2}) - t + i(x_1^2 + x_2^2)/4, \qquad K_{-2} = \partial_t,$$

$$P_j = \partial_{x_j}, \qquad B_j = -t \partial_{x_j} + i x_j/2, \qquad j = 1, 2, \tag{5.3}$$

$$M = x_1 \partial_{x_2} - x_2 \partial_{x_1}, \qquad E = i, \qquad D = x_1 \partial_{x_1} + x_2 \partial_{x_2} + 2t \partial_t + 1$$

and commutation relations

$$[D, K_{\pm 2}] = \pm 2 K_{\pm 2}, \qquad [D, B_j] = B_j, \quad [D, P_j] = -P_j,$$

$$[D, M] = 0, \qquad [M, K_{\pm 2}] = 0, \quad [P_j, M] = (-1)^{j+1} P_l,$$

$$[B_j, M] = (-1)^{j+1} B_l, \quad [K_2, K_{-2}] = D, \quad [K_2, B_j] = 0,$$

$$[K_{-2}, B_j] = -P_j, \qquad [K_{-2}, P_j] = 0, \quad [P_j, K_2] = B_j,$$

$$[P_j, B_j] = \tfrac{1}{2} E, \qquad [P_j, B_l] = 0, \quad j, l = 1, 2, \; j \neq l, \tag{5.4}$$

with E in the center of $\mathcal{G}_3^c$. In the following we will study only the *Schrödinger algebra* $\mathcal{G}_3$, the real Lie algebra with basis (5.3).

A second useful basis for $\mathcal{G}_3$ is given by the operators B_j, P_j, E, which generate the five-dimensional Weyl algebra $\mathcal{W}_2$, the operator M, and the three operators L_1, L_2, L_3 where

$$L_1 = D, \qquad L_2 = K_2 + K_{-2}, \qquad L_3 = K_{-2} - K_2. \tag{5.5}$$

Here,

$$[L_1, L_2] = -2L_3, \qquad [L_3, L_1] = 2L_2, \qquad [L_2, L_3] = 2L_1, \tag{5.6}$$

so that the L_ν form a basis for the Lie algebra $sl(2, R)$; compare with (1.11). It follows that $\mathcal{G}_3$ is a semidirect product of $sl(2, R) \oplus o(2)$ and $\mathcal{W}_2$. Here $o(2)$ is the one-dimensional Lie algebra spanned by M.

Using standard results from Lie theory, we can exponentiate the operators (5.3) to obtain a local Lie group G_3 (the *Schrödinger group*) of operators acting on the space $\mathcal{F}$ of locally analytic functions of the real variables t, x_j and mapping solutions of (5.2) into solutions. The required computations can be carried out in simple analogy with expressions (1.15)–(1.19).

The action of the Weyl group W_2 is given by operators

$$\mathbf{T}(\mathbf{w}, \mathbf{z}, \rho) = \exp(w_1 B_1) \exp(z_1 P_1) \exp(w_2 B_2) \exp(z_2 P_2) \exp(\rho E),$$

$$\mathbf{w} = (w_1, w_2), \quad \mathbf{z} = (z_1, z_2),$$

ISBN-0-201-13503-5

such that

$$T(w,z,\rho)T(w',z',\rho') = T\left(w+w',z+z',\rho+\rho'+\frac{1}{2}w' \cdot z\right), \qquad (5.7)$$

where

$$T(w,z,\rho)\Phi(t,x) = \exp[i(2x \cdot w - tw \cdot w + 4\rho)/4]\Phi(t,x-tw+z),$$

$$\Phi \in \mathcal{F}.$$

Here $x \cdot w = x_1 w_1 + x_2 w_2$. The action of $SO(2)$ is given by $T(\theta) = \exp(\theta M)$,

$$T(\theta)T(\theta') = T(\theta + \theta')$$

where

$$T(\theta)\Phi(t,x) = \Phi(t,x\Theta), \qquad \Theta = \begin{pmatrix} \cos\theta & \sin\theta \\ -\sin\theta & \cos\theta \end{pmatrix}. \qquad (5.8)$$

Finally, the action of $SL(2,R)$ is given by operators $T(A)$,

$$T(A)\Phi(t,x) = \exp\left[\frac{i\beta x \cdot x}{4(\delta + t\beta)}\right](\delta + t\beta)^{-1}\Phi\left[\frac{\gamma + t\alpha}{\delta + t\beta}, (\delta + t\beta)^{-1}x\right],$$

$$A = \begin{pmatrix} \alpha & \beta \\ \gamma & \delta \end{pmatrix} \in SL(2,R), \qquad (5.9)$$

where

$$T(A)T(B) = T(AB), \qquad A,B \in SL(2,R).$$

The one-parameter subgroups of $SL(2,R)$ generated by $K_{\pm 2}, L_1, L_2, L_3$, respectively, are given by expressions (1.17). The adjoint actions of $SO(2)$ and $SL(2,R)$ on W_2 are

$$T^{-1}(A)T(w,z,\rho)T(A) = T(w',z',\rho'),$$

$$\rho' = \rho + (w' \cdot z' - w \cdot z)/4, w' = \delta w + \beta z, z' = \alpha z + \gamma w$$

$$T^{-1}(\theta)T(w,z,\rho)T(\theta) = T(w\Theta, z\Theta, \rho). \qquad (5.10)$$

These relations define G_3 as a semidirect product of $SL(2,R) \oplus SO(2)$ and W_2:

$$g = (A,\theta,v) \in G_3, \quad A \in SL(2,R), \quad \theta \in SO(2), \quad v = (w,z,\rho) \in W_2;$$

$$T(g) = T(A)T(\theta)T(v). \qquad (5.11)$$

ISBN-0-201-13503-5

The group G_3 acts on the Lie algebra $\mathcal{G}_3$ of differential operators via the adjoint representation

$$K \to K^g = \mathbf{T}(g)K\mathbf{T}^{-1}(g)$$

and this action splits $\mathcal{G}_3$ into G_3 orbits. We will classify the orbit structure of the factor algebra $\tilde{\mathcal{G}} \cong \mathcal{G}_3 / \{E\}$ where $\{E\}$ is the center of $\mathcal{G}_3$. Let $K \in \mathcal{G}_3$ and let a_2, a_0, a_{-2}, respectively, be the coefficients corresponding to K_2, D, K_{-2} in the expansion of $K \neq 0$ in terms of the basis (5.3). Setting $\alpha = a_2 a_{-2} + a_0^2$, we find that α is invariant under the adjoint representation.

The following list is a complete set of orbit representatives in the sense that any $K \neq 0$ lies on the same G_3 orbit as exactly one of the operators in this list:

Case 1 $(\alpha < 0)$ $K_{-2} - K_2 + \beta^2 M, |\beta| \neq 1, K_{-2} - K_2 + M + \gamma B_1,$

Case 2 $(\alpha > 0)$ $D + \beta M,$ (5.12)

Case 3 $(\alpha = 0)$ $K_2 + M, K_2 + P_1, K_2, M, P_1 + B_2, P_1.$

We next consider the problem of determining higher-order differential operators S that are symmetries of (5.2). The special structure of (5.2) enables us to simplify this problem somewhat. Since we will only apply S to solutions Ψ of $Q\Psi = 0$, without loss of generality we can require that S contain no derivatives in t. In other words, wherever ∂_t appears in S, we can replace it by $i(\partial_{x_1 x_1} + \partial_{x_2 x_2})$. Another way to view this is to note that if S is a symmetry operator, then so is $S' = S + XQ$ where X is an arbitrary differential operator. Moreover, $S'\Psi = S\Psi$ for any solution Ψ of (5.2). There is a unique choice of X such that S' contains no derivatives with respect to t.

With this in mind we see that only the operators P_j, B_j, E, generating the Weyl algebra, and M are first order or less in the x_j. The elements $K_2 = -i(B_1^2 + B_2^2)$, $K_{-2} = i(P_1^2 + P_2^2)$, and $D = -i(\{B_1, P_1\} + \{B_2, P_2\})$ are second order. (These equalities are valid modulo the replacement of ∂_t by $i(\partial_{x_1 x_1} + \partial_{x_2 x_2})$.) More generally, we can compute all symmetries S_2 that are second order or less in x_1 and x_2:

$$S = \sum_{i,j=1}^{2} a_{ij}(\mathbf{x},t)\partial_{x_i x_j} + \sum_{j=1}^{2} b_j(\mathbf{x},t)\partial_{x_j} + c(\mathbf{x},t).$$

A tedious computation shows that such S form a 20-dimensional vector space. A basis for this space is provided by the zeroth-order operator E, the five first-order operators P_j, B_j, M, and the three second-order operators $iK_{\pm 2}, iD$ listed earlier, plus the eleven second-order operators

$$B_1^2 - B_2^2, \quad B_1 P_1 - B_2 P_2, \quad P_1^2 - P_2^2, \quad \{B_1, M\}, \quad \{B_2, M\}, \quad \{P_1, M\},$$

$$\{P_2, M\}, \quad B_1 B_2, \quad P_1 P_2, \quad B_1 P_2 + B_2 P_1, \quad M^2.$$

ISBN-0-201-13503-5

It follows that all second-order symmetries are symmetric quadratic forms in B_j, P_j, E, and M.

The problem of R-separation of variables for equation (5.2) was solved in [20]. In the following we will concern ourselves only with those systems which, in addition to admitting R-separable solutions, have the property that the separated factors satisfy three ordinary differential equations, one in each of the separation variables. Since (5.2) is an equation in three variables, there are now two separation constants associated with each separable system.

The relationship between orbits of first-order symmetries (5.12) and separable systems is now rather tenuous. It is true that corresponding to any first-order symmetry K we can find a new system of coordinates $\{u,v,w\}$, not unique, such that the variable u can be separated out of equation (5.2). (See the analogous discussion for the Helmholtz equation in Section 1.2.) However, the resulting equation in v,w may not permit separation of variables. Thus, diagonalization of a symmetry operator K may correspond to a partial, but not total, separation of variables.

The results of [20] are as follows. Corresponding to every R-separation of variables for (5.2) we can find a pair of differential operators K,S such that:

1. K and S are symmetries of (5.2) and $[K,S]=0$.
2. $K \in \tilde{\mathcal{G}}_3$; that is, K is first order in x_1, x_2, and t.
3. S is second order in x_1, x_2 and contains no term in ∂_t.

The R-separation of variables is characterized by the simultaneous equations

$$Q\Psi = 0, \qquad K\Psi = i\lambda\Psi, \qquad S\Psi = \mu\Psi. \tag{5.14}$$

In particular, the eigenvalues λ, μ are the usual separation constants for the R-separable solutions Ψ.

It follows that K lies in the symmetry algebra $\tilde{\mathcal{G}}_3$ while S can be expressed as a symmetric quadratic form in B_j, P_j, E, and M. Thus the possible coordinate systems in which (5.2) R-separates can always be characterized by eigenfunction equations for operators at most second order in the enveloping algebra of $\mathcal{G}_3$. The possible commuting operators K,S, R-separable coordinates $\{u,v,w\}$, and separated solutions are listed in Table 12.

The notation for the coordinate systems that we introduce in Table 12 requires some comment. Coordinate systems 13–17 are not of much interest to us because they result from the fact that if P_1 is diagonalized, the free-particle Schrödinger equation (5.2) essentially collapses to the free-particle equation (1.2). However, the remaining coordinate systems are associated with the Hamiltonians for the free-particle, linear potential, harmonic oscillator, and repulsive oscillator in exactly the same manner as

ISBN-0-201-13503-5

Table 12 Operators and R-Separable Coordinates for the Equation $(i\partial_t + \partial_{xx} + \partial_{yy})\Psi = 0$.

Operators K, S	Coordinates $\{u,v,w\}$	Multiplier $e^{i\Re}$	Separated solutions
1a Fc^1 K_2, B^2	$x = uw$ $y = vw$	$\Re = (u^2+v^2)w/4$	Exponential Exponential
1b Fc^2 K_{-2}, P_1^2	$x = u$ $y = v$	0	Exponential Exponential
2a Fr^1 K_2, M^2	$x = uw\cos v$ $y = uw\sin v$	$u^2 w/4$	Bessel Exponential
2b Fr^2 K_{-2}, M^2	$x = u\cos v$ $y = u\sin v$	0	Bessel Exponential
3a Fp^1 $K_2, \{B_2, M\}$	$x = w(u^2-v^2)/2$ $y = uvw$	$(u^2+v^2)^2 w/16$	Parabolic cylinder Parabolic cylinder
3b Fp^2 $K_{-2}, \{P_2, M\}$	$x = (u^2-v^2)/2$ $y = uv$	0	Parabolic cylinder Parabolic cylinder
4a Fe^1 $K_2, M^2 - B_2^2$	$x = w\cosh u\cos v$ $y = w\sinh u\sin v$	$(\sinh^2 u + \cos^2 v)w/4$	Modified Mathieu Mathieu
4b Fe^2 $K_{-2}, M^2 - P_2^2$	$x = \cosh u\cos v$ $y = \sinh u\sin v$	0	Modified Mathieu Mathieu
5a Lc^1 $K_2 - 2aP_1 - 2bP_2,$ $B_2^2 + 2bEP_2$	$x = uw + a/w$ $y = vw + b/w$	$(u^2+v^2)w/4$ $-(au+bv)/2w$	Airy Airy
5b Lc^2 $K_{-2} + 2aB_1 + 2bB_2,$ $P_1^2 - 2aEB_1$	$x = u + aw^2$ $y = v + bw^2$	$(au + bv)w$	Airy Airy
6a Lp^1 $K_2 - aP_1,$ $\{B_2, M\} - aP_2^2$	$x = (u^2-v^2)w/2$ $\quad + a/w$ $y = uvw$	$(u^2+v^2)^2 w/16$ $-(u^2-v^2)a/4w$	Anharmonic oscillator Anharmonic oscillator
6b Lp^2 $K_{-2} - aB_1,$ $\{P_2, M\} + aB_2^2$	$x = (u^2-v^2)/2 + aw^2/2$ $y = uv$	$(u^2-v^2)aw/4$	Anharmonic oscillator Anharmonic oscillator
7 Oc $K_{-2} - K_2,$ $P_1^2 + B_1^2$	$x = u(1+w^2)^{1/2}$ $y = v(1+w^2)^{1/2}$	$(u^2+v^2)w/4$	Hermite Hermite
8 Or $K_{-2} - K_2,$ M^2	$x = u(1+w^2)^{1/2}\cos v$ $y = u(1+w^2)^{1/2}\sin v$	$u^2 w/4$	Laguerre Exponential
9 Oe $K_{-2} - K_2,$ $M^2 - P_2^2 - B_2^2$	$x = (1+w^2)^{1/2}\cosh u\cos v$ $y = (1+w^2)^{1/2}\sinh u\sin v$	$(\sinh^2 u + \cos^2 v)w/4$	Ince Ince

ISBN-0-201-13503-5

Table 12 (*Continued*)

Operators K, S	Coordinates $\{u,v,w\}$	Multiplier $e^{i\mathfrak{R}}$	Separated solutions
10a Rc^1 $D, \{B_1, P_1\}$	$x = u\|w\|^{1/2}$ $y = v\|w\|^{1/2}$	0	Parabolic cylinder Parabolic cylinder
10b Rc^2 $K_{-2} + K_2,$ $P_1^2 - B_1^2$	$x = u\|w^2 - 1\|^{1/2}$ $y = v\|w^2 - 1\|^{1/2}$	$\varepsilon(u^2 + v^2)w/4$	Parabolic cylinder Parabolic cylinder
11a Rr^1 D, M^2	$x = u\|w\|^{1/2}\cos v$ $y = u\|w\|^{1/2}\sin v$	0	Whittaker Exponential
11b Rr^2 $K_{-2} + K_2,$ M^2	$x = \|w^2 - 1\|^{1/2} u \cos v$ $y = \|w^2 - 1\|^{1/2} u \sin v$	$\varepsilon u^2\, w/4$	Whittaker Exponential
12a Re^1 $D,$ $M^2 - \frac{1}{2}\{B_2, P_2\}$	$x = \|w\|^{1/2}\cosh u \cos v$ $y = \|w\|^{1/2}\sinh u \sin v$	0	Ince Ince
12b Re^2 $K_{-2} + K_2,$ $M^2 - P_2^2 + B_2^2$	$x = \|w^2 - 1\|^{1/2}\cosh u \cos v$ $y = \|w^2 - 1\|^{1/2}\sinh u \sin v$	$\varepsilon(\sinh^2 u + \cos^2 v)/4$	Ince Ince
13 $L1$ $P_1, B_2^2 - 2bEP_2$	$x = u$ $y = vw + b/w$	$wv^2/4 - bv/2w$	Exponential Airy
14 $L2$ $P_1, P_2^2 - 2aEB_2$	$x = u$ $y = v + aw^2$	avw	Exponential Airy
15 $O1$ $P_1, P_2^2 + B_2^2$	$x = u$ $y = v(1 + w^2)^{1/2}$	$wu^2/4$	Exponential Hermite
16 $R1$ $P_1, \{B_2, P_2\}$	$x = u$ $y = v\|w\|^{1/2}$	0	Exponential Parabolic cylinder
17 $R2$ $P_1, P_2^2 - B_2^2$	$x = u$ $y = v\|w^2 - 1\|^{1/2}$	$\varepsilon v^2 w/4$	Exponential Parabolic cylinder

described in Section 2.1. We denote each system in the form Ab^j. The capital letter corresponds to the type of Hamiltonian; that is, $F\leftrightarrow$free particle, $L\leftrightarrow$linear potential, $O\leftrightarrow$harmonic oscillator, $R\leftrightarrow$repulsive oscillator. The small letter indicates the type of coordinate used in each of these Hamiltonians; that is, $c\leftrightarrow$Cartesian, $r\leftrightarrow$radial (polar) coordinates, $p\leftrightarrow$ parabolic, and $e\leftrightarrow$elliptic coordinates. The superscript j is used to distinguish two systems on the same G_3 orbit.

In each case $w = t$ and the separated solution in the variable w is an exponential function. In the last column of Table 12 we list first the form of the separated solution in u followed by the separated solution in v. The symbol $\varepsilon = \pm 1$ denotes the sign of $1 - w^2$ and the anharmonic oscillator

ISBN-0-201-13503-5

functions are solutions of a differential equation of the form

$$f''(u)+(\lambda u^2+\alpha u^4-\beta)f(u)=0, \qquad \alpha,\beta\in R. \qquad (5.15)$$

From the viewpoint of Galilean and dilatation symmetry alone there are 26 inequivalent coordinate systems. (As indicated in the remarks following Table 6, this list of 26 Galilean–dilatation inequivalent coordinate systems is not precisely a list of orbits because certain pairs of Galilean–dilatation orbits yield separable coordinates that differ only in the sign of a parameter.) However, we can also regard two coordinate systems as equivalent if the first can be transformed to the second under the action of some $g\in G_3$. In terms of operators, the system described by K,S is equivalent to the system described by K',S' if, under the adjoint action of G_3 on the enveloping algebra of $\mathcal{G}_3$, the two-dimensional space spanned by K,S can be mapped onto the two-dimensional space spanned by K',S'. Under this more general equivalence relation not all of the 26 systems are inequivalent. Indeed, the systems denoted Ab^1 and Ab^2 lie on the same two-dimensional orbits, so that there are only 17 equivalence classes of orbits. (For convenience in applications, the representatives of orbits $5,6,13,14$ contain parameters a,b. Some of these parameters can be normalized to ±1 with the dilatation symmetry.)

We can describe these equivalences in terms of the operator $J=\exp[\pi(K_2-K_{-2})/4]$:

$$J\Phi(t,\mathbf{x})=\frac{\sqrt{2}}{(1+t)}\exp\left[i(1+t)^{-1}\frac{\mathbf{x}\cdot\mathbf{x}}{4}\right]\Phi\left(\frac{t-1}{t+1},\sqrt{2}\,(t+1)^{-1}\mathbf{x}\right), \qquad \Phi\in\mathcal{F}. \qquad (5.16)$$

Note that $J^2=\exp[\pi(K_2-K_{-2})/2]$, and

$$\begin{aligned}
&J^2\Phi(t,\mathbf{x})=t^{-1}\exp\left[i\mathbf{x}\cdot\mathbf{x}/4t\right]\Phi(-t^{-1},t^{-1}\mathbf{x}),\\
&J^4\Phi(t,\mathbf{x})=-\Phi(t,-\mathbf{x}), \qquad J^8\Phi(t,\mathbf{x})=\Phi(t,\mathbf{x}).
\end{aligned} \qquad (5.17)$$

It is easy to show that $J(K_{-2}+K_2)J^{-1}=D$, and, checking the adjoint action of J on second-order operators, we can verify that the three coordinate systems Rc^2,Rr^2,Re^2 are equivalent under J to the three systems Rc^1,Rr^1,Re^1, respectively.

Denoting the adjoint action of J^2 on $K\in\mathcal{G}_3$ by $K'=J^2KJ^{-2}$, we find $P'_j=-B_j, B'_j=P_j, K'_{-2}=-K_2, K'_2=-K_{-2}, D'=-D, M'=M, E'=E$, so that the six pairs of the form Fa^1,Fa^2 or La^1,La^2 are equivalent under J^2.

We next demonstrate that the operators (5.3) can be interpreted as a Lie algebra of skew-symmetric operators on the Hilbert space $L_2(R_2)$ of complex-valued Lebesgue square-integrable functions on the plane. This is accomplished by considering t as a fixed parameter and replacing ∂_t by

ISBN-0-201-13503-5

$i(\partial_{x_1 x_1} + \partial_{x_2 x_2})$ in expressions (5.3). It is then straightforward to show that the resulting operators, multiplied by i and restricted to the domain of infinitely differentiable functions on R_2 with compact support, have unique self-adjoint extensions. In fact, these operators are real linear combinations of the operators

$$\mathcal{K}_2 = i\left(x_1^2 + x_2^2\right)/4, \qquad \mathcal{K}_{-2} = i\left(\partial_{x_1 x_1} + \partial_{x_2 x_2}\right), \qquad \mathcal{P}_j = \partial_{x_j},$$

$$\mathcal{B}_j = ix_j/2, \qquad \mathcal{M} = x_1\partial_{x_2} - x_2\partial_{x_1}, \qquad \mathcal{E} = i, \tag{5.18}$$

$$\mathcal{D} = x_1\partial_{x_1} + x_2\partial_{x_2} + 1, \qquad j = 1,2.$$

Note that when the parameter $t = 0$, the operators (5.3) reduce to (5.18). Thus the script operators (5.18) satisfy the same commutation relations (5.4) as do the italic operators (5.3). More specifically, we have the general identity

$$\exp(t\mathcal{K}_{-2})\mathcal{K}\exp(-t\mathcal{K}_{-2}) = K \tag{5.19}$$

relating corresponding script $(\mathcal{K})$ and italic (K) operators. Here $\exp(t\mathcal{K}_{-2})$ is a unitary operator on $L_2(R_2)$ which corresponds to time translation for the free-particle system. It is shown in [67, p. 493] that

$$\exp(a\mathcal{K}_{-2})f(\mathbf{x}) = \text{l.i.m.}(4\pi ia)^{-1}\int\int_{-\infty}^{\infty}\exp\left[-\frac{(\mathbf{x}-\mathbf{y})^2}{4ia}\right]f(\mathbf{y})\,dy_1\,dy_2, \tag{5.20}$$

$$f \in L_2(R_2).$$

If $f \in L_2(R_2)$, then we can show that $\Psi(t,\mathbf{x}) = \exp(t\mathcal{K}_{-2})f(\mathbf{x})$ satisfies $\partial_t\Psi = \mathcal{K}_{-2}\Psi$ or $i\partial_t\Psi = -\Delta_2\Psi$ (for almost every t) whenever f is in the domain of $\mathcal{K}_{-2}$, and $\Psi(0,\mathbf{x}) = f(\mathbf{x})$. Also, the unitary operators $\exp(\alpha K) = \exp(t\mathcal{K}_{-2})\exp(\alpha\mathcal{K})\exp(-t\mathcal{K}_{-2})$ map Ψ into $\Phi = \exp(\alpha K)\Psi$, which also satisfies $i\partial_t\Phi = -\Delta_2\Phi$ for each linear combination $\mathcal{K}$ of the operators (5.18). Thus the operators $\exp(\alpha K)$ are symmetries of (5.2).

We will see later that the operators (5.18) generate a global unitary irreducible representation of G_3 on $L_2(R_2)$. Assuming this here, we let $U(g), g \in G_3$, be the corresponding unitary operators and set $T(g) = \exp(t\mathcal{K}_{-2})U(g)\exp(-t\mathcal{K}_{-2})$. It then follows that the $T(g)$ are unitary symmetries of (5.2) with associated infinitesimal operators $K = \exp(t\mathcal{K}_{-2})\mathcal{K}\exp(-t\mathcal{K}_{-2})$.

Next consider the operator $\mathcal{L}_3 = \mathcal{K}_{-2} - \mathcal{K}_2 = i\left[\Delta_2 - \frac{1}{4}\left(x_1^2 + x_2^2\right)\right]$. If $f \in L_2(R_2)$, then $\Psi(t,\mathbf{x}) = \exp(t\mathcal{L}_3)f(\mathbf{x})$ satisfies $\partial_t\Psi = \mathcal{L}_3\Psi$ or

$$i\partial_t\Psi = -\Delta_2\Psi + \frac{1}{4}\left(x_1^2 + x_2^2\right)\Psi \tag{5.21}$$

ISBN-0-201-13503-5

and $\Psi(0, \mathbf{x}) = f(\mathbf{x})$. Similarly, the unitary operators $\mathbf{V}(g) = \exp(t\mathcal{L}_3) \times \mathbf{U}(g)\exp(-t\mathcal{L}_3)$ are symmetries of (5.21), the Schrödinger equation for the harmonic oscillator, and the associated infinitesimal operators $\exp(t\mathcal{L}_3) \times \mathcal{K} \exp(-t\mathcal{L}_3)$ can be expressed as first-order differential operators in t and $\mathbf{x}$. Analogous statements hold for the operator $\mathcal{L}_2 = \mathcal{K}_{-2} + \mathcal{K}_2 = i\left(\Delta_2 + \frac{1}{4}(x_1^2 + x_2^2)\right)$ with associated equation $\partial_t \Psi = \mathcal{L}_2 \Psi$ or

$$i\,\partial_t \Psi = -\Delta_2 \Psi - \frac{1}{4}(x_1^2 + x_2^2)\Psi \tag{5.22}$$

(Schrödinger equation for the repulsive oscillator) and the operator $\mathcal{K}_{-2} - \mathcal{B}_1 = i(\Delta_2 - x_1/2)$ with associated equation $\partial_t \Psi = (\mathcal{K}_{-2} - \mathcal{B}_1)\Psi$,

$$i\,\partial_t \Psi = -\Delta_2 \Psi + \frac{1}{2}x_1 \Psi \tag{5.23}$$

(linear potential).

These remarks show explicitly the equivalence of equations (5.2), (5.21)–(5.23). Though we have chosen to start with equation (5.2), an analysis of any of the other equations would have led to the same (script) symmetry algebra (5.18).

From Table 12 we see that, except for coordinates 13–17, which are essentially the same as those discussed in Section 2.1, every R-separable coordinate system corresponds to a G_3 orbit that contains exactly one of the Hamiltonian operators $i\mathcal{K}_{-2}, i\mathcal{L}_3, i\mathcal{L}_2$, or $i(\mathcal{K}_{-2} - \mathcal{B}_1)$. Thus each coordinate system is naturally associated with one of these four Hamiltonians. Moreover, the remarks accompanying expressions (1.29), (1.30) also hold here: an R-separable coordinate system for the free-particle equation corresponds to a truly separable coordinate system for one of the other three Schrödinger equations, namely, that equation whose Hamiltonian is diagonalized by the system.

Consider a pair of commuting self-adjoint operators $i\mathcal{K}, \mathcal{S}$ where $\mathcal{K} \in G_3$ and $\mathcal{S}$ is a symmetric quadratic operator in the enveloping algebra of G_3. These operators have a common spectral resolution; that is, there is a complete set of (generalized) eigenfunctions $f_{\lambda,\mu}(\mathbf{x})$ in $L_2(R_2)$ with

$$i\mathcal{K}f_{\lambda,\mu} = \lambda f_{\lambda,\mu}, \qquad \mathcal{S}f_{\lambda,\mu} = \mu f_{\lambda,\mu}, \qquad \langle f_{\lambda,\mu}, f_{\lambda',\mu'}\rangle = \delta_{\lambda\lambda'}\delta_{\mu\mu'}, \tag{5.24}$$

where

$$\langle h_1, h_2\rangle = \int\int_{-\infty}^{\infty} h_1(\mathbf{x})\bar{h}_2(\mathbf{x})\,dx_1, dx_2, \qquad h_j \in L_2(R_2) \tag{5.25}$$

(see [77, p. 76]). Now suppose $i\mathcal{K}', \mathcal{S}'$ are another pair of commuting self-adjoint operators on the same G_3 orbit as $i\mathcal{K}, \mathcal{S}$. Then by renormaliza-

ISBN-0-201-13503-5

tion of these operators if necessary, it follows that there is a $g \in G_3$ such that

$$\mathcal{K}' = \mathbf{U}(g)\mathcal{K}\mathbf{U}(g^{-1}), \qquad \mathcal{S}' = \mathbf{U}(g)\mathcal{S}\mathbf{U}(g^{-1}).$$

Thus the spectral resolution of the primed pair is identical to that for the unprimed pair. Indeed for $f'_{\lambda,\mu} = \mathbf{U}(g)f_{\lambda,\mu}$ we have

$$i\mathcal{K}'f'_{\lambda,\mu} = \lambda f'_{\lambda,\mu}, \qquad \mathcal{S}'f'_{\lambda,\mu} = \mu f'_{\lambda,\mu}, \qquad \langle f'_{\lambda,\mu}, f'_{\lambda',\mu'} \rangle = \delta_{\lambda\lambda'}\delta_{\mu\mu'}, \quad (5.26)$$

and the $f'_{\lambda,\mu}$ form a complete ON set in $L_2(R_2)$.

In the following we will frequently need the spectral resolution of a pair $i\mathcal{K}, \mathcal{S}$ where $i\mathcal{K}$ is one of the four Hamiltonians listed earlier. However, in many cases we will be able to use the unitary symmetry operators $\mathbf{U}(g)$ to construct an equivalent pair $i\mathcal{K}', \mathcal{S}'$ whose spectral resolution is much simpler to compute. This information will then provide the spectral resolution of the original pair.

As a special case of these remarks, consider the operator $\mathcal{K}_{-2} = i\Delta_2$. If $\{f_{\lambda,\mu}\}$ is the basis (5.24) of generalized eigenfunctions for the pair $\mathcal{K}, \mathcal{S}$, then $\{f'_{\lambda,\mu}(t,\mathbf{x}) = \exp(t\mathcal{K}_{-2})f_{\lambda,\mu}(\mathbf{x})\}$ is the corresponding basis of generalized eigenvectors for the italic operators

$$K = \exp(t\mathcal{K}_{-2})\mathcal{K}\exp(-t\mathcal{K}_{-2}), \qquad S = \exp(t\mathcal{K}_{-2})\mathcal{S}\exp(-t\mathcal{K}_{-2})$$

and the $f'_{\lambda,\mu}(t,\mathbf{x})$ are also solutions of the free-particle Schrödinger equation (5.2). Similar remarks hold for the other Hamiltonians. This clarifies the relationship between the two $(\mathbf{x})$ and three $(\mathbf{x},t)$ variable models of G_3.

We now explicitly compute the spectral resolutions of the pairs of commuting operators listed in Table 12. We begin with the Oc orbit and the two-variable model; that is, we determine the spectral resolution of the pair $\mathcal{L}_3 = \mathcal{K}_{-2} - \mathcal{K}_2, \mathcal{P}_1^2 + \mathcal{B}_1^2$. Equations (5.24) are

$$\left[-\Delta_2 + (x_1^2 + x_2^2)/4 \right] f = \lambda f, \qquad \left(\partial_{x_1 x_1} - x_1^2/4 \right) f = \mu f.$$

Note that these equations are separable in the variables x_1, x_2. Comparing with (1.34) we find the well-known ON basis of eigenfunctions

$$f_{\lambda,\mu} = oc_{n,m}(\mathbf{x}) = (2^{m+n}\pi n! m!)^{-1/2}\exp(-\mathbf{x} \cdot \mathbf{x}/4)H_n(x_1/\sqrt{2})$$

$$\times H_n(x_2/\sqrt{2}), \qquad \mu = -n - \tfrac{1}{2}, \quad \lambda + \mu = m + \tfrac{1}{2}, \quad n,m = 0,1,2,\ldots,$$

$$\langle oc_{n,m}, oc_{n',m'} \rangle = \delta_{nn'}\delta_{mm'}, \qquad\qquad\qquad (5.27)$$

where $H_n(x)$ is a Hermite polynomial.

ISBN-0-201-13503-5

At this point we can show directly that the operators (5.18) exponentiate to yield a global unitary irreducible representation of G_3. Indeed, from the recurrence relations (2.28), (2.29), (2.33) for the Hermite polynomials we can see that the operators $\mathcal{L}_1, \mathcal{L}_2, \mathcal{L}_3$ acting on the oc basis define a unitary representation of $sl(2,R)$ that is a direct sum of representations from the discrete series, and the $\mathcal{W}_2$ operators define a unitary irreducible representation of $\mathcal{W}_2$. As follows from the work of Bargmann [10, 115], this Lie algebra representation extends to a global representation of G_3, irreducible since its restriction to W_2 is already irreducible.

We now compute the unitary operators $U(g)$ on $L_2(R_2)$. The operators

$$U(\mathbf{w},\mathbf{z},\rho)=\exp(w_1\mathcal{B}_1)\exp(z_1\mathcal{P}_1)\exp(w_2\mathcal{B}_2)\exp(z_2\mathcal{P}_2)\exp(\rho\mathcal{E})$$

defining the irreducible representation of W_2 are

$$U(\mathbf{w},\mathbf{z},\rho)h(\mathbf{x})=\exp(i\rho+i\mathbf{w}\cdot\mathbf{x}/2)h(\mathbf{x}+\mathbf{z}), \qquad h\in L_2(R_2). \quad (5.28)$$

The operator $U(\theta)=\exp(\theta\,\mathfrak{M})$ is

$$U(\theta)h(\mathbf{x})=h(\mathbf{x}\Theta) \qquad\qquad\qquad (5.29)$$

where Θ is given by (5.8). The operators $U(A), A\in SL(2,R)$, are more difficult to compute. We have the integral operator $\exp(a\mathcal{K}_{-2})$ from (5.20). Also,

$$\exp(b\mathcal{K}_2)h(\mathbf{x})=\exp(i b\mathbf{x}\cdot\mathbf{x}/4)h(\mathbf{x}), \qquad \exp(c\mathcal{D})h(\mathbf{x})=e^c h(e^c\mathbf{x}). \quad (5.30)$$

Using group multiplication in $SL(2,R)$, we find

$$\exp(\varphi\mathcal{L}_2)=\exp(\tanh(\varphi)\mathcal{K}_2)\exp(\sinh(\varphi)\cosh(\varphi)\mathcal{K}_{-2})\exp(-\ln(\cosh\varphi)\mathcal{D})$$

so that

$$\exp(\varphi\mathcal{L}_2)h(\mathbf{x})=\frac{\exp(i\coth(\varphi)\mathbf{x}\cdot\mathbf{x}/4)}{4\pi i\sinh\varphi}$$

$$\cdot\int\!\!\int_{-\infty}^{\infty}\exp\left[\frac{i}{4}\left(-\frac{2}{\sinh\varphi}\mathbf{x}\cdot\mathbf{y}+\coth(\varphi)\mathbf{y}\cdot\mathbf{y}\right)\right]$$

$$\cdot h(\mathbf{y})\,dy_1\,dy_2, \qquad \varphi\neq 0 \qquad\qquad (5.31)$$

(In this and the following two integrals, use of the short form l.i.m. is to be

ISBN-0-201-13503-5

understood.) Similar computations yield

$$\exp(\theta \mathcal{L}_3)h(\mathbf{x}) = \frac{\exp(i\cot(\theta)\mathbf{x}\cdot\mathbf{x}/4)}{4\pi i \sin\theta}$$

$$\cdot \int\int_{-\infty}^{\infty} \exp\left[\frac{i}{4}\left(-\frac{2}{\sin\theta}\mathbf{x}\cdot\mathbf{y}+\cot(\theta)\mathbf{y}\cdot\mathbf{y}\right)\right]$$

$$\cdot h(\mathbf{y})\,dy_1\,dy_2, \qquad \theta \neq n\pi, \tag{5.32}$$

$$\exp\left[\rho(\mathcal{K}_{-2}+a\mathcal{B}_1)\right]h(\mathbf{x}) = \frac{\exp\left[i(a\rho x_1/2 - a^2\rho^3/12)\right]}{4\pi i \rho}$$

$$\cdot \int\int_{-\infty}^{\infty} \exp\left\{\frac{i}{4\rho}\left[(x_1 - a\rho^2 - y_1)^2 + (x_2 - y_2)^2\right]\right\}$$

$$\cdot h(\mathbf{y})\,dy_1\,dy_2, \qquad \rho \neq 0. \tag{5.33}$$

2.6 Bases and Overlaps for the Schrödinger Equation

From (5.20) and (5.27) it follows that the basis functions $oc_{n,m}(\mathbf{x})$ map to the ON basis functions $Oc_{n,m}(t,\mathbf{x}) = \exp(t\mathcal{K}_{-2})oc_{n,m}(\mathbf{x})$, solutions of the Schrödinger equation, where

$$Oc_{n,m}(t,\mathbf{x}) = (2^{m+n+1}\pi n!\,m!)^{1/2}\exp\left[\frac{i\pi(m+n-1)}{2} - \frac{(u^2+v^2)(1-iw)}{4}\right]$$

$$\times \left(\frac{w+i}{w-i}\right)^{(m+n)/2}(w-i)^{-1}H_m(u/\sqrt{2})H_n(v/\sqrt{2}) \tag{6.1}$$

with

$$x_1 = u(1+w^2)^{1/2}, \qquad x_2 = v(1+w^2)^{1/2}, \qquad t = w.$$

The functions (6.1) correspond to the separable coordinate system Oc in Table 12.

Next we compute the spectral resolution for the system Or:

$$i(\mathcal{K}_{-2} - \mathcal{K}_2)f = \lambda f, \qquad \mathfrak{M}^2 f = \mu f.$$

The basis of eigenfunctions is

$$or_{n,m}^+(\mathbf{x}) = \left[m!/2^m\pi(n+m)!\right]^{1/2}\exp(-r^2/4)r^m L_n^{(m)}(r^2/2)\cos m\theta,$$

$$or_{n,m}^-(\mathbf{x}) = \tan(n\theta)or_{n,m}^+(\mathbf{x}), \tag{6.2}$$

ISBN-0-201-13503-5

where n,m are integers with $m \geqslant 1, n \geqslant 0$ and $x_1 = r\cos\theta, x_2 = r\sin\theta$. The eigenvalues λ, μ are related to m, n via $\lambda = 2n + m + 1, \mu = -m^2$. For $m = 0$ there is one additional eigenvector

$$or^+_{\lambda,0}(\mathbf{x}) = (2/\pi n!)^{1/2} \exp(-r^2/4) L^{(0)}_n(r^2/2). \tag{6.3}$$

Here, the $L^{(\alpha)}_n(r)$ are Laguerre polynomials. The orthogonality relations are

$$\langle or^\varepsilon_{n,m}, or^{\varepsilon'}_{n',m'} \rangle = \delta_{\varepsilon\varepsilon'}\delta_{nn'}\delta_{mm'}, \qquad \varepsilon, \varepsilon' = \pm.$$

The three-variable basis functions $Or_{n,m}(t,\mathbf{x}) = \exp(t\mathcal{K}_{-2})or_{n,m}(\mathbf{x})$ are

$$Or^+_{n,m}(t,\mathbf{x}) = K\left(\frac{m!}{\pi^3 2^m (n+m)!}\right)^{1/2} \frac{(-1)^{m+n}}{2^{2m}} \frac{(w+i)^{m/2+n}}{(w-i)^{m/2+n+1}}$$

$$\cdot \exp\left[\frac{u^2(iw-1)}{4}\right] L^{(m)}_n\left(\frac{u^2}{2}\right)\cos mv,$$

$$Or^-_{n,m}(t,\mathbf{x}) = \tan(mv)Or^+_{n,m}(t,\mathbf{x}), \qquad m \geqslant 1. \tag{6.4}$$

For $m = 0$ we have $K = \sqrt{2}$; otherwise $K = 1$. Also,

$$x_1 = (1 + w^2)^{1/2} u\cos v, \qquad x_2 = (1 + w^2)^{1/2} u\sin v, \qquad t = w.$$

The equations for the system Oe,

$$i(\mathcal{K}_{-2} - \mathcal{K}_2)f = \lambda f, \qquad (\mathcal{M}^2 - \mathcal{P}^2_2 - \mathcal{B}^2_2)f = \mu f,$$

separate in elliptic coordinates $x_1 = \cosh\zeta\cos\eta, x_2 = \sinh\zeta\sin\eta$. We obtain the ON basis

$$oe^+_{p,m}(\mathbf{x}) = \pi^{-1}hc^m_p(i\zeta, \tfrac{1}{2})hc^m_p(\eta, \tfrac{1}{2}),$$

$$oe^-_{p,m}(\mathbf{x}) = \pi^{-1}hs^m_p(i\zeta, \tfrac{1}{2})hs^m_p(\eta, \tfrac{1}{2}), \tag{6.5}$$

where

$$hc^m_p(\eta, \tfrac{1}{2}) = \exp(-\tfrac{1}{4}\cos 2\eta)C^m_p(\eta, \tfrac{1}{2})$$

$$hs^m_p(\eta, \tfrac{1}{2}) = \exp(-\tfrac{1}{4}\cos 2\eta)S^m_p(\eta, \tfrac{1}{2}),$$

and m, p are integers with $o \leqslant m \leqslant p, (-1)^{m-p} = 1$. The eigenvalues λ, μ are related to p, m via $\lambda = p + 1, \mu = \lambda/2 + a^m_p(\tfrac{1}{2})$ or $\mu = \lambda/2 + b^m_p(\tfrac{1}{2})$ and the

ISBN-0-201-13503-5

orthogonality relations are

$$\langle oe_{p,m}^{\varepsilon}, oe_{p',m'}^{\varepsilon'} \rangle = \delta_{\varepsilon\varepsilon'} \delta_{pp'} \delta_{mm'}, \qquad \varepsilon, \varepsilon' = \pm.$$

The functions $C_p^m(\eta,\zeta), S_p^m(\eta,\zeta)$ are Ince polynomials [7], that is, poly-nomial solutions of period 2π for the Whittaker–Hill equation

$$\frac{d^2 v}{d\eta^2} + \zeta \sin 2\eta \, \frac{dv}{d\eta} + (a - p\zeta \cos 2\eta) v = 0. \qquad (6.6)$$

This equation has been investigated in detail by Arscott [8], and it is his notation for the solutions and eigenvalues that we use. The $C_p^m(\eta,\zeta)$ are polynomials of order p in $\cos\eta$ and correspond to the eigenvalues $a = a_p^m(\zeta)$, while the $S_p^m(\eta,\zeta)$ are polynomials of order p in $\sin\eta$ and corre-spond to the eigenvalues $a = b_p^m(\zeta)$.

The three-variable basis functions $Oe_{p,m}(t,\mathbf{x}) = \exp(t\mathcal{K}_{-2})oe_{p,m}(\mathbf{x})$ are

$$Oe_{p,m}^+(t,\mathbf{x}) = (\lambda_p^{m+}/\pi)\exp\left[iw(\sinh^2 u + \cos^2 v)/4 \right]$$

$$\cdot (w - i)^{(p/2)+1}(w + i)^{-p/2} hc_p^m\left(iu, \tfrac{1}{2} \right) hc_p^m\left(v, \tfrac{1}{2} \right) \qquad (6.7)$$

where

$$x_1 = (1 + w^2)^{1/2} \cosh u \cos v, \qquad x_2 = (1 + w^2)^{1/2} \sinh u \sin v, \qquad t = w.$$

The expression for $Oe_{p,m}^-(t,\mathbf{x})$ is as above except the phase factor $\lambda_p^{m+}, |\lambda_p^{m+}| = 1$, is replaced by λ_p^{m-} and the functions $hc_p^m(\eta,\zeta)$ are replaced by $hs_p^m(\eta,\zeta)$. The constants $\lambda_p^{m\pm}$ are calculable in principle from a knowl-edge of the explicit form of the Ince polynomials. Note that the expression $Oe_{p,m} = \exp(t\mathcal{K}_{-2})oe_{p,m}$ is a nontrivial relation satisfied by products of Ince polynomials. We are able to evaluate this integral (in a manner analogous to the evaluation of (3.38), Section 1.3) because we know in advance that the integral is an R-separable solution of the Schrödinger equation in the variables u, v, w.

For the remaining cases in Table 12 there are always two coordinate systems associated with each orbit. For simplicity we shall always treat the coordinate system with superscript 1. The corresponding results for system 2 follow immediately upon application of the unitary operators J or J^2, (5.16), (5.17).

The Fc system is defined by equations

$$i\mathcal{K}_2 f = -\tfrac{1}{4}\gamma^2 f, \qquad \mathcal{B}_1 f = \tfrac{1}{2} i\gamma \cos(\alpha) f,$$

and has a basis of generalized eigenfunctions

$$fc_{\gamma,\alpha}(\mathbf{x}) = r^{-1/2}\delta(r-\gamma)\delta(\theta-\alpha), \qquad 0 \le \alpha < 2\pi, \quad 0 \le \gamma,$$
$$\langle fc_{\gamma,\alpha}, fc_{\gamma',\alpha'}\rangle = \delta(\gamma-\gamma')\delta(\alpha-\alpha'), \qquad x_1 = r\cos\theta, \qquad x_2 = r\sin\theta. \tag{6.8}$$

The three-variable basis functions $Fc_{\gamma,\alpha}(t,\mathbf{x}) = \exp(t\mathcal{K}_{-2})fc_{\gamma,\alpha}(\mathbf{x})$ are

$$Fc_{\gamma,\alpha}(t,\mathbf{x}) = \frac{\gamma^{1/2}}{4\pi it}\exp\left\{\frac{i\left[(x_1-\gamma\cos\alpha)^2+(x_2-\gamma\sin\alpha)^2\right]}{4t}\right\}. \tag{6.9}$$

The Fr system is defined by

$$i\mathcal{K}_2 f = -\tfrac{1}{4}\gamma^2 f, \qquad i\mathfrak{M}f = -mf$$

with basis

$$fr_{\gamma,m}(\mathbf{x}) = (2\pi r)^{-1/2}\delta(r-\gamma)e^{im\theta}, \qquad \langle fr_{\gamma,m}, fr_{\gamma',m'}\rangle = \delta(\gamma-\gamma')\delta_{mm'}. \tag{6.10}$$

Here $0 \le \gamma$, $m=0, \pm 1,\ldots$ and r,θ are polar coordinates. The three-variable basis functions are

$$Fr_{\gamma,m}(t,\mathbf{x}) = \left(\frac{\gamma}{2\pi}\right)^{1/2}\frac{i^{m-1}}{2t}\exp\left[\frac{i(r^2+\gamma^2)}{4t}\right]\exp(im\theta)J_m\left(\frac{-r\gamma}{2t}\right) \tag{6.11}$$

where $J_m(z)$ is a Bessel function.

The F_p system is determined by equations

$$i\mathcal{K}_2 f = -\tfrac{1}{4}\gamma^2 f, \qquad \{\mathcal{B}_2, \mathfrak{M}\}f = -\mu\gamma f.$$

with eigenbasis

$$fp_{\gamma,\mu}^+(\mathbf{x}) = (2\pi r)^{-1/2}(1+\cos\theta)^{-i\mu/2-\frac{1}{4}}(1-\cos\theta)^{i\mu/2-\frac{1}{4}}\delta(r-\gamma), \quad -\pi \le \theta < 0,$$
$$= 0, \qquad 0 \le \theta \le \pi,$$
$$fp_{\gamma,\mu}^-(\mathbf{x}) = fp_{\gamma,\mu}^-(r,\theta) = fp_{\gamma,\mu}^+(r,-\theta). \tag{6.12}$$

Here r,θ are polar coordinates, $0 \le \gamma$, $-\infty < \mu < \infty$, and the spectrum is continuous with multiplicity two. The orthogonality relations are

$$\langle fp_{\gamma,\mu}^\pm, fp_{\gamma',\mu'}^\pm\rangle = \delta(\gamma-\gamma')\delta(\mu-\mu'), \qquad \langle fp_{\gamma,\mu}^\pm, fp_{\gamma',\mu'}^\mp\rangle = 0.$$

ISBN-0-201-13503-5

The three-variable basis functions are

$$Fp_{\gamma,\mu}^{+}(t,\mathbf{x}) = \frac{i\gamma^{1/2}\exp(i\gamma^2/4t)}{2^3\pi t \cos(i\mu\pi)} \exp\left(\frac{i(\xi^2+\eta^2)^2}{16t}\right)$$

$$\cdot[D_{-i\mu/2-\frac{1}{2}}(\sigma\xi t^{-1/2})D_{i\mu/2-\frac{1}{2}}(\sigma\eta t^{-1/2}) + D_{-i\mu/2-\frac{1}{2}}(-\sigma\xi t^{-1/2})$$

$$\cdot D_{i\mu/2-\frac{1}{2}}(-\sigma\eta t^{-1/2})], \qquad t>0,$$

$$Fp_{\gamma,\mu}^{+}(-t,\mathbf{x}) = \overline{Fp}\,_{\gamma,-\mu}^{+}(t,\mathbf{x}),$$

$$Fp_{\gamma,\mu}^{-}(t,x_1,x_2) = Fp_{\gamma,\mu}^{+}(t,x_1,-x_2), \qquad (6.13)$$

where $\sigma = \gamma^{1/2}\exp(i\pi/4)$ and ξ,η are parabolic coordinates

$$2x_1 = \xi^2 - \eta^2, \qquad x_2 = \xi\eta.$$

The *Fe* system is defined by equations

$$i\mathcal{K}_2 f = -\tfrac{1}{4}\gamma^2 f, \qquad \left(\mathcal{M}^2 + 4\mathcal{B}_1^2 - 4\mathcal{B}_2^2\right)f = -\mu f$$

(equivalent to 4a in Table 12 since $\mathcal{K}_2 = -i(\mathcal{B}_1^2 + \mathcal{B}_2^2)$). The basis functions are

$$fe_{\gamma,n}(\mathbf{x}) = (r\pi)^{-1/2}\delta(r-\gamma)\begin{cases} ce_n(\theta,\gamma^2/2), & n=0,1,2,\ldots, \\ se_{-n}(\theta,\gamma^2/2), & n=-1,-2,\ldots, \end{cases}$$

$$\gamma \geqslant 0, \qquad \langle fe_{\gamma,n}, fe_{\gamma',n'}\rangle = \delta(\gamma-\gamma')\delta_{nn'}, \qquad (6.14)$$

where $ce_n(\theta,q), se_n(\theta,q)$ are the periodic Mathieu functions (B.26) and r,θ are polar coordinates. The eigenvalues $\mu = \mu_n$ are discrete and all of multiplicity one. The basis functions $Fe_{\gamma,n}(t,\mathbf{x}) = \exp(t\mathcal{K}_{-2})fe_{\gamma,n}(\mathbf{x})$ are

$$Fe_{\gamma,n}(t,\mathbf{x}) = \frac{A_{\gamma,n}}{4\pi i\tau}\left(\frac{\gamma}{\pi}\right)^{1/2}\exp\left[i\tau(\cos^2\sigma + \sinh^2\rho + \gamma^2)\right]$$

$$\cdot\begin{cases} ce_n(\sigma,\gamma^2/2)Ce_n(\rho,\gamma^2/2), & n=0,1,2,\ldots, \\ se_{-n}(\sigma,\gamma^2/2)Se_{-n}(\rho,\gamma^2/2), & n=-1,-2,\ldots, \end{cases} \qquad (6.15)$$

where $A_{\gamma,n}$ is a normalization constant, $Se_n(\rho,q)$ and $Ce_n(\rho,q)$ are modified Mathieu functions (3.40), Section 1.3, and

$$x_1 = -2\tau\cosh\rho\cos\sigma, \qquad x_2 = -2\tau\sinh\rho\sin\sigma, \qquad t=\tau.$$

The Lc system (transformed so that $b=0$) can be defined by equations

$$i(\mathcal{K}_2+a\mathcal{P}_1)f=\lambda f, \qquad \mathcal{B}_2^2 f=-\tfrac{1}{4}\rho^2 f, \qquad a\neq 0,$$

with basis functions

$$lc_{\lambda,\rho}(\mathbf{x})=\frac{\delta(x_2-\rho)}{(2\pi|a|)^{1/2}}\exp\left[-ia^{-1}\left(\lambda x_1+\frac{\rho^2 x_1}{4}+\frac{x_1^3}{12}\right)\right],$$

$$\langle lc_{\lambda,\rho},lc_{\lambda',\rho'}\rangle=\delta(\lambda-\lambda')\delta(\rho-\rho'), \qquad -\infty<\lambda,\rho<\infty. \tag{6.16}$$

The three-variable basis functions are

$$Lc_{\lambda,\rho}(t,\mathbf{x})=\frac{(9a)^{-1/3}}{8iw(2\pi|a|)^{1/2}}\exp\left[i\left((u^2+v^2)\frac{w}{4}-\frac{au}{w}-\frac{\rho v}{2}-\frac{a}{3w^3}-\lambda/w\right)\right]$$

$$\cdot\mathrm{Ai}\left[(36a)^{-1/3}\left(\frac{u}{a}+\frac{\lambda}{a}+\frac{\rho^2}{4a}\right)\right] \tag{6.17}$$

where $\mathrm{Ai}(z)$ is an Airy function ((1.52), Section 2.1). Here

$$x_1=uw+a/w, \qquad x_2=vw, \qquad t=w.$$

The Lp system is defined by

$$i(\mathcal{K}_2+a\mathcal{P}_1)f=\lambda f, \qquad (\{\mathcal{B}_2,\mathcal{M}\}+a\mathcal{P}_2^2)f=\mu f$$

with basis functions

$$lp_{\lambda,n}(\mathbf{x})=(2\pi|a|)^{-1/2}h_n(x_2)\exp\left[-i(\lambda x_1+x_1 x_2^2/4+x_1^3/12)/a\right],$$

$$\langle lp_{\lambda,n},lp_{\lambda',n'}\rangle=\delta(\lambda-\lambda')\delta_{nn'}, \qquad -\infty<\lambda<\infty, \qquad n=0,1,2,\ldots. \tag{6.18}$$

Here the anharmonic oscillator function $h_n(x)$ is a solution of

$$\frac{d^2h(x)}{dx^2}-\left(\frac{\mu}{a}+\frac{\lambda x^2}{a^2}+\frac{x^4}{4a^2}\right)h(x)=0, \qquad \lambda,a \text{ fixed}, \tag{6.19}$$

such that

$$\int_{-\infty}^{\infty}|h_n(x)|^2\,dx=1. \tag{6.20}$$

The eigenvalues $\mu=\mu_n(\lambda)$ of (6.19) subject to condition (6.20) are discrete [100, p. 250] with multiplicity one, and we assume them ordered so that $\mu_0<\mu_1<\mu_2<\ldots$. Here $h_n(x)$ is either even or odd for each value of n.

ISBN-0-201-13503-5

Denote a general solution of (6.19) by $h_{\mu,\lambda,a}(x)$. Then using separation of variables it is straightforward to show that the basis functions $Lp_{\lambda,n}(t,\mathbf{x}) = \exp(t\mathcal{K}_{-2})lp_{\lambda,n}(\mathbf{x})$ are

$$Lp_{\lambda,n}(t,\mathbf{x}) = C_{\lambda,n}w^{-1}\exp\left\{i\left[\frac{(u^2+v^2)^2 w}{16} - \frac{a(u^2-v^2)}{4w} - \frac{a^2}{12w^2} - \frac{\lambda}{w}\right]\right\}$$

$$\cdot h_{2\mu_n,\lambda,a/2}(u)h_{2\mu_n,\lambda,a/2}(iv) \qquad (6.21)$$

where the two h functions have the same parity as $h_n(x)$ and $C_{\lambda,n}$ is a normalization constant. (Note that since (6.19) is invariant under the replacement $x \to -x$, for each μ,λ,a this equation has a single even solution in x and a single odd solution to within multiplication by a normalization constant.) Also,

$$x_1 = (u^2 - v^2)w/2 + a/w, \qquad x_2 = uvw, \qquad t = w.$$

The Rc system is defined by the equations

$$i\mathcal{D}f = \rho f, \qquad \{\mathcal{B}_1, \mathcal{P}_1\}f = \mu f$$

with basis eigenfunctions

$$rc_{\lambda\mu}^{\varepsilon\varepsilon'}(\mathbf{x}) = (2\pi)^{-1}(x_1)_\varepsilon^{-i\lambda-\frac{1}{2}}(x_2)_{\varepsilon'}^{-i\mu-\frac{1}{2}}, \qquad (6.22)$$

$$-\infty < \lambda, \mu < \infty, \ \varepsilon, \varepsilon' = \pm, \ \lambda = \rho - \mu;$$

see (1.46). The orthogonality relations are

$$\langle rc_{\lambda\mu}^{\varepsilon\varepsilon'}, rc_{\bar{\lambda}\bar{\mu}}^{\bar{\varepsilon}\bar{\varepsilon}'}\rangle = \delta_{\varepsilon\bar{\varepsilon}}\delta_{\varepsilon'\bar{\varepsilon}'}\delta(\lambda - \bar{\lambda})\delta(\mu - \bar{\mu}).$$

The three-variable eigenfunctions are

$$Rc_{\lambda\mu}^{++}(t,\mathbf{x}) = (8\pi^2 iw)^{-1}\left[\exp(i\pi/4)(2w)^{1/2}\right]^{-i(\lambda+\mu)+1}$$

$$\cdot \Gamma\left(\tfrac{1}{2} - i\lambda\right)\Gamma\left(\tfrac{1}{2} - i\mu\right)\exp\left[\frac{i(u^2+v^2)}{8}\right]$$

$$\cdot D_{i\lambda-\frac{1}{2}}\left(\frac{-u}{(2i)^{1/2}}\right)D_{i\mu-\frac{1}{2}}\left(\frac{-v}{(2i)^{1/2}}\right), \qquad t > 0, \quad (6.23)$$

where $x_1 = |w|^{1/2}u$, $x_2 = |w|^{1/2}v$, $t = w$. The remaining three-variable basis

ISBN-0-201-13503-5

functions are given by

$$Rc_{\lambda\mu}^{++}(u,v) = \exp\left[\pi(i+\lambda+\mu)\right]Rc_{\lambda\mu}^{--}(-u,-v)$$
$$= \exp\left[\pi(i/2+\lambda)\right]Rc_{\lambda\mu}^{-+}(-u,v)$$
$$= \exp\left[\pi(i/2+\mu)\right]Rc_{\lambda\mu}^{+-}(u,-v). \qquad (6.24)$$

The Rr system is defined by the equations

$$\mathcal{D}f = i\rho f, \qquad \mathcal{M}f = imf.$$

The eigenfunctions are

$$rr_{\rho,m}(\mathbf{x}) = (2\pi)^{-1}r^{i\rho-1}e^{im\theta}, \qquad -\infty < \rho < \infty, \quad m = 0, \pm 1, \dots,$$
$$x_1 = r\cos\theta, \qquad x_2 = r\sin\theta, \qquad (6.25)$$

satisfying the orthogonality relations

$$\langle rr_{\rho,m}, rr_{\rho',m'} \rangle = \delta_{mm'}\delta(\rho-\rho').$$

The three-variable basis functions are

$$Rr_{\rho,m}(t,\mathbf{x}) = 2^{-m+i\rho-2}\exp\left[i\pi\frac{(3m-1+i\rho)}{4}\right]\pi^{-1}w^{i\rho/2-\frac{1}{2}}$$

$$\cdot \Gamma\left(\frac{m+i\rho+1}{2}\right)\frac{u^m}{\Gamma(m+1)}\,{}_1F_1\left(\begin{array}{c}(m+1-i\rho)/2\\ m+1\end{array}\bigg|\frac{iu^2}{4}\right)\exp(imv) \qquad (6.26)$$

where

$$x_1 = \sqrt{w}\,u\cos v, \qquad x_2 = \sqrt{w}\,u\sin v, \qquad t = w > 0.$$

The Re system is defined by equations

$$\mathcal{D}f = i\lambda f, \qquad \left(\mathcal{M}^2 + \tfrac{1}{2}\{\mathcal{B}_2, \mathcal{P}_2\}\right)f = \mu f.$$

The ON basis of eigenfunctions is

$$re_{\lambda m}^+(\mathbf{x}) = (2\pi)^{-1/2}r^{i\lambda-1}Gc_m(\theta, \tfrac{1}{4}, -\lambda), \qquad (6.27a)$$

$$re_{\lambda m}^-(\mathbf{x}) = (2\pi)^{-1/2}r^{i\lambda-1}Gs_m(\theta, \tfrac{1}{4}, -\lambda), \qquad (6.27b)$$

$$m = 0, 1, 2, \dots, \quad -\infty < \lambda < \infty,$$

ISBN-0-201-13503-5

where $x_1 = r\cos\theta$, $x_2 = r\sin\theta$. Here we have introduced the notation

$$Gc_m\left(\theta, \tfrac{1}{4}, -\lambda\right) = \exp\left[i\cos(2\theta)/16\right] gc_m\left(\theta, \tfrac{1}{4}, -\lambda\right),$$

$$Gs_m\left(\theta, \tfrac{1}{4}, -\lambda\right) = \exp\left[i\cos(2\theta)/16\right] gs_m\left(\theta, \tfrac{1}{4}, -\lambda\right).$$

The functions $gc_m(\theta, \alpha, \beta)$, $gs_m(\theta, \alpha, \beta)$ are even and odd nonpolynomial solutions of the Whittaker–Hill equation

$$\frac{d^2 g}{d\theta^2} + \left(\mu + \frac{\alpha^2}{8} + \alpha\beta\cos 2\theta - \frac{\alpha^2}{8}\cos 4\theta\right) g = 0 \qquad (6.28)$$

with period 2π. The subscript m (the number of zeros in the interval $[0, 2\pi]$) labels the discrete eigenvalues $\mu = \mu_m$ of the operator $\mathfrak{M}^2 + \tfrac{1}{2}\{\mathcal{B}_2, \mathcal{P}_2\}$. This notation is due to Urwin and Arscott [127]. Each of the solutions Gc_m, Gs_m can be written as an infinite trigonometric series in $\cos n\theta$, $\sin n\theta$, respectively, which converges for the discrete eigenvalues μ_m. The orthogonality relations are

$$\langle re^{\pm}_{\lambda m}, re^{\pm}_{\lambda' m'}\rangle = \delta(\lambda - \lambda')\delta_{mm'}, \qquad \langle re^{\pm}_{\lambda m}, re^{\mp}_{\lambda' m'}\rangle = 0$$

and the three-variable basis functions are

$$Re^{+}_{\lambda m}(t, \mathbf{x}) = K^{\lambda+}_m w^{(i\lambda - 1)/2} Gc_m\left(iu, \tfrac{1}{4}, -\lambda\right) Gc_m\left(v, \tfrac{1}{4}, -\lambda\right),$$

$$Re^{-}_{\lambda m}(t, \mathbf{x}) = K^{\lambda-}_m w^{(i\lambda - 1)/2} Gs_m\left(iu, \tfrac{1}{4}, -\lambda\right) Gs_m\left(v, \tfrac{1}{4}, -\lambda\right), \qquad (6.29)$$

where

$$x_1 = \sqrt{w}\ \cosh u\cos v, \qquad x_2 = \sqrt{w}\ \sinh u\sin v, \qquad t = w > 0.$$

The constants $K^{\lambda\pm}_m$ are in principle calculable by choosing special values of the parameters u, v, w. In fact, in the process of calculating the functions $Re^{\pm}$ by separation of variables, we obtain relations

$$K^{\lambda+}_m Gc_m\left(iu, \tfrac{1}{4}, -\lambda\right) Gc_m\left(v, \tfrac{1}{4}, -\lambda\right) = \exp\left[i(\sinh^2 u + \cos^2 v)/4\right] \int_{-\pi}^{\pi} d\theta$$

$$\times Gc_m\left(\theta, \tfrac{1}{4}, -\lambda\right)\exp\left[-i(\cosh u\cos v\cos\theta + \sinh u\sin v\sin\theta)^2/8\right]$$

$$\times D_{i\lambda - 1}\left(-[\cosh u\cos v\cos\theta + \sinh u\sin v\sin\theta]/(2i)^{1/2}\right)$$

with a similar expression for the functions $Gs_n(\theta, \tfrac{1}{4}, -\lambda)$. The constants $K^{\lambda\pm}_m$ can be calculated for particular values of the arguments; for example,

ISBN-0-201-13503-5

if $Gc_m(\theta, \frac{1}{4}, -\lambda) = \sum_{k=0}^{\infty} A_k^m \cos 2k\theta$, then

$$K_m^{\lambda+} = 2\pi D_{i\lambda-1}(0) A_0^m \left[Gc_m(\pi/2, \tfrac{1}{4}, -\lambda) Gc_m(0, \tfrac{1}{4}, -\lambda) \right]^{-1}.$$

This completes our determination of bases for the solution space of the Schrödinger equation (5.2).

Exactly as in Section 2.1 we can show that these results lead to a number of Hilbert space expansion theorems. Indeed, if $\{f_{\lambda\mu}\}$ is an ON basis for $L_2(R_2)$, then $\{U(g)f_{\lambda\mu}\}$ for any $g \in G_3$ is also an ON basis. In particular, each of the three-variable models constructed above provides a basis for $L_2(R_2)$ (as well as a basis of solutions for (5.2)). Furthermore, we can derive discrete and continuous generating functions for each basis.

Now we compute some overlap functions $\langle Aa_{\lambda\mu}, Bb_{\lambda'\mu'} \rangle$ that allow us to expand eigenfunctions $Aa_{\lambda\mu}$ in terms of eigenfunctions $Bb_{\lambda'\mu'}$. The utility of these formulas is that they are invariant under the action of G, so the same expressions allow us to expand $U(g)Aa_{\lambda\mu}$ in terms of $U(g)Bb_{\lambda'\mu'}$, where the results may be much less obvious. In the following we use the two-variable bases to compute some overlaps of interest. Because of G_3 invariance, identical results hold for the three-variable bases.

Here we omit overlaps involving the discrete basis oe. This basis is of special interest but the overlap computation involves use of the Bargmann–Segal Hilbert space of analytic functions, which we will not discuss here. Detailed results and an interesting connection between Ince polynomials and the representation theory of $SU(2)$ are presented in [21]. For most of the other bases we give an overlap with either of the discrete bases oc or or. The principle behind these computations should now be obvious, so the interested reader can derive for himself any of the other overlaps.

$$\langle fc_{\gamma,\alpha}, or_{nm}^{\pm} \rangle = \gamma^{1/2} or_{nm}^{\pm}(\gamma\cos\alpha, \gamma\sin\alpha); \qquad (6.30)$$

$$\langle fr_{\gamma,p}, or_{nm}^{\pm} \rangle = \begin{cases} 0 \\ \text{if} \quad p \neq \pm m, \\[2ex] \left[\dfrac{\gamma m!}{2^{m+1}(n+m)!} \right]^{1/2} \exp\left(\dfrac{-\gamma^2}{4} \right) \gamma^m L_n^{(m)}\left(\dfrac{\gamma^2}{2} \right) \\ \text{if} + \text{and } p = \pm m \neq 0, \\[2ex] \dfrac{ip}{m} \left[\dfrac{\gamma m!}{2^{m+1}(n+m)!} \right]^{1/2} \exp\left(\dfrac{-\gamma^2}{4} \right) \gamma^m L_n^{(m)}\left(\dfrac{\gamma^2}{2} \right) \\ \text{if} - \text{and } p = \pm m \neq 0 \\[2ex] \left(\dfrac{4\gamma}{n!} \right)^{1/2} \exp\left(\dfrac{-\gamma^2}{4} \right) L_n^{(o)}\left(\dfrac{\gamma^2}{2} \right) \\ \text{if} \quad p = m = 0; \end{cases} \qquad (6.31)$$

ISBN-0-201-13503-5

$$\langle fp^{+}_{\gamma\mu}, or^{\pm}_{nm}\rangle = \left[\frac{\gamma m!}{2^{m}\pi(n+m)!}\right]^{1/2} \exp\left(\frac{-\gamma^2}{4}\right)\gamma^{m}L_n^{(m)}\left(\frac{\gamma^2}{2}\right)$$

$$\times \exp\left[-\pi i\frac{(1\mp 1)}{4}\right](a_m \pm a_{-m});$$

$$\langle f\bar{p}_{\gamma\mu}, or^{\pm}_{nm}\rangle = \langle fp^{+}_{\gamma\mu}, or^{\pm}_{n,-m}\rangle;$$

$$a_m = \exp\left[\pi\frac{(i/2-\mu)}{2}\right]\Gamma\left(m+\tfrac{1}{2}\right)\left[\frac{(-1)^{m}\Gamma\left(\tfrac{1}{2}+i\mu\right)}{\Gamma\left(m+i\mu+\tfrac{1}{2}\right)}\right.$$

$$\times {}_2F_1\left[\begin{matrix}\tfrac{1}{2}+i\mu,\tfrac{1}{2}+m\\m+i\mu+1\end{matrix}\middle| -1\right] - \frac{i\Gamma\left(\tfrac{1}{2}-i\mu\right)}{\Gamma(m-i\mu+1)}$$

$$\left.\times {}_2F_1\left[\begin{matrix}\tfrac{1}{2}-i\mu,\tfrac{1}{2}+m\\m-i\mu+1\end{matrix}\middle| -1\right]\right]; \tag{6.32}$$

$$\langle fe_{\gamma,p}, or^{+}_{nm}\rangle = \theta(p)\left(1+(-1)^{m-p}\right)A_m^p\left[\frac{\gamma m!}{2^{m+2}\pi^2(n+m)!}\right]^{1/2}$$

$$\times \exp\left(\frac{-\gamma^2}{4}\right)\gamma^{m}L_n^{(m)}\left(\frac{\gamma^2}{2}\right); \tag{6.33}$$

where $\theta(x)=1$ for $x\geq 0$, and $\theta(x)=0$ otherwise. A similar expression for $\langle fe_{\gamma,p}, or_{nm}\rangle$ can be obtained by replacing $\theta(p)$ by $\theta(-p)$ and A_m^p by B_m^p in expression (6.33). Here A_m^p, B_m^p are the coefficients in the trigonometric expansions of the even and odd Mathieu functions, respectively. Also,

$$\langle lc_{\lambda,\rho}, oc_{n,m}\rangle = \exp(-\rho^2/4)(2^{m-1}\pi m!)^{-1/2}H_m(\rho/\sqrt{2})C_n \tag{6.34}$$

where

$$2^{2/3}\exp\left[-i\left(^{1/6}+\lambda+\rho^2/4+\sqrt{2y}\,\right)\right]\text{Ai}\left[2^{2/3}\left(1/4-i\lambda-i\rho^2/4-i(2y)^{1/2}\right)\right]$$

$$= \sum_{n=0}^{\infty}\left[\left((2i)^{1/2}y\right)^{n}/n!\right]C_n,$$

and we have normalized so that $a=-1$,

$$\langle lc_{\lambda,\rho}, lp_{\mu,n}\rangle = (2\pi|a|)^{-1}\bar{h}_n(\rho)\delta[(\lambda-\mu)/a], \tag{6.35}$$

$$\langle rc^{++}_{\lambda\mu}, oc_{n,m}\rangle = \pi^{-2}(2^{m+n+3}n!m!)^{-1/2}\mathcal{C}_m^{\lambda}\mathcal{C}_n^{\mu} \tag{6.36}$$

where

$$
\ell_m^\lambda = \begin{cases}
2^{m+i\lambda-\frac12}\Gamma(i\lambda/2+1/4)\Gamma((m+1)/2)\,{}_2F_1\left(\begin{matrix}-m/2,i\lambda/2+1/4\\1/2\end{matrix}\Big|2\right),\\
\text{m even,}\\
2^{m+i\lambda+1}\Gamma(i\lambda/2+1/4)\Gamma(m/2)\,{}_2F_1\left(\begin{matrix}(1-m)/2,i\lambda/2+3/4\\3/2\end{matrix}\Big|2\right),\\
\text{m odd.}
\end{cases}
$$

The remaining overlaps for rc^{+-}, rc^{-+}, and rc^{--} can be calculated by using relations (6.24).

$$
\langle rr_{\lambda m}^+, or_{nm'}\rangle = \delta_{mm'}(2/n!)^{m/2-i\lambda}\big[(m+n)!/m!\big]^{1/2}
$$

$$
\times \Gamma((m+1-i\lambda)/2)\,{}_2F_1\left(\begin{matrix}-n,(m+1-i\lambda)/2\\m+1\end{matrix}\Big|2\right), \qquad (6.37)
$$

$$
\langle rr_{\lambda m}^-, or_{nm'}\rangle = -i(-1)^{\operatorname{sign}m}\langle rr_{\lambda m}^+, or_{nm'}\rangle, \qquad (6.38)
$$

$$
\langle rr_{\lambda 0}^+, or_{n,m'}\rangle = \delta_{om'}\left(2^{-\frac12-i\lambda}/(n!)^{1/2}\right)\Gamma((1-i\lambda)/2)
$$

$$
\times {}_2F_1\left(\begin{matrix}-n,(1-i\lambda)/2\\1\end{matrix}\Big|2\right). \qquad (6.39)
$$

$$
\langle oc_{n_1,n_2}, or_{nm}^\pm\rangle = K\delta_{n_1+n_2,2n+m}\,i^{n_1}\left(2^{2n+m+1}n!n_2!(n+m)!/n_1!\right)^{1/2}
$$

$$
\times\binom{1}{-i}\Big[i^m\{\Gamma((n_1+n_2-m)/2)\Gamma((n_2-n_1+m)/2)\}^{-1}
$$

$$
\times {}_2F_1\left[\begin{matrix}-n_1,1-(n_1+n_2+m)/2\\(n_2-n_1-m)/2\end{matrix}\Big|-1\right]\pm i^{-m}\{\Gamma((n_1+n_2-m)/2)
$$

$$
\times \Gamma((n_2-n_1+m)/2)\}^{-1}\,{}_2F_1\left[\begin{matrix}-n_1,1-(n_1+n_2-m)/2\\(n_2-n_1+m)/2\end{matrix}\Big|-1\right]\Big]. \qquad (6.40)
$$

ISBN-0-201-13503-5

For the basis *re* we have

$$\langle re^+_{\lambda m}, or^+_{nm'}\rangle = \tfrac{1}{2}\left(1 + (-1)^{m-m'}\right)\overline{A}^m_{m'}(2\pi)^{1/2}\langle rr^+_{\lambda m'}, or_{n,m'}\rangle, \qquad (6.41)$$

$$\langle re^-_{\lambda m}, or^-_{nm'}\rangle = \tfrac{1}{2}\left(1 + (-1)^{m-m'}\right)\overline{B}^m_{m'}(2\pi)^{1/2}\langle rr^-_{\lambda m'}, or_{n,m'}\rangle, \qquad (6.42)$$

where $\overline{A}^m_{m'}$ and $\overline{B}^m_{m'}$ are the coefficients for the expansion of the functions $gc_m(\theta, \tfrac{1}{4}, -\lambda)$ and $gs_m(\theta, \tfrac{1}{4}, -\lambda)$, respectively, in trigonometric series analogous to (B.26) [127].

2.7 The Real and Complex Heat Equations $(\partial_t - \partial_{xx} - \partial_{yy})\Phi = 0$

The heat equation in three-dimensional space-time (suitably normalized) is

$$Q\Phi = 0, \qquad Q = \partial_t - \partial_{x_1 x_1} - \partial_{x_2 x_2} \qquad (7.1)$$

where t, x_1, x_2 are the real time and space variables, respectively. Since this equation can be obtained from the Schrödinger equation by replacing t in (5.2) with $-it$, the symmetry algebras of these two equations are closely related. The symmetry algebra of (7.1) is nine dimensional with basis

$$H_2 = t^2\partial_t + tx_1\partial_{x_1} + tx_2\partial_{x_2} + t + (x_1^2 + x_2^2)/4, \qquad H_{-2} = \partial_t,$$

$$P_j = \partial_{x_j}, \qquad B_j = t\partial_{x_j} + x_j/2, \qquad M = x_1\partial_{x_2} - x_2\partial_{x_1}, \qquad (7.2)$$

$$H^0 = x_1\partial_{x_1} + x_2\partial_{x_2} + 2t\partial_t + 1, \qquad H_0 = 1, \qquad j = 1,2,$$

and commutation relations

$$[H^0, H_{\pm 2}] = \pm 2H_{\pm 2}, \qquad [H^0, B_j] = B_j, \qquad [H^0, P_j] = -P_j,$$

$$[P_j, H_2] = B_j, \qquad [P_j, B_j] = \tfrac{1}{2}H_0, \qquad [P_j, B_l] = 0,$$

$$[H_{-2}, H_2] = H^0, \qquad [H_{\pm 2}, M] = [H_2, B_j] = [H_{-2}, P_j] = [H^0, M] = 0,$$

$$[B_j, M] = (-1)^{j+1}B_l, \qquad [H_{-2}, B_j] = P_j, \qquad [P_j, M] = (-1)^{j+1}P_l, \qquad (7.3)$$

$$j,l = 1,2, j \neq l,$$

where H_0 is in the center of the algebra. We denote by $\mathcal{G}_3'$ the real Lie algebra with basis (7.2). The operators B_j, P_j, H_0 span the five-dimensional Weyl subalgebra $\mathcal{W}_2$ of $\mathcal{G}_3'$ and local Lie theory yields the associated local

ISBN-0-201-13503-5

group action

$$\mathbf{T}(\mathbf{w},\mathbf{z},\rho)\Psi(t,\mathbf{x})=\exp\left[\frac{1}{2}\mathbf{x}\cdot\mathbf{w}+\frac{t}{4}\mathbf{w}\cdot\mathbf{w}+\rho\right]\Psi(t,\mathbf{x}+t\mathbf{w}+\mathbf{z}). \qquad (7.4)$$

Here, $\mathbf{w}=(w_1,w_2)$, $\mathbf{z}=(z_1,z_2)$, $\mathbf{x}\cdot\mathbf{w}=x_1w_1+x_2w_2$, and $w_j,z_j,\rho\in R$. The operators act on the space $\mathcal{F}$ of functions $\Psi(t,\mathbf{x})$, analytic in some given domain $\mathcal{D}$ in three-space. Furthermore, these operators map solutions of the heat equation into solutions.

Similarly, the operators $H_{\pm2}$, H_0 span the three-dimensional subalgebra $sl(2,R)$ and determine operators

$$\mathbf{T}(A)\Psi(t,\mathbf{x})=\exp\left[-\frac{\beta}{4}(\delta+t\beta)^{-1}\mathbf{x}\cdot\mathbf{x}\right](\delta+t\beta)^{-1}\Psi\left(\frac{\gamma+t\alpha}{\delta+t\beta},(\delta+t\beta)^{-1}\mathbf{x}\right),$$

$$A=\begin{pmatrix}\alpha & \beta\\ \gamma & \delta\end{pmatrix}\in SL(2,R), \qquad \Psi\in\mathcal{F}, \qquad (7.5)$$

which define a local representation of $SL(2,R)$. The operator M determines a local representation of $SO(2)$:

$$\mathbf{T}(\theta)\Psi(t,\mathbf{x})=\Psi(t,\mathbf{x}\Theta), \qquad \Theta=\begin{pmatrix}\cos\theta & \sin\theta\\ -\sin\theta & \cos\theta\end{pmatrix}. \qquad (7.6)$$

The local Lie group G_3' of symmetry operators $\mathbf{T}$ can be represented as a semidirect product of W_2 and $SL(2,R)\times SO(2)$ by means of expressions analogous to (5.10) and (5.11).

For future use we point out explicitly the special case of (7.5) where

$$A_0=2^{-1/2}\begin{pmatrix}1 & 1\\ -1 & 1\end{pmatrix}, \qquad A_0^8=\begin{pmatrix}1 & 0\\ 0 & 1\end{pmatrix}:$$

$$\mathbf{T}(A_0)\Psi(t,\mathbf{x})=\exp\left[-\tfrac{1}{4}(1+t)^{-1}\mathbf{x}\cdot\mathbf{x}\right]\frac{\sqrt{2}}{1+t}\Psi\left(\frac{t-1}{t+1},\frac{\sqrt{2}}{t+1}\mathbf{x}\right),$$

$$\mathbf{T}(A_0^2)\Psi(t,\mathbf{x})=\exp\left[-\frac{1}{4t}\mathbf{x}\cdot\mathbf{x}\right]t^{-1}\Psi(-t^{-1},t^{-1}\mathbf{x}),$$

$$\mathbf{T}(A_0^4)\Psi(t,\mathbf{x})=-\Psi(t,-\mathbf{x}). \qquad (7.7)$$

Here, $\mathbf{T}(A_0^2)$ is the *Appell transform* [4, 13].

The problem of R-separation of variables for the heat equation (7.1) is analogous to that for the free-particle Schrödinger equation (5.2) and the results are similar [56]. Here the R-separable solutions of (7.1) take the

ISBN-0-201-13503-5

form

$$\Phi(t,\mathbf{x}) = \exp\left[\,\mathfrak{R}(u,v,w)\right] U(u)V(v)W(w), \qquad \mathfrak{R} \text{ real}, \qquad (7.8)$$

where either $\mathfrak{R} \equiv 0$ or $\mathfrak{R} \neq 0$ cannot be written as a sum $\mathfrak{R} = A(u) + B(v) + C(w)$. We require that $\{u,v,w\}$ be a real analytic coordinate system such that substitution of (7.8) into (7.1) reduces the partial differential equation to three ordinary differential equations, one for each of the factors U, V, W. Two coordinate systems are considered equivalent if they can be obtained from one another under the adjoint action of G_3'.

The results announced in [56] are as follows. Corresponding to every R-separation of variables for (7.1) we can find a pair of differential operators H, S such that:

1. H and S are symmetries of (7.1) and $[H,S]=0$.
2. $H \in \mathcal{G}_3'$; that is, H is first order in x_1, x_2, and t.
3. S is second order in x_1, x_2, and contains no term in ∂_t.

The R-separation of variables is characterized by the simultaneous equations

$$Q\Phi = 0, \qquad H\Phi = i\lambda\Phi, \qquad S\Phi = \mu\Phi. \qquad (7.9)$$

The eigenvalues λ, μ are the usual separation constants for R-separable solutions Φ.

It follows from these remarks that S can always be expressed as a symmetric quadratic form in B_j, P_j, E, and M. The possible coordinates and their characterizations are listed in Table 13.

For each system in Table 13 we have $w = t$ and the separated solution in the variable w is exponential. In the last column of the table we list first the form of the separated solution in u followed by the separated solution in v. The anharmonic oscillator functions are solutions of a differential equation of the form

$$f''(u) + (\lambda u^2 + \alpha u^4 - \beta)f(u) = 0. \qquad (7.10)$$

From the viewpoint of Galilean and dilatation symmetry alone there are 26 distinct coordinate systems. However, from the viewpoint of G_3' symmetry there are only 17 systems. It is easy to show that two systems whose labels differ only in the superscripts lie on the same G_3' orbit. Indeed, systems of the form Fa^1, Fa^2 or La^1, La^2 are related by $\mathbf{T}(A_0^2)$, (7.7), and systems of the form Ra^1, Ra^2 are related by $\mathbf{T}(A^0)$. These are the only G_3' equivalences.

The eigenfunctions of the commuting pair H^0, M^2 are of special interest for this equation. From Table 13, the corresponding eigenfunctions separate in the variables $u = [(x^2+y^2)/t]^{1/2} = rt^{-1/2}$, $v = \theta$, $w = t$, where $x =$

Table 13 Operators and R-Separable Coordinates for the Equation $(\partial_t - \partial_{xx} - \partial_{yy})\Phi = 0$

Operators H,S	Coordinates $\{u,v,w\}$	Multiplier $e^{\mathcal{R}}$	Separated Solutions
1a $\;Fc^1$ H_2, B_1^2	$x = uw$ $y = vw$	$\mathcal{R} = -(u^2+v^2)w/4$	Exponential Exponential
1b $\;Fc^2$ H_{-2}, P_1^2	$x = u$ $y = v$	0	Exponential Exponential
2a $\;Fr^1$ H_2, M^2	$x = uw\cos v$ $y = uw\sin v$	$-u^2 w/4$	Bessel Exponential
2b $\;Fr^2$ H_{-2}, M^2	$x = u\cos v$ $y = u\sin v$	0	Bessel Exponential
3a $\;Fp^1$ $H_2, \{B_2, M\}$	$x = (u^2-v^2)w/2$ $y = uvw$	$-(u^2+v^2)^2 w/16$	Parabolic cylinder Parabolic cylinder
3b $\;Fp^2$ $H_{-2}, \{P_2, M\}$	$x = (u^2-v^2)/2$ $y = uv$	0	Parabolic cylinder Parabolic cylinder
4a $\;Fe^1$ $H_2, M^2 - B_2^2$	$x = w\cosh u\cos v$ $y = w\sinh u\sin v$	$-(\sinh^2 u + \cos^2 v)w/4$	Modified Mathieu Mathieu
4b $\;Fe^2$ $H_{-2}, M^2 - P_2^2$	$x = \cosh u\cos v$ $y = \sinh u\sin v$	0	Modified Mathieu Mathieu
5a $\;Lc^1$ $H_2 - 2aP_1 - 2bP_2,$ $B_2^2 - 2bP_2 H_0$	$x = uw + a/w$ $y = vw + b/w$	$-(u^2+v^2)w/4$ $+(au+bv)/2w$	Airy Airy
5b $\;Lc^2$ $H_{-2} + 2aB_1 + 2bB_2$ $P_1^2 + 2aB_1 H_0$	$x = u + aw^2$ $y = v + bw^2$	$-(au+bv)w$	Airy Airy
6a $\;Lp^1$ $H_2 - aP_1,$ $\{B_2, M\} - aP_2^2$	$x = (u^2-v^2)w/2 + a/w$ $y = uvw$	$-(u^2+v^2)^2 w/16$ $+a(u^2-v^2)/4w$	Anharmonic oscillator Anharmonic oscillator
6b $\;Lp^2$ $H_{-2} - 2aB_1,$ $\{P_2, M\} + 2aB_2^2$	$x = (u^2-v^2)/2 + aw^2$ $y = uv$	$-a(u^2-v^2)w/2$	Anharmonic oscillator Anharmonic oscillator
7 $\;\;Oc$ $H_{-2} + H_2$ $P_1^2 + B_1^2$	$x = u(1+w^2)^{1/2}$ $y = v(1+w^2)^{1/2}$	$-(u^2+v^2)w/4$	Parabolic cylinder Parabolic cylinder
8 $\;\;Or$ $H_{-2} + H_2, M^2$	$x = (1+w^2)^{1/2}u\cos v$ $y = (1+w^2)^{1/2}u\sin v$	$-u^2 w/4$	Whittaker Exponential
9 $\;\;Oe$ $H_{-2} + H_2,$ $M^2 - P_2^2 - B_2^2$	$x = (1+w^2)^{1/2}\cosh u\cos v$ $y = (1+w^2)^{1/2}\sinh u\sin v$	$-(\sinh^2 u + \cos^2 v)$ $\cdot w/4$	Ince Ince

ISBN-0-201-13503-5

Table 13 (Continued)

Operators H, S	Coordinates $\{u, v, w\}$	Multiplier $e^{\mathfrak{R}}$	Separated Solutions
10a Rc^1 $H^0, \{B_1, P_1\}$	$x = |w|^{1/2} u$ $y = |w|^{1/2} v$	0	Hermite Hermite
10b Rc^2 $H_{-2} - H_2,$ $P_1^2 - B_1^2$	$x = u|1 - w^2|^{1/2}$ $y = v|1 - w^2|^{1/2}$	$-\varepsilon(u^2 + v^2)w/4$ $\varepsilon = \mathrm{sign}(1 - w^2)$	Hermite Hermite
11a Rr^1 H^0, M^2	$x = |w|^{1/2} u \cos v$ $y = |w|^{1/2} u \sin v$	0	Laguerre Exponential
11b Rr^2 $H_{-2} - H_2, M^2$	$x = |1 - w^2|^{1/2} u \cos v$ $y = |1 - w^2|^{1/2} u \sin v$	$-\varepsilon u^2 w/4$	Laguerre Exponential
12a Re^1 $H^0,$ $M^2 - \frac{1}{2}\{B_2, P_2\}$	$x = |w|^{1/2} \cosh u \cos v$ $y = |w|^{1/2} \sinh u \sin v$	0	Finite Ince Finite Ince
12b Re^2 $H_{-2} - H_2,$ $M^2 - P_2^2 + B_2^2$	$x = |1 - w^2|^{1/2} \cosh u \cos v$ $y = |1 - w^2|^{1/2} \sinh u \sin v$	$-\varepsilon(\sinh^2 u$ $+ \cos^2 v)w/4$	Finite Ince Finite Ince
13 $L1$ $P_1, B_2^2 + 2bP_2 H_0$	$x = u$ $y = vw + b/w$	$-v^2 w/4 + bv/2w$	Exponential Airy
14 $L2$ $P_1, P_2^2 + 2aB_2 H_0$	$x = u$ $y = v + aw^2$	$-avw$	Exponential Airy
15 $O1$ $P_1, P_2^2 + B_2^2$	$x = u$ $y = v(1 + w^2)^{1/2}$	$-v^2 w/4$	Exponential Parabolic cylinder
16 $R1$ $P_1, \{B_2, P_2\}$	$x = u$ $y = v|w|^{1/2}$	0	Exponential Hermite
17 $R2$ $P_1, P_2^2 - B_2^2$	$x = u$ $y = v|1 - w^2|^{1/2}$	$-\varepsilon v^2 w/4$	Exponential Hermite

$r\cos\theta$, $y = r\sin\theta$. Moreover, the solutions $\Phi_{m,n}(t, \mathbf{x})$ of the heat equation (bounded at $\mathbf{x} = \mathbf{0}$) that satisfy

$$H^0 \Phi_{m,n} = (m + 2n + 1)\Phi_{m,n}, \qquad M\Phi_{m,n} = im\Phi_{m,n}, \qquad (7.11)$$

$$n = 0, 1, 2, \ldots, m = n, n - 1, \ldots, -n$$

are expressible in terms of Laguerre polynomials

$$\Phi_{m,n}(t, \mathbf{x}) = t^n \left(re^{i\theta}\right)^m L_n^{(m)}\left(-r^2/4t\right). \qquad (7.12)$$

Studies of expansions of solutions of the heat equation in terms of these polynomials can be found in [25] and [29].

ISBN-0-201-13503-5

It is well known that if $f(x)$ is a bounded continuous function defined in the plane R_2, then there is a unique solution $\Phi(t,x)$ of the heat equation, bounded and continuous in (t,x) for all $x \in R_2$, $t \geqslant 0$, and continuously differentiable in t, twice continuously differentiable in x_1,x_2 for all $x \in R, t > 0$, such that $\Phi(0,x) = f(x)$ [107]. This solution is

$$\Phi(t,x) = (4\pi t)^{-1} \iint_{-\infty}^{\infty} \exp\left[-(x-y)\cdot(x-y)/4t\right] f(y)\, dy_1\, dy_2$$
$$= I^t(f), \qquad t > 0. \tag{7.13}$$

In analogy with our work in Section 2.2 we can construct another model of the symmetry algebra (7.2). First we restrict the operators (7.2) to the solution space of the heat equation. This allows us to replace ∂_t by Δ_2 in the expressions for these operators and to consider $t \geqslant 0$ as a fixed parameter. With this interpretation the operators (7.2) are the symmetry operators at a fixed time t. At time $t = 0$ they become

$$\mathcal{K}_2 = (x_1^2 + x_2^2)/4, \qquad \mathcal{K}_{-2} = \Delta_2, \qquad \mathcal{P}_j = \partial_{x_j}, \qquad \mathcal{B}_j = x_j/2,$$
$$\mathcal{M} = x_1 \partial_{x_2} - x_2 \partial_{x_1}, \qquad \mathcal{K}^0 = x_1 \partial_{x_1} + x_2 \partial_{x_2} + 1, \tag{7.14}$$
$$\mathcal{K}_0 = 1, \qquad j = 1,2,$$

and when acting on, say, the space $\mathcal{F}_0$ of infinitely differentiable functions $f(x)$ on R_2 with compact support, these operators satisfy the usual commutation relations (7.3).

Exactly as in Section 2.2 we can interpret (7.13) in the form

$$\Phi(t,x) = I^t(f) = \exp(t\Delta_2) f(x) = \exp(t\mathcal{K}_{-2}) f(x), \qquad f \in \mathcal{F}_0, t > 0, \tag{7.15}$$

and show that the connection between the italic operators H, (7.2), and the corresponding script operators $\mathcal{K}$, (7.14), is

$$H \exp(t\mathcal{K}_{-2}) = \exp(t\mathcal{K}_{-2})\mathcal{K} \tag{7.16}$$

where $H \in \mathcal{G}_3'$ and $\mathcal{K}$ is obtained from H by setting $t = 0$. In addition, we can obtain results of the form

$$\exp(aH)\exp(t\mathcal{K}_{-2}) = \exp(t\mathcal{K}_{-2})\exp(a\mathcal{K}) \tag{7.17}$$

and show that the equations

$$\partial_t \Phi = \left(\Delta_2 + a_1 x\cdot x + a_2 \partial_{x_1} + a_3 \partial_{x_2} + a_4(x_1 \partial_{x_2} - x_2 \partial_{x_1})\right.$$
$$\left. + a_5 x_1 + a_6 x_2 + a_7(x_1 \partial_{x_1} + x_2 \partial_{x_2}) + a_8\right)\Phi, \qquad a_j \in R, \tag{7.18}$$

have isomorphic symmetry algebras and are equivalent to (7.1). The

ISBN-0-201-13503-5

techniques for solving the Cauchy problem, discussed in Section 2.2, work for all of the equations (7.18).

At this point it is useful to discuss a method for constructing explicit solutions of the heat equation, which applies equally well to many other equations studied in this book. Every R-separable coordinate system for (7.1) is associated with a pair of commuting operators, one of which is first order. By diagonalizing this first-order operator, we can separate out the corresponding coordinate and reduce the heat equation to an equation with one less variable. For example, diagonalizing the symmetry operator ∂_t, we can separate out the t variable and obtain solutions

$$\Phi(t,\mathbf{x}) = \exp(-k^2 t)F(\mathbf{x})$$

where F is any solution of the Helmholtz equation

$$\Delta_2 F + k^2 F = 0. \tag{7.19}$$

This rather obvious remark becomes less trivial when we realize that each of the symmetry operators $\mathbf{T}(g)$, (7.4)–(7.6), maps Φ into another solution $\mathbf{T}(g)\Phi$. For example, if $g = A_0$, (7.7), we have the result that

$$\mathbf{T}(A_0)\Phi(t,\mathbf{x}) = \exp\left[\left(-k^2(t-1) - \tfrac{1}{2}\mathbf{x}\cdot\mathbf{x}\right)/(t+1)\right]$$
$$\times 2^{1/2}(t+1)^{-1}F\left(2^{1/2}\mathbf{x}/(t+1)\right) \tag{7.20}$$

is a solution of the heat equation for any solution F of the Helmholtz equation (7.19). By choosing appropriate group elements g and solutions F, we can construct solutions of the heat equation satisfying a wide variety of initial and boundary conditions. Some examples are given by Bateman [13, p. 340].

Now we proceed to a study of the complex heat equation. This is equation (7.1) where now t, x_1, x_2 are complex variables. It is obvious that the symmetry algebra $\mathcal{G}_3^c$ of this equation is (complex) nine dimensional with basis (7.2). The basis operators can be exponentiated to yield the local Lie group G_3^c of symmetry operators acting on the space $\mathcal{F}$ of functions $\Psi(t,\mathbf{x})$ analytic in some domain $\mathcal{D}$ in complex (t, x_1, x_2) space. The group action is given by (7.4)–(7.6) where now the parameters $\mathbf{w}, \mathbf{z}, \rho$ are allowed to take arbitrary complex values and the matrices $A = \begin{pmatrix} \alpha & \beta \\ \gamma & \delta \end{pmatrix}$ range over the group $SL(2,\mathcal{C})$. Of course these operators map solutions of the complex heat equation into solutions.

The problem of R-separation of variables for this equation can be formulated in a manner analogous to that of the complex heat equation $(\partial_t - \partial_{xx})\Phi = 0$ in Section 2.2. We expect that all such R-separable systems will correspond to a pair of commuting symmetry operators in the envelop-

ISBN-0-201-13503-5

ing algebra of $\mathcal{G}_3^c$. It is clear that all the real R-separable systems listed in Tables 12 and 13 can be analytically continued to yield R-separable systems for the complex heat equation. However, each system Aa in Table 12 is complex equivalent to the system Aa in Table 13. Furthermore, the systems Oc, Or, Oe are complex equivalent to the systems Rc, Rr, Re, respectively.

There exist other R-separable systems complex inequivalent to these. For example, if we diagonalize the operator ∂_t, we can reduce (7.1) to the complex Helmholtz equation and, from Table 3, find separable solutions that are products of Bessel functions, clearly inequivalent to any of the entries in Tables 12 and 13.

The separation of variables problem for (7.1) has been completely solved by E. G. Kalnins (private communication), who finds 38 nontrivial separable systems, each system characterized by a pair of commuting symmetry operators. Rather than discuss these results here, we will simply make use of the separable systems at hand and use them to apply Weisner's method.

As suggested by (7.12), for Laguerre polynomial solutions of (7.1) it is appropriate to introduce new coordinates

$$z = -\left(x_1^2 + x_2^2\right)/4t, \qquad s = i(x_1 + ix_2)/2, \qquad \tau = t. \qquad (7.21)$$

In terms of these coordinates the basis functions (7.12) become (suitably normalized)

$$\Phi_{m,n}(t,s,z) = \tau^n s^m L_n^{(m)}(z), \qquad H^0 \Phi_{m,n} = (m+2n+1)\Phi_{m,n},$$
$$M\Phi_{m,n} = im\Phi_{m,n}, \qquad n = 0,1,2,\ldots. \qquad (7.22)$$

These expressions make sense for any $m \in \mathbb{C}$ such that m is not a negative integer. Since the Laguerre polynomial $L_n^{(m)}(z)$ can be expressed as a confluent hypergeometric function (see (B.9i)), we can choose another set of eigenfunctions

$$\Psi_{m,n}(\tau,s,z) = \tau^n s^m {}_1F_1\left({-n \atop m+1}\Big|z\right), \qquad \Phi_{m,n} = \binom{m+n}{n}\Psi_{m,n} \qquad (7.23)$$

and a linearly independent set of eigenfunctions

$$\Psi'_{m,n}(\tau,s,z) = \tau^n s^m z^{-m} {}_1F_1\left({-n-m \atop -m+1}\Big|z\right). \qquad (7.24)$$

That is, $\Psi_{m,n}$ and $\Psi'_{m,n}$ form a basis for the space of solutions of the eigenvalue equations (7.22) for fixed n, m.

ISBN-0-201-13503-5

In terms of the τ, s, z coordinates, the operators (7.2) become

$$H_2 = \tau^2 \partial_\tau + \tau s \partial_s + \tau z \partial_z + \tau (1 - z), \quad H_{-2} = \tau^{-1}(\tau \partial_\tau - z \partial_z),$$

$$H_{-1} = \partial_s + zs^{-1}\partial_z, \quad H^+_{-1} = s\tau^{-1}\partial_z, \quad H^-_1 = \tau \partial_s + \tau zs^{-1}\partial_z - \tau zs^{-1} \quad (7.25)$$

$$H^+_1 = s\partial_z - s, \qquad \tilde{H}^0 = s\partial_s, \qquad H^0 = s\partial_s + 2\tau \partial_\tau + 1, \qquad H_0 = 1,$$

where

$$H^-_{-1} = -iP_1 - P_2, \quad H^+_{-1} = -iP_1 + P_2, \quad H^-_1 = -iB_1 - B_2,$$
$$H^+_1 = -iB_1 + B_2, \quad \tilde{H}^0 = iM. \tag{7.26}$$

Note that

$$\left[H^0, H^\alpha_j\right] = jH^\alpha_j, \qquad \left[\tilde{H}^0_j, H^\alpha_j\right] = \alpha H^\alpha_j. \tag{7.27}$$

It is clear from the explicit expressions (7.25) that each of the Lie algebra operators maps a polynomial in z to another such polynomial. It follows from this and the commutation relations (7.27) that $H^\alpha_j \Psi_{m,n}$ must be a constant times $\Psi_{m+(\alpha)1, n+[j-(\alpha)1]/2}$. Differentiating the power series (7.23) term by term, we can verify

$$H_2\Psi_{m,n} = (m - n + 1)\Psi_{m, n+1}, \quad H_{-2}\Psi_{m,n} = n\Psi_{m, n-1},$$

$$H^-_{-1}\Psi_{m,n} = m\Psi_{m-1, n}, \qquad H^+_{-1}\Psi_{m,n} = -n(m+1)^{-1}\Psi_{m+1, n-1},$$

$$H^-_1\Psi_{m,n} = m\Psi_{m-1, n+1}, \qquad H^+_1\Psi_{m,n} = -(n+m+1)(m+1)^{-1}\Psi_{m+1, n},$$

$$\tilde{H}^0\Psi_{m,n} = m\Psi_{m,n}, \qquad H^0\Psi_{m,n} = (m + 2n + 1)\Psi_{m,n}.$$

$$\tag{7.28}$$

Note that the first six of these relations agree exactly with the six differential recurrence formulas (B.8) for the functions ${}_1F_1$. Thus we have an interpretation of the recurrence formulas in terms of the action of the symmetry algebra of the complex heat equation.

Note also that the operators $H_{\pm 2}, H^0$, which form a basis for an $sl(2, ¢)$ subalgebra of $\mathcal{G}^c_3$, yield the same recurrence formulas for Laguerre functions as did the operators $J^\pm, J^0$ in Section 2.4 (see (4.9)). This is due to the fact that equation (4.1) can be obtained from (7.1) by introducing polar coordinates and separating out the angular variable. Thus all the results of Section 2.4 can be obtained as special cases of results concerning solutions of (7.1).

Moreover, most of Chapter 4 in the author's book [82] is concerned with identities for Laguerre functions that can be obtained from a study of the subalgebra of $\mathcal{G}_3^c$ with basis $\{H_{-1}^-, H_1^+, \tilde{H}^0, H_0\}$ and commutation relations $(H_{\pm 1}^{\pm} = H^{\pm}, \tilde{H}^0 = H^0)$,

$$[H^0, H^{\pm}] = \pm H^{\pm}, \qquad [H^+, H^-] = H_0, \qquad [H_0, H] = 0. \quad (7.29)$$

(See also [78].) Thus it is clear that the special function theory associated with the heat equation is rich in useful results. Here, we will present only a few examples illustrating the interplay between the symmetry of this equation and identities obeyed by the separated solutions.

It is easy to see that the fundamental generating function (4.11) for Laguerre polynomials arises when the solution $\exp(\alpha H_2)\Phi_{-2l-1,0}$ of the complex heat equation is evaluated in two different ways. Similarly, if we apply the operator $\exp(\alpha H_1^-)$ to the basis function $\Phi_{m,0}(\tau, s, z) = s^m$ and make use of the recurrence formula $H_1^- \Phi_{l,n} = (n+1)\Phi_{l-1,n+1}$ and the Lie theory relation

$$\exp(\alpha H_1^-)\Psi(\tau, s, z) = \exp(-\tau z\alpha/s)\Psi(\tau, s + \alpha\tau, z[1 + \alpha\tau/s]),$$

we obtain the generating function

$$e^{-\alpha z}(1+\alpha)^m = \sum_{n=0}^{\infty} \alpha^n L_n^{(m-n)}(z), \qquad m \in \mathcal{C}, \quad |\alpha| < 1. \quad (7.30)$$

(Here we have set $\tau = s$ and factored s^m out of both sides of this expression.)

We can obtain the action of the local symmetry group G_3^c in terms of the coordinates t, s, z by combining expressions (7.21) with (7.4)–(7.7). In particular, the Appell transform has the simple appearance

$$\mathbf{T}(A_0^2)\Phi(\tau, s, z) = \tau^{-1} e^z \Phi(-\tau^{-1}, s\tau^{-1}, -z). \quad (7.31)$$

Applying this operator to the basis function $\Psi_{m,n}$, (7.23), with $m, n \in \mathcal{C}$ such that m is not a negative integer, we obtain

$$\mathbf{T}(A_0^2)\Psi_{m,n} = (-1)^n \tau^{-m-n-1} s^m e^z {}_1F_1\left(\begin{array}{c} -n \\ m+1 \end{array}\bigg| -z\right).$$

This expression is a simultaneous eigenfunction of H^0 and M again, with eigenvalues $-m - 2n - 1$ and im, respectively. Furthermore, it is analytic in z at $z = 0$. Hence, there exists a constant $c_{m,n}$ such that

$$\mathbf{T}(A_0^2)\Psi_{m,n} = c_{m,n}\Psi_{m,-m-n-1}.$$

ISBN-0-201-13503-5

Setting $z = 0$ on both sides of this expression, we obtain $c_{m,n} = (-1)^n$ or

$$e^z {}_1F_1\left(\begin{array}{c} -n \\ m+1 \end{array}\bigg| -z\right) = {}_1F_1\left(\begin{array}{c} m+n+1 \\ m+1 \end{array}\bigg| z\right). \tag{7.32}$$

This is the important transformation formula for the ${}_1F_1$ listed in Appendix B.

The heat equation can be written in the form $(H_{-2} - P_1^2 - P_2^2)\Phi = 0$ or, what is the same thing, $(H_{-2} + H_{-1}^+ H_{-1}^-)\Phi = 0$. It follows from (7.25) that in terms of coordinates $\{\tau, s, z\}$ this equation reads

$$(z\,\partial_{zz} + (s\,\partial_s - z + 1)\,\partial_z + \tau\,\partial_\tau)\Phi = 0. \tag{7.33}$$

A straightforward application of Weisner's method shows that any solution Φ of (7.33) that is analytic in τ, s, z in a suitable region, such that Φ can be expanded in a Laurent series in τ, s about $\tau = 0$, $s = 0$ and such that $\Phi(\tau, s, 0)$ is bounded in this region, must satisfy an identity of the form

$$\Phi(\tau, s, z) = \sum_{m,n} c_{m,n} L_n^{(m)}(z)\tau^n s^m \tag{7.34}$$

where the $c_{m,n}$ are complex constants. Conversely, a uniformly convergent series of the form (7.34) in some region of τ, s, z space defines a solution of the complex heat equation. We conclude that all generating functions of the form (7.34) are obtainable as solutions of the heat equation. One way to find such functions Φ is to characterize them as simultaneous eigenfunctions of a pair of commuting operators in the enveloping algebra of $\mathcal{G}_3^c$. For example, the equations

$$\{B_1, P_1\}\Phi = (4\alpha + 2)\Phi, \qquad H^0\Phi = (\lambda + 1)\Phi, \qquad \alpha, \lambda \in \mathbb{C}, \tag{7.35}$$

correspond to the coordinates u, v, w where

$$u = \tau^{-1/2}(s + \tau z/s), \qquad v = \tau^{-1/2}(-s + \tau z/s), \qquad w = \tau; \tag{7.36}$$

see 10a in Table 13. In terms of the new coordinates we have

$$H^0 = 2w\,\partial_w + 1, \qquad \{B_1, P_1\} = -8\left(\partial_{uu} - \tfrac{1}{2}u\,\partial_u - \tfrac{1}{4}\right)$$

and solutions $\Phi^{\alpha,\lambda}$ of (7.35) can be written in the form $\Phi^{\alpha,\lambda} = w^{\lambda/2}U(u)V(v)$ where

$$2U'' - uU' + \alpha U = 0, \qquad 2V'' + vV' + (\alpha - \lambda)V = 0. \tag{7.37}$$

ISBN-0-201-13503-5

Comparing these equations with (2.24) and (2.25), we find the independent solutions $H_\alpha(u/2)$ and $\exp(u^2/4)H_{-\alpha-1}(iu/2)$ for U and $H_{\lambda-\alpha}(iv/2)$, $\exp(-v^2/4)H_{\alpha-\lambda-1}(v/2)$ for V. To be definite we choose the solutions

$$\Phi^{\alpha,\lambda}(u,v,w) = w^{\lambda/2}H_\alpha(u/2)H_{\lambda-\alpha}(iv/2). \tag{7.38}$$

By changing to u,v,w coordinates in the expressions (7.25) for the $\mathcal{G}_3^c$ symmetry operators and applying these operators to the functions (7.38), the reader can obtain a family of simple recurrence formulas obeyed by products of Hermite functions. Let us apply Weisner's method to the generating function (7.38) in the case where λ and α are positive integers with $\alpha \leqslant \lambda$. In this case the Hermite functions appearing in (7.38) are Hermite polynomials. From (7.34) and (7.36) we obtain

$$\tau^{\lambda/2}H_\alpha\big[2^{-1}(\tau^{-1/2}s+\tau^{1/2}z/s)\big]H_{\lambda-\alpha}\big[i2^{-1}(-\tau^{-1/2}s+\tau^{1/2}z/s)\big]$$
$$= \sum_{k=0}^{\lambda} s^{\lambda-2k}\tau^k c_k L_k^{(\lambda-2k)}(z). \tag{7.39}$$

(We have used the facts that $H_\alpha(x)$ is a polynomial of order α and $H_\alpha(-x)=(-1)^\alpha H_\alpha(x)$ to obtain this result.) Setting $x=s\tau^{-1/2}$, we find

$$H_\alpha\big[2^{-1}(x+z/x)\big]H_{\lambda-\alpha}\big[i2^{-1}(-x+z/x)\big] \tag{7.40}$$
$$= \sum_{k=0}^{\lambda} x^{\lambda-2k}c_k L_k^{(\lambda-2k)}(z).$$

To obtain a simple generating function for the coefficients c_k, we set $z=0$ and use the fact that $L_n^{(m)}(0)=\binom{m+n}{n}$ where $\binom{m+n}{n}$ is a binomial coefficient (B.1):

$$H_\alpha(x/2)H_{\lambda-\alpha}(-ix/2) = \sum_{k=0}^{\lambda}\binom{\lambda-k}{k}c_k x^{\lambda-2k}. \tag{7.41}$$

By explicitly computing the coefficient of $x^{\lambda-2k}$ on the left-hand side of this equation, we can express c_k as a terminating hypergeometric series $_3F_2$.

The polynomial functions (7.38) can be used as an alternative (but less useful) basis for solutions of the heat equation. Thus, one can compute matrix elements of the group operators $T(g)$ in this basis, expand an arbitrary solution Φ in terms of the basis and so on.

For λ and α complex numbers, we can derive infinite series identities that are similar to (7.40) but slightly more complicated.

ISBN-0-201-13503-5

2.8 Concluding Remarks

We close this chapter by pointing out several important research results closely related to our subject but which will not be treated in detail here.

In [140], Winternitz, Smorodinsky, Uhlir, and Fris determined all potentials $V(x,y)$ such that the time-independent Schrödinger equation

$$(-\Delta_2 + V(x,y))\Phi = \lambda\Phi \qquad (8.1)$$

admits a first- or second-order symmetry operator. They showed that the possible symmetry operators are of the form $L + f(x,y)$ where $L \in \mathcal{E}(2)$ ((1.6), (1.7) in Section 1.1) for first-order symmetries, and of the form $S + f(x,y)$ where S is a second-order symmetric operator in the enveloping algebra of $\mathcal{E}(2)$ for second-order symmetries. The equations that admit first-order symmetries separate in corresponding coordinate systems (2.31) or (2.32), Section 1.2. Equations that admit no first-order symmetries but do admit second-order symmetries separate in one of the four coordinate systems listed in Table 1. The latter equations are class II. Note that the Lie algebra $\mathcal{E}(2)$ appears in this study even though it is *not* the symmetry algebra of equation (8.1) except in the trivial case in which $V(x,y)$ is constant. The reason that $\mathcal{E}(2)$ appears here is that a first- or second-order symmetry operator for (8.1) has the property that its derivative terms L or S must necessarily commute with the Laplace operator Δ_2; hence L and S must belong to the enveloping algebra of $\mathcal{E}(2)$, as follows from the results of Section 1.1. Usually however, the complete symmetry operator will not belong to the enveloping algebra of $\mathcal{E}(2)$ because the functional part $f(x,y)$ of the operator will not be zero or even a constant. Here, $f(x,y)$ will depend on the potential.

Just as in Section 1.1, separation of variables corresponding to first-order symmetries turns out to be rather trivial. The interesting cases are the class II equations that admit second-order symmetries but no nontrivial first-order symmetries. Such equations admit separation of variables in one or more of the four coordinate systems listed in Table 1. Each separable coordinate system is determined by the pure differential part of the symmetry operator, that is, the part which belongs to the enveloping algebra of $\mathcal{E}(2)$. Thus, the occurrence of the operator Δ_2 in (8.1) limits the number of separable coordinate systems to four at most. Whether or not a given equation (8.1) separates in one of these coordinate systems depends on the explicit form of the potential V. One finds that (8.1) separates in one of the coordinate systems if and only if this equation admits the second-order symmetry $S + f(x,y)$ where the pure differential operator S corresponds to the coordinate system.

Although the interesting cases of (8.1) as treated in [140] are all class II, it frequently happens that such cases arise from class I equations by a partial separation of variables. For example, (8.1) arises from the time-dependent Schrödinger equation (5.1) if we split off the time variable by the assumption $\Psi(x,y,t) = e^{-i\lambda t}\Phi(x,y)$. The class I harmonic oscillator, repulsive oscillator, and linear potential equations which we have studied in the time-dependent case become class II equations in [140] when the t variable is separated. This is because separation of the t variable drastically reduces the symmetry of the Schrödinger equation.

An especially interesting class II equation that appears in [140] corresponds to the potential $v(x,y) = -\alpha/x^2 - \beta/y^2$ where α, β are real constants such that $\alpha^2 + \beta^2 > 0$. In [18] and [103], Boyer and Niederer have both listed this potential in their classifications of all potentials $V(x,y)$ such that the time-dependent Schrödinger equation admits nontrivial first-order symmetry operators. In [19], Boyer studied the time-dependent Schrödinger equation

$$\left(i\,\partial_t + \partial_{xx} + \partial_{yy} - \alpha/x^2 - \beta/y^2\right)\Psi = 0 \qquad (8.2)$$

from the point of view presented in this book. He showed that this equation is still class II. However, it is highly tractable since it can be obtained from the free-particle Schrödinger equation (class I), $(i\,\partial_t + \Delta_4)\Psi = 0$ by a partial separation of variables. Boyer showed that (8.2) R-separates in 25 coordinate systems for $\alpha = 0$, $\beta \neq 0$ and in 15 coordinate systems for $\alpha \neq 0$, $\beta \neq 0$. Moreover, he found that each separable coordinate system corresponded to a pair of commuting second-order symmetry operators of (8.2). The special function identities that he obtained from this study are similar to, but not the same as, those obtained in Section 2.5.

In [5, 6], Armstrong used methods due to the author and the Wigner–Echart theorem to study the quantum-mechanical systems of Section 2.3, all of which admit $SL(2,R)$ as a dynamical symmetry group. He considered infinite families of self-adjoint operators on $L_2(R+)$ that transform irreducibly under the adjoint action of $SL(2,R)$ and used group theory to compute the matrix elements of these operators with respect to a basis of eigenvectors of L_3. See also [99]. An extension of the theory which is similar in viewpoint to the methods of this book is contained in [86] and [87].

Finally, in [23] group theory and separation of variables are used to determine all possible first- and second-order raising operators for Hamiltonians of the form $H = -\Delta_2 + V(x,y)$. (A *raising operator* R for H satisfies the commutation relation $[H,R] = \mu R$, $\mu > 0$. If the eigenvector Ψ of H satisfies $H\Psi = \lambda\Psi$, then formally $H(R\Psi) = (\lambda + \mu)R\Psi$, so R maps an eigenvector corresponding to eigenvalue λ to an eigenvector corresponding to eigenvalue $\lambda + \mu$.) In [23] it is shown that a necessary condition for H to admit a second-order raising operator is that equation (8.1) separate in one of the four coordinate systems listed in Table 1. A complete list of possible raising operators is given.

Exercises

1. Compute the symmetry algebra of the free-particle Schrödinger equation (1.2).

2. Determine the decomposition of the Schrödinger algebra $\mathcal{G}_2$ into orbits under the adjoint action of G_2.

3. Show that under the adjoint action of $SL(2,R)$, the Lie algebra $sl(2,R)$ decomposes into three orbits.

4. Expression (1.30) shows explicitly the equivalence between the Schrödinger equations for the free particle and the harmonic oscillator. Derive the corresponding expression giving the equivalence between the free-particle and linear potential equations.

5. Derive the bilinear expansions (1.55) for the fundamental solution of the Schrödinger equation

$$k(t,x-y) = (4\pi it)^{-1/2}\exp\left[-(x-y)^2/4it\right]$$

with respect to the bases $\{F_\lambda^{(j)}\}$, $j=2$, 4. (See [35, 136] for detailed discussions of such continuous generating functions.)

6. Use the methods of Section 2.2 to solve the Cauchy problem for

$$\partial_t\Phi = \partial_{xx}\Phi + x\Phi.$$

That is, find a bounded solution $\Phi(t,x)$ of this equation for $t>0$, continuous for $t \geqslant 0$, such that $\Phi(0,x)=f(x)$ where $f(x)$ is bounded and continuous on the real line.

7. The Hermite functions $H_n(z)$, (2.26), are polynomials for $n=0,1,2,\ldots$ and for $n=-1,-2,\ldots$ they are called Hermite functions of the second kind. Show that the second-kind functions can be expressed in terms of the error function and its derivatives [37]. Verify that the functions $\Phi_n(z,s)= H_n(z)s^n$, $n=0, \pm 1, \pm 2,\ldots$, satisfy the recurrence relations

$$H_1\Phi_n = \Phi_{n+1}, \qquad H^0\Phi_n = \left(n+\tfrac{1}{2}\right)\Phi_n, \qquad H_{-1}\Phi_n = (n/2)\Phi_{n-1},$$

$$H_2\Phi_n = \Phi_{n+2}, \qquad H_{-2}\Phi_n = \tfrac{1}{2}n(n-1)\Phi_{n-2}.$$

where the operators H_j, (2.23), form a basis for the symmetry algebra of the complex heat equation. Show that this representation is not irreducible. Use the simple models constructed in Section 2.2 to compute the matrix elements of this representation and obtain the corresponding special function identities. In particular, derive the identity associated with the expression $\exp(\alpha H_1)\Phi_{-1}$.

8. Compute the bilinear expansion for the fundamental solution $k(t,x,y)$, (3.19), of the radial free-particle Schrödinger equation in terms of the Laguerre polynomial basis. Show that the expansion is a special case of the Hille–Hardy formula (4.27). Determine the bilinear expansion of $k(t,x,y)$ in terms of the continuum basis $\{\Psi_\lambda^{(2)}\}$, (3.16).

9. Compute the symmetry algebra of the complex heat equation $\partial_t\Phi - \partial_{xx}\Phi - \partial_{yy}\Phi = 0$.

CHAPTER 3

The Three-Variable Helmholtz and Laplace Equations

3.1 The Helmholtz equation $(\Delta_3 + \omega^2)\Psi = 0$

The Helmholtz or reduced wave equation in three variables

$$(\Delta_3 + \omega^2)\Psi(x_1, x_2, x_3) = 0, \qquad \Delta_3 = \partial_{x_1 x_1} + \partial_{x_2 x_2} + \partial_{x_3 x_3}, \qquad \omega > 0, \quad (1.1)$$

has been widely studied from the point of view of separation of variables, and the possible separable coordinate systems for this equation are well known [97, 98]. The connection between the separable systems and the Euclidean symmetry group $E(3)$ of (1.1) was first pointed out in [76]. However, it is only recently that this connection with group theory has been employed systematically to derive properties of the separable solutions of the Helmholtz equation.

Applying our usual methods, we find that (apart from the trivial symmetry E) the symmetry algebra of (1.1) is six dimensional with basis

$$
\begin{aligned}
P_j &= \partial_j = \partial_{x_j}, \qquad j = 1, 2, 3; \\
J_1 &= x_3 \partial_2 - x_2 \partial_3, \qquad J_2 = x_1 \partial_3 - x_3 \partial_1, \qquad J_3 = x_2 \partial_1 - x_1 \partial_2,
\end{aligned}
\qquad (1.2)
$$

and commutation relations

$$[J_l, J_m] = \sum_n \varepsilon_{lmn} J_n, \quad [J_l, P_m] = \sum_n \varepsilon_{lmn} P_n, \quad [P_l, P_m] = 0, \qquad (1.3)$$

$$l, m, n = 1, 2, 3,$$

where ε_{lmn} is the tensor such that $\varepsilon_{123} = \varepsilon_{312} = \varepsilon_{231} = 1$, $\varepsilon_{132} = \varepsilon_{213} = \varepsilon_{321} = -1$, with all other components zero. We take the real Lie algebra $\mathcal{E}(3)$ with basis (1.2) as the symmetry algebra of (1.1). In terms of the P operators,

ENCYCLOPEDIA OF MATHEMATICS and Its Applications, Gian-Carlo Rota (ed.).
Vol. 4: Willard Miller, Jr., Symmetry and Separation of Variables

ISBN-0-201-13503-5

the Helmholtz equation reads

$$(P_1^2 + P_2^2 + P_3^2)\Psi = -\omega^2\Psi. \qquad (1.4)$$

Here $\mathcal{E}(3)$ is isomorphic to the Lie algebra of the Euclidean group in three-space $E(3)$ and the subalgebra $so(3)$ with basis $\{J_1, J_2, J_3\}$ is isomorphic to the Lie algebra of the proper rotation group $SO(3)$. To show this explicitly we first consider the well-known realization of $SO(3)$ as the group of real 3×3 matrices A such that $A'A = E_3$ and $\det A = 1$ (see, e.g., [45, 85]). Here E_3 is the 3×3 identity matrix $(E_3)_{jl} = \delta_{jl}$ and $(A')_{jl} = A_{lj}$, $j, l = 1, 2, 3$. The Lie algebra of $SO(3)$ in this realization is the space of 3×3 skew-symmetric matrices $\mathcal{Q}$ ($\mathcal{Q}' = -\mathcal{Q}$). A basis for this Lie algebra is provided by the matrices

$$\mathcal{J}_1' = \begin{bmatrix} 0 & 0 & 0 \\ 0 & 0 & -1 \\ 0 & 1 & 0 \end{bmatrix}, \quad \mathcal{J}_2' = \begin{bmatrix} 0 & 0 & 1 \\ 0 & 0 & 0 \\ -1 & 0 & 0 \end{bmatrix}, \quad \mathcal{J}_3' = \begin{bmatrix} 0 & -1 & 0 \\ 1 & 0 & 0 \\ 0 & 0 & 0 \end{bmatrix}, \qquad (1.5)$$

with commutation relations $[\mathcal{J}_l', \mathcal{J}_m'] = \sum_n \varepsilon_{lmn} \mathcal{J}_n'$, in agreement with (1.3). A convenient parametrization of $SO(3)$ is that in terms of *Euler angles* (φ, θ, ψ):

$$A(\varphi, \theta, \psi) = \exp(\varphi \mathcal{J}_3') \exp(\theta \mathcal{J}_1') \exp(\psi \mathcal{J}_3'), \qquad (1.6)$$

$$0 \leqslant \varphi < 2\pi, \quad 0 \leqslant \theta \leqslant \pi, \quad 0 \leqslant \psi < 2\pi.$$

As the Euler angles run over their full domain of values, $A(\varphi, \theta, \psi)$ runs over all elements of $SO(3)$. The coordinates are one to one on the group manifold except for those elements for which $\theta = 0, \pi$, in which cases only the sum $\varphi + \psi$ is uniquely determined. More detailed discussions of these coordinates can be found in many references (e.g., [45, 85, 124]).

The Euclidean group in three-space $E(3)$ can be realized as a group of 4×4 real matrices. The elements of $E(3)$ are

$$g(A, \mathbf{a}) = \begin{bmatrix} & A & & 0 \\ & & & 0 \\ & & & 0 \\ a_1 & a_2 & a_3 & 1 \end{bmatrix}, \quad A \in SO(3), \quad \mathbf{a} = (a_1, a_2, a_3) \in R^3, \qquad (1.7)$$

and the group product is given by matrix multiplication

$$g(A, \mathbf{a}) g(A', \mathbf{a}') = g(AA', \mathbf{a}A' + \mathbf{a}'). \qquad (1.8)$$

$E(3)$ acts as a transformation group in three-space R^3. The group element $g(A, \mathbf{a})$ maps the point $\mathbf{x} \in R^3$ to the point

$$\mathbf{x}g = \mathbf{x}A + \mathbf{a} \in R^3. \qquad (1.9)$$

ISBN-0-201-13503-5

It follows easily from this definition that $\mathbf{x}(gg')=(\mathbf{x}g)g'$ for all $\mathbf{x}\in R^3$, $g,g'\in E(3)$, and that $\mathbf{x}g(E_3,\mathbf{0})=\mathbf{x}$ where $g(E_3,\mathbf{0})$ is the identity element of $E(3)$. Geometrically, g corresponds to a rotation A about the origin $(0,0,0)\in R^3$ followed by a translation $\mathbf{a}$ [85].

A basis for the Lie algebra of the matrix group $E(3)$ is provided by the matrices

$$
\mathcal{J}_l = \begin{bmatrix} & & & 0 \\ & \mathcal{J}_l' & & 0 \\ & & & 0 \\ 0 & 0 & 0 & 0 \end{bmatrix}, \quad l=1,2,3; \qquad
\mathcal{P}_1 = \begin{bmatrix} & & & 0 \\ & 0 & & 0 \\ & & & 0 \\ 1 & 0 & 0 & 0 \end{bmatrix},
$$

$$
\mathcal{P}_2 = \begin{bmatrix} & & & 0 \\ & 0 & & 0 \\ & & & 0 \\ 0 & 1 & 0 & 0 \end{bmatrix}, \qquad
\mathcal{P}_3 = \begin{bmatrix} & & & 0 \\ & 0 & & 0 \\ & & & 0 \\ 0 & 0 & 1 & 0 \end{bmatrix}, \quad (1.10)
$$

with commutation relations identical to (1.3). This shows that the Lie algebra $\mathcal{E}(3)$ with basis (1.2) is isomorphic to the Lie algebra of $E(3)$. The explicit relation between the Lie algebra generators (1.10) and the group elements (1.7) is

$$g(\varphi,\theta,\psi,\mathbf{a})\equiv g(A(\varphi,\theta,\psi),\mathbf{a})$$
$$=\exp(\varphi\mathcal{J}_3)\exp(\theta\mathcal{J}_1)\exp(\psi\mathcal{J}_3)\exp(a_1\mathcal{P}_1+a_2\mathcal{P}_2+a_3\mathcal{P}_3). \quad (1.11)$$

Using standard Lie theory, we can extend the action of $\mathcal{E}(3)$ by Lie derivatives (1.2) on the space $\mathcal{F}$ of analytic functions defined on some open connected set $\mathcal{D}\subseteq R^3$ to a local representation $\mathbf{T}$ of $E(3)$ on $\mathcal{F}$. We find

$$\mathbf{T}(g)\Phi(\mathbf{x})=\{\exp(\varphi J_3)\exp(\theta J_1)\exp(\psi J_3)$$
$$\times\exp(a_1P_1+a_2P_2+a_3P_3)\}\Phi(\mathbf{x})=\Phi(\mathbf{x}g) \quad (1.12)$$

where $\mathbf{x}g$ is given by (1.9). Thus the action (1.9) of $E(3)$, as a transformation group is exactly that induced by the Lie derivatives (1.2). As usual,

$$\mathbf{T}(gg')=\mathbf{T}(g)\mathbf{T}(g'), \qquad g,g'\in E(3), \quad (1.13)$$

and the operators $\mathbf{T}(g)$ map solutions of the Helmholtz equation into solutions.

Computing the space $\mathcal{S}$ of second-order symmetries of (1.1), we find that this equation is class I. Indeed, factoring out the space q of trivial symmetries RQ, $R\in\mathcal{F}$, $Q=P_1^2+P_2^2+P_3^2+\omega^2$ (recall that RQ is the zero operator on the solution space of (1.1)), we find that the factor space $\mathcal{S}/q$ is 41 dimensional, with a basis consisting of the identity operator E, the 6

ISBN-0-201-13503-5

first-order operators J_l, P_l, and 34 purely second-order symmetrized operators. The space $\mathcal{E}(3)^2$ of second-order symmetrized operators is spanned by the elements $\{J_l, J_m\}$, $\{J_l, P_m\}$, $\{P_l, P_m\} \Rightarrow 2P_l P_m$, and these elements are subject only to the relations $\mathbf{J} \cdot \mathbf{P} = J_1 P_1 + J_2 P_2 + J_3 P_3 \equiv 0$ and $\mathbf{P} \cdot \mathbf{P} = P_1^2 + P_2^2 + P_3^2 = -\omega^2$, the latter relation holding on the solution space of (1.1) (see [76]).

The group $E(3)$ acts on $\mathcal{E}(3)$ via the adjoint representation and decomposes $\mathcal{E}(3)$ into three orbit types with representatives

$$P_3, \quad J_3, \quad J_3 + aP_3, \quad a \neq 0. \tag{1.14}$$

Note that $\exp(aP_3)$ is a translation along the three-axis, $\exp(\varphi J_3)$ is a rotation about this axis, and $\exp(\varphi J_3 + \varphi aP_3) = \exp(\varphi J_3)\exp(\varphi aP_3)$ is a rotation about the three-axis followed by a translation along the axis (a *screw-displacement*). Thus we have the Lie algebra version of the theorem that every Euclidean transformation is a translation, a rotation, or a screw-displacement (see [85]).

Since (1.1) is an equation in three variables, two separation constants are associated with each separable coordinate system. Thus we expect the separated solutions to be characterized as common eigenfunctions of a pair of commuting symmetry operators in the enveloping algebra of $\mathcal{E}(3)$. This turns out to be the case. Just as for the two-variable Helmholtz equation in Section 1.2, we find a number of rather trivial nonorthogonal coordinate systems which correspond to the diagonalization of first-order operators. In addition to these, there are eleven types of orthogonal separable coordinate systems, each of which corresponds to a pair of independent commuting operators S_1, S_2 in $\mathcal{E}(3)^2$. The associated separable solutions $\Psi = U(u)V(v)W(w)$ are characterized by the eigenvalue equations

$$\left(\Delta_3 + \omega^2\right)\Psi = 0, \quad S_1\Psi = \omega_1^2\Psi, \quad S_2\Psi = \omega_2^2\Psi \tag{1.15}$$

where ω_1^2, ω_2^2 are the separation constants [121, 76]. (It can be shown that there are no nontrivial R-separable solutions.)

Put another way, a separable coordinate system is associated with a two-dimensional subspace of commuting operators in $\mathcal{E}(3)^2$ and S_1, S_2 is a basis (nonunique) for this subspace. The group $E(3)$ acts on the set of all two-dimensional subspaces of commuting operators in $\mathcal{E}(3)^2$ via the adjoint representation and decomposes this set into orbits of equivalent subspaces. As usual, one regards separable coordinates associated with equivalent subspaces as equivalent, since one can obtain any such system from any other by a Euclidean transformation. As proved in [76], there are eleven types of distinct (nontrivial) orbits, and they match exactly the eleven types of orthogonal separable coordinates. Representative operators from each orbit and the associated coordinate systems are listed in Table 14.

ISBN-0-201-13503-5

Table 14 Operators and Separable Coordinates for
$$(\Delta_3 + \omega^2)\Psi = 0 \;((x_1, x_2, x_3) = (x, y, z))$$

Commuting operators S_1, S_2	Separable coordinates
1 P_2^2, P_3^2	Cartesian x, y, z
2 J_3^2, P_3^2	Cylindrical $x = r\cos\varphi,$ $y = r\sin\varphi,\; z = z$
3 $\{J_3, P_2\}, P_3^2$	Parabolic cylindrical $x = (\xi^2 - \eta^2)/2,$ $y = \xi\eta,\; z = z$
4 $J_3^2 + d^2 P_1^2, P_3^2,$ $d > 0$	Elliptic cylindrical $x = d\cosh\alpha\cos\beta,$ $y = d\sinh\alpha\sin\beta,\; z = z$
5 $\mathbf{J}\cdot\mathbf{J}, J_3^2$	Spherical $x = \rho\sin\theta\cos\varphi,$ $y = \rho\sin\theta\sin\varphi,\; z = \rho\cos\theta$
6 $\mathbf{J}\cdot\mathbf{J} - a^2(P_1^2 + P_2^2), J_3^2,$ $a > 0$	Prolate spheroidal $x = a\sinh\eta\sin\alpha\cos\varphi$ $y = a\sinh\eta\sin\alpha\sin\varphi$ $z = a\cosh\eta\cos\alpha$
7 $\mathbf{J}\cdot\mathbf{J} + a^2(P_1^2 + P_2^2), J_3^2,$ $a > 0$	Oblate spheroidal $x = a\cosh\eta\sin\alpha\cos\varphi$ $y = a\cosh\eta\sin\alpha\sin\varphi$ $z = a\sinh\eta\cos\alpha$
8 $\{J_1, P_2\} - \{J_2, P_1\}, J_3^2$	Parabolic $x = \xi\eta\cos\varphi,$ $y = \xi\eta\sin\varphi,\; z = (\xi^2 - \eta^2)/2$
9 $J_3^2 - c^2 P_3^2 + c(\{J_2, P_1\} + \{J_1, P_2\}),$ $c(P_2^2 - P_1^2) + \{J_2, P_1\} - \{J_1, P_2\}$	Paraboloidal $x = 2c\cosh\alpha\cos\beta\sinh\gamma$ $y = 2c\sinh\alpha\sin\beta\cosh\gamma$ $z = c(\cosh 2\alpha + \cos 2\beta - \cosh 2\gamma)/2$

10   $P_1^2 + aP_2^2 + (a+1)P_3^2 + \mathbf{J}\cdot\mathbf{J},$
 	
     $J_2^2 + a(J_1^2 + P_3^2),$
 	
     $a > 1$
	Ellipsoidal
	$x = \left[\dfrac{(\mu-a)(\nu-a)(\rho-a)}{a(a-1)}\right]^{1/2}$
	$y = \left[\dfrac{(\mu-1)(\nu-1)(\rho-1)}{1-a}\right]^{1/2}$
	$z = \left[\dfrac{\mu\nu\rho}{a}\right]^{1/2}$

11 $\mathbf{J}\cdot\mathbf{J}, J_1^2 + bJ_2^2,$ $1 > b > 0$	Conical $x = r\left[\dfrac{(b\mu-1)(b\nu-1)}{1-b}\right]^{1/2}$ $y = r\left[\dfrac{b(\mu-1)(\nu-1)}{b-1}\right]^{1/2},\; z = r[b\mu\nu]^{1/2}$

ISBN-0-201-13503-5

We will briefly study each of these systems to determine the form of the separated solutions and the significance of the eigenvalues of the commuting symmetry operators. We begin by considering solutions Ψ of the Helmholtz equation that are eigenfunctions of the operator P_3:

$$P_3\Psi = i\lambda\Psi, \qquad \Psi(x,y,z) = c^{i\lambda z}\Phi(x,y).$$

In this case we can split off the variable z, and equation (1.1) reduces to

$$\left(\Delta_2 + \left[\omega^2 - \lambda^2\right]\right)\Phi(x,y) = 0, \tag{1.16}$$

The Helmholtz equation in two variables. It follows from the results of Section 1.2 (see Table 1) that this reduced equation permits separation of variables in exactly four orthogonal coordinate systems. The corresponding systems for the full equation (1.1) are 1–4 in Table 14.

Next we consider solutions Ψ of (1.1) that are eigenfunctions of J_3:

$$iJ_3\Psi = m\Psi, \qquad \Psi(x,y,z) = e^{im\varphi}\Phi(r,z).$$

Here r,φ,z are cylindrical coordinates 2 and $J_3 = -\partial_\varphi$. We now split off the variable φ, and equation (1.1) reduces to

$$\left(\partial_{rr} + r^{-1}\partial_r - m^2/r^2 + \partial_{zz} + \omega^2\right)\Phi = 0. \tag{1.17}$$

This equation is class II, though it arises from a class I equation via partial separation of variables. The reduced equation separates in five coordinate systems, corresponding to systems 2, 5–8.

For spherical coordinates 5 the separated equations in ρ, θ are

$$P'' + \frac{2}{\rho}P' + \left(\omega^2 - \frac{l(l+1)}{\rho^2}\right)P = 0, \tag{1.18a}$$

$$\Theta'' + \cot\theta\,\Theta' + \left(l(l+1) - \frac{m^2}{\sin^2\theta}\right)\Theta = 0, \tag{1.18b}$$

$$\mathbf{J}\cdot\mathbf{J}\Psi = -l(l+1)\Psi.$$

The separated solutions take the form

$$P(\rho) = \rho^{-1/2}J_{\pm(l+\frac{1}{2})}(\omega\rho), \qquad \Theta(\theta) = P_l^{\pm m}(\cos\theta) \tag{1.19}$$

where $J_\nu(z)$ is a Bessel function and $P_l^m(\cos\theta)$ is a Legendre function (see

ISBN-0-201-13503-5

(B.6iv)). The coordinates ρ, θ, φ vary in the ranges

$$0 \leqslant \rho, \qquad 0 \leqslant \theta \leqslant \pi, \qquad 0 \leqslant \varphi < 2\pi$$

to cover the full space R^3.

For prolate spheroidal (or ellipsoidal) coordinates 6 (Table 14) the separated equations in η, α are

$$H'' + \coth(\eta)H' + (-\lambda + a^2\omega^2 \sinh^2\eta - m^2/\sinh^2\eta)H = 0,$$
$$A'' + \cot(\alpha)A' + (\lambda + a^2\omega^2 \sin^2\alpha - m^2/\sin^2\alpha)A = 0, \qquad (1.20)$$
$$(\mathbf{J} \cdot \mathbf{J} - a^2P_1^2 - a^2P_2^2)\Psi = -\lambda\Psi.$$

Equations (1.20) are two forms of the *spheroidal wave equation* [7, 79]. The corresponding solutions Ψ of (1.1) that are bounded and single valued in R^3 are of the form

$$H(\eta)A(\alpha)e^{im\varphi} = Ps_n^{|m|}(\cosh\eta, a^2\omega^2) Ps_n^{|m|}(\cos\alpha, a^2\omega^2)e^{im\varphi}, \qquad (1.21)$$

$$m \text{ integer}, \ n = 0, 1, 2, \ldots, \ -n \leqslant m \leqslant n,$$

where $Ps_n^m(z, \gamma)$ is a spheroidal wave function. The discrete eigenvalues $\lambda_n^{|m|}(a^2\omega^2)$ are analytic functions of $a^2\omega^2$. For $a = 0$ the spheroidal wave equation reduces to the equation for Legendre functions (1.18b) and $Ps_n^{|m|}(\cos\alpha, 0) = P_n^{|m|}(\cos\alpha)$. Furthermore, $\lambda_n^{|m|}(0) = n(n+1)$. The coordinates vary in the range $0 \leqslant \alpha < 2\pi$, $\eta \geqslant 0$, $0 \leqslant \varphi < 2\pi$.

For oblate spheroidal (or ellipsoidal) coordinates 7 the separated equations in η, α are

$$H'' + \tanh(\eta)H' + (-\lambda + a^2\omega^2 \cosh^2\eta + m^2/\cosh^2\eta)H = 0,$$
$$A'' + \cot(\alpha)A' + (\lambda - a^2\omega^2 \sin^2\alpha - m^2/\sin^2\alpha)A = 0, \qquad (1.22)$$
$$(\mathbf{J} \cdot \mathbf{J} + a^2P_1^2 + a^2P_2^2)\Psi = -\lambda\Psi.$$

Again these equations are forms of the spheroidal wave equation. The corresponding solutions Ψ of (1.1) that are bounded and single valued in R^3 take the form

$$Ps_n^{|m|}(-i\sinh\eta, a^2\omega^2) Ps_n^{|m|}(\cos\alpha, -a^2\omega^2)e^{im\varphi}, \qquad (1.23)$$

$$m \text{ integer}, \ n = 0, 1, 2, \ldots, \ -n \leqslant m \leqslant n,$$

with eigenvalues $\lambda_n^{|m|}(-a^2\omega^2)$.

ISBN-0-201-13503-5

For parabolic coordinates 8 the separated equations in ξ, η are

$$\Xi'' + \xi^{-1}\Xi' + (\omega^2\xi^2 - m^2/\xi^2 - \lambda)\Xi = 0,$$
$$H'' + \eta^{-1}H' + (\omega^2\eta^2 - m^2/\eta^2 + \lambda)H = 0, \tag{1.24}$$
$$(\{J_1, P_2\} - \{J_2, P_1\})\Psi = \lambda\Psi,$$

and the separated solutions take the form

$$\Xi(\xi) = \xi^m \exp(\pm i\omega\xi^2/2) \,_1F_1\left(\begin{matrix} i\lambda/4\omega + (m+1)/2 \\ m+1 \end{matrix} \middle| \mp i\omega\xi^2\right),$$

$$H(\eta) = \eta^m \exp(\pm i\omega\eta^2/2) \,_1F_1\left(\begin{matrix} -i\lambda/4\omega + (m+1)/2 \\ m+1 \end{matrix} \middle| \mp i\omega\eta^2\right). \tag{1.25}$$

The foregoing eight systems are the only ones whose separated solutions are eigenfunctions of a second-order operator that is the square of a first-order symmetry operator. The remaining three systems are somewhat less tractable.

For paraboloidal coordinates 9 the separated equations in α, β, γ are

$$A'' + \left(-q - \lambda c \cosh 2\alpha + \frac{\omega^2 c^2}{2} \cosh 4\alpha\right)A = 0,$$

$$B'' + \left(q + \lambda c \cos 2\beta - \frac{\omega^2 c^2}{2} \cos 4\beta\right)B = 0, \tag{1.26}$$

$$\Gamma'' + \left(-q + \lambda c \cosh 2\gamma + \frac{\omega^2 c^2}{2} \cosh 4\gamma\right)\Gamma = 0, \qquad q = \mu - c^2\omega^2/2,$$

where

$$(J_3^2 - c^2 P_3^2 + c\{J_2, P_1\} + c\{J_1, P_2\})\Psi = -\mu\Psi,$$
$$(cP_2^2 - cP_1^2 + \{J_2, P_1\} - \{J_1, P_2\})\Psi = \lambda\Psi. \tag{1.27}$$

Each of the equations (1.26) can be transformed to the Whittaker–Hill equation (6.28), Section 2.6 [127]. Single-valued solutions of (1.1) take the form

$$\Psi(\alpha, \beta, \gamma) = gc_n(i\alpha; 2c\omega, \lambda/2\omega) \, gc_n(\beta; 2c\omega, \lambda/2\omega)$$
$$\times gc_n(i\gamma + \pi/2; 2c\omega, \lambda/2\omega), \qquad n = 0, 1, 2, \ldots, \mu = \mu_n, \tag{1.28}$$

or the same form with gc_n replaced by gs_n.

ISBN-0-201-13503-5

For ellipsoidal coordinates μ, ν, ρ where $0 < \rho < 1 < \nu < a < \mu < \infty$ for single-valued coordinates, the separation equations all take the form

$$\left(4(h(\xi))^{1/2} \frac{d}{d\xi} (h(\xi))^{1/2} \frac{d}{d\xi} + \lambda_1 \xi + \lambda_2 + \omega^2 \xi^2 \right) E(\xi) = 0,$$

$$h(\xi) = (\xi - a)(\xi - 1)\xi, \qquad \xi = \mu, \nu, \rho, \tag{1.29}$$

with

$$\left(\mathbf{J} \cdot \mathbf{J} + P_1^2 + a P_2^2 + (a+1) P_3^2 \right) \Psi = \lambda_1 \Psi,$$

$$\left(J_2^2 + a J_1^2 + a P_3^2 \right) \Psi = \lambda_2 \Psi. \tag{1.30}$$

For computational purposes it is more convenient to introduce the equivalent separable coordinates α, β, γ defined by

$$\rho = \mathrm{sn}^2(\alpha, k), \qquad \nu = \mathrm{sn}^2(\beta, k), \qquad \mu = \mathrm{sn}^2(\gamma, k), \qquad k = a^{-1/2}, \tag{1.31}$$

where $\mathrm{sn}(z, k)$ is a Jacobi elliptic function (see Appendix C). The relationship between α, β, γ and x, y, z is

$$x = ik^{-1}k'^{-1} \mathrm{dn}\,\alpha\,\mathrm{dn}\,\beta\,\mathrm{dn}\,\gamma, \qquad y = -kk'^{-1} \mathrm{cn}\,\alpha\,\mathrm{cn}\,\beta\,\mathrm{cn}\,\gamma,$$

$$z = k\,\mathrm{sn}\,\alpha\,\mathrm{sn}\,\beta\,\mathrm{sn}\,\gamma \tag{1.32}$$

where $\mathrm{cn}\,\alpha$, $\mathrm{dn}\,\alpha$ are elliptic functions and $k' = (1 - k^2)^{.1/2}$ To obtain real values for x, y, z we choose α real, β complex such that $\mathrm{Re}\,\beta = K$, and γ complex such that $\mathrm{Im}\,\gamma = K'$ where $K(k)$ is defined by (C.3) and $K' = K(k')$. To cover all real values of x, y, z once, it is sufficient to let α vary in the interval $[-K, K]$, β vary in $[K - iK', K + iK']$ (parallel to the imaginary axis), and γ vary in $[-K + iK', K + iK']$ (parallel to the real axis). In these new variables the separation equations take the form of the *ellipsoidal wave equation*

$$\left\{ \frac{d^2}{d\xi^2} + k^2\lambda_2 + k^2\lambda_1 \mathrm{sn}^2 \xi + k^2\omega^2 \mathrm{sn}^4 \xi \right\} E(\xi) = 0, \qquad \xi = \alpha, \beta, \gamma. \tag{1.33}$$

From the periodicity properties of the elliptic functions it follows that if ξ is replaced by $\xi + 4Kn + 4iK'm$ in (1.32), where n, m are integers and ξ is any one of α, β, γ, then x, y, z remain unchanged. Thus only those solutions $E(\xi)$ of (1.33) that are doubly periodic and single valued in ξ with real period $4K$ and imaginary period $4iK'$ are single-valued functions of x, y, z. The doubly periodic single-valued solutions of (1.33) are called *ellipsoidal wave functions* and are denoted by the symbol el(ξ) in Arscott's notation [7, Chapter X]. There are eight types of such functions, each expressible in the

ISBN-0-201-13503-5

form

$$\text{sn}^s z \, \text{cn}^c z \, \text{dn}^d z \, F(\text{sn}^2 z), \qquad s,c,d=0,1,$$

where F is a convergent power series in its argument. The eigenvalues are countable and discrete.

For conical coordinates r, μ, ν (System 11, Table 14) it is convenient to set $\mu = \text{sn}^2(\alpha, k)$, $\nu = \text{sn}^2(\beta, k)$ where $k = b^{1/2} > 0$. Then

$$x = rk'^{-1} \text{dn}(\alpha, k) \text{dn}(\beta, k), \qquad y = irkk'^{-1} \text{cn}(\alpha, k) \text{cn}(\beta, k)$$
$$z = rk \, \text{sn}(\alpha, k) \text{sn}(\beta, k), \tag{1.34}$$

and the variables have the range $0 \leqslant r$, $-2K < \alpha < 2K$, $K \leqslant \beta < K + 2iK'$ (see [7, p..24]). The separation equations are

$$R'' + 2r^{-1} R' + \left(\omega - l(l+1)r^{-2}\right)R = 0,$$
$$A'' + \left(\lambda - l(l+1)k^2 \text{sn}^2 \alpha\right)A = 0,$$
$$B'' + \left(\lambda - l(l+1)k^2 \text{sn}^2 \beta\right)B = 0,$$
$$\mathbf{J} \cdot \mathbf{J}\Psi = -l(l+1)\Psi, \qquad \left(J_1^2 + bJ_2^2\right)\Psi = \lambda\Psi. \tag{1.35}$$

The first equation has solutions of the form $R(r) = r^{-1/2} J_{\pm(l+\frac{1}{2})}(\omega r)$, in agreement with (1.18a). The latter two equations are examples of the *Lamé equation*. If α or β is increased by integral multiples of $4K$ or $4iK'$, it follows from (1.34) that x, y, and z are unchanged. Thus only those solutions $A(\alpha), B(\beta)$ of (1.35) that are doubly periodic and single valued in α, β, respectively, lead to single-valued functions of x,y,z. It is known (see [7]) that doubly periodic solutions of Lamé's equation exist only in the cases where $l=0,1,2,\ldots$. Furthermore, for positive integer l there exist exactly $2l+1$ such solutions corresponding to $2l+1$ distinct eigenvalues λ. The solutions, exactly one for each pair of eigenvalues λ,l, can be expressed as finite series called *Lamé polynomials*. There are eight types of Lamé polynomials, each expressible in the form

$$\text{sn}^s \alpha \, \text{cn}^c \alpha \, \text{dn}^d \alpha Fp(\text{sn}^2 \alpha), \qquad s,c,d=0,1, \qquad s+c+d+2p=l,$$

where $Fp(z)$ is a polynomial of order p in z. In Section 3.3 we shall study these functions in more detail.

3.2 A Hilbert Space Model: The Sphere S_2

In analogy with the methods of Chapter 1 we can introduce a Hilbert space structure on the solution space of (1.1) in such a way that the separated solutions can be interpreted as eigenfunctions of self-adjoint

ISBN-0-201-13503-5

operators in the enveloping algebra of $\mathcal{E}(3)$. By an obvious extension of arguments in Section 1.3 we can show that $\Psi(\mathbf{x})$ satisfies $(\Delta_3 + \omega^2)\Psi(\mathbf{x}) = 0$ if it can be represented in the form

$$\Psi(\mathbf{x}) = \iint_{S_2} e^{i\omega\mathbf{x}\cdot\hat{\mathbf{k}}} h(\hat{\mathbf{k}}) \, d\Omega(\hat{\mathbf{k}}) = I(h), \tag{2.1}$$

$$\mathbf{x} = (x_1, x_2, x_3), \quad \hat{\mathbf{k}} = (k_1, k_2, k_3).$$

Here $\hat{\mathbf{k}}$ is a unit vector ($\hat{\mathbf{k}} \cdot \hat{\mathbf{k}} = 1$) that runs over the unit sphere S_2: $k_1^2 + k_2^2 + k_3^2 = 1$, $d\Omega$ is the usual solid-angle measure on the sphere, and h is an arbitrary complex-valued measurable function on S_2 (with respect to $d\Omega$) such that

$$\iint_{S_2} |h(\hat{\mathbf{k}})|^2 \, d\Omega(\hat{\mathbf{k}}) < \infty.$$

The set $L_2(S_2)$ of such functions h is a Hilbert space with inner product

$$\langle h_1, h_2 \rangle = \iint_{S_2} h_1(\hat{\mathbf{k}}) \bar{h}_2(\hat{\mathbf{k}}) \, d\Omega(\hat{\mathbf{k}}), \tag{2.2}$$

or, in terms of spherical coordinates on S_2

$$\hat{\mathbf{k}} = (\sin\theta\cos\varphi, \sin\theta\sin\varphi, \cos\theta), \quad 0 \leqslant \theta \leqslant \pi, \ -\pi \leqslant \varphi < \pi,$$

$$d\Omega(\hat{\mathbf{k}}) = \sin\theta \, d\theta \, d\varphi \tag{2.3}$$

and

$$\langle h_1, h_2 \rangle = \int_{-\pi}^{\pi} d\varphi \int_0^{\pi} h_1(\theta, \varphi) \bar{h}_2(\theta, \varphi) \sin\theta \, d\theta.$$

The elements $g(A, \mathbf{a})$ of $E(3)$ act on the solutions of the Helmholtz equation via the operators $\mathbf{T}(g)$, (1.9), (1.12). Using (2.1) we find

$$\mathbf{T}(g)\Psi(\mathbf{x}) = I(\mathbf{T}(g)h) \tag{2.4}$$

whenever $\Psi = I(h)$, where the operators $\mathbf{T}(g)$ on $L_2(S_2)$ are defined by

$$\mathbf{T}(g)h(\hat{\mathbf{k}}) = \exp(i\omega\mathbf{a}\cdot\hat{\mathbf{k}}A)h(\hat{\mathbf{k}}A), \tag{2.5}$$

$$g = (A, \mathbf{a}), \quad A \in SO(3), \quad \mathbf{a} \in R^3.$$

Thus the $\mathbf{T}(g)$ acting on Ψ induce operators (which we also call $\mathbf{T}(g)$) acting on h. It is easy to verify directly that the operators (2.5) satisfy the

ISBN-0-201-13503-5

group homomorphism property $\mathbf{T}(g_1 g_2) = \mathbf{T}(g_1)\mathbf{T}(g_2)$. Moreover, these operators are unitary on $L_2(S_2)$:

$$\langle \mathbf{T}(g)h_1, \mathbf{T}(g)h_2 \rangle = \langle h_1, h_2 \rangle, \qquad h_j \in L_2(S_2).$$

This result and (2.5) itself depend on the invariance of the measure under rotations: $d\Omega(\hat{\mathbf{k}}A) = d\Omega(\hat{\mathbf{k}})$.

A similar computation shows that the Lie algebra generators on $L_2(S_2)$ induced by the generators (1.2) on the solution space are

$$P_1 = i\omega k_1 = i\omega \sin\theta \cos\varphi, \qquad P_2 = i\omega k_2 = i\omega \sin\theta \sin\varphi, \qquad P_3 = i\omega k_3 = i\omega \cos\theta,$$

$$J_1 = k_3 \partial_{k_2} - k_2 \partial_{k_3} = \sin\varphi\, \partial_\theta + \cos\varphi \cot\theta\, \partial_\varphi,$$

$$J_2 = k_1 \partial_{k_3} - k_3 \partial_{k_1} = -\cos\varphi\, \partial_\theta + \sin\varphi \cot\theta\, \partial_\varphi,$$

$$J_3 = k_2 \partial_{k_1} - k_1 \partial_{k_2} = -\partial_\varphi. \tag{2.6}$$

In analogy with (1.12) these operators are related to the group operators (2.5) by

$$\mathbf{T}(g) = \exp(\varphi' J_3)\exp(\theta' J_1)\exp(\psi' J_3)\exp(a_1 P_1 + a_2 P_2 + a_3 P_3)$$

where φ', θ', ψ' are the Euler angles for A. Furthermore, the operators (2.6) are skew-Hermitian on the dense subspace $\mathfrak{D}$ of $L_2(S_2)$ consisting of infinitely differentiable functions on S_2.

We have shown that the $\mathbf{T}(g)$ define a unitary (irreducible) representation of $E(3)$ on $L_2(S_2)$. The elements of $\mathcal{E}(3)^2$ are easily seen to be symmetric on $\mathfrak{D}$ and we shall show explicitly that their domains can be extended to define self-adjoint operators in dense subspaces of $L_2(S_2)$. Corresponding to each pair of commuting operators listed in Table 14 we shall find a pair of commuting self-adjoint operators S, S' on $L_2(S_2)$ and determine the spectral resolution of this pair. These results will then be used to obtain information about the space $\mathcal{H}$ consisting of solutions Ψ of the Helmholtz equation such that $\Psi = I(h)$ for some $h \in L_2(S_2)$, (2.1). Here $\mathcal{H}$ is a Hilbert space with inner product

$$(\Psi_1, \Psi_2) \equiv \langle h_1, h_2 \rangle, \qquad \Psi_j = I(h_j). \tag{2.7}$$

(It is not hard to show that no nonzero $h \in L_2(S_2)$ can be mapped by I to the zero solution of the Helmholtz equation). It follows that I is a unitary transformation from $L_2(S_2)$ to $\mathcal{H}$. Also, the operators $\mathbf{T}(g)$ on $\mathcal{H}$ defined by (1.9), (1.12) are now seen to be unitary.

We can also interpret each function $\Psi(\mathbf{x})$ in $\mathcal{H}$ as an inner product

$$\Psi(\mathbf{x}) = I(h) = \langle h, H(\mathbf{x}, \cdot)\rangle, \qquad H(\mathbf{x}, \hat{\mathbf{k}}) = e^{-i\omega\mathbf{x}\cdot\hat{\mathbf{k}}} \in L_2(S_2). \tag{2.8}$$

Just as we saw in Section 1.3, the existence of the unitary mapping I allows us to transform problems involving $\mathcal{H}$ to problems involving $L_2(S_2)$. In particular, if S, S' are a pair of commuting operators from Table 14, we can interpret them as a pair of commuting self-adjoint operators on $L_2(S_2)$ and compute a basis of eigenfunctions for $L_2(S_2)$:

$$S f_{\lambda\mu} = \lambda f_{\lambda\mu}, \qquad S' f_{\lambda\mu} = \mu f_{\lambda\mu}, \qquad \langle f_{\lambda\mu}, f_{\lambda'\mu'} \rangle = \delta(\lambda - \lambda')\delta(\mu - \mu'). \quad (2.9)$$

Then the functions $\Psi_{\lambda\mu}(\mathbf{x}) = I(f_{\lambda\mu})$ will form a corresponding basis in $\mathcal{H}$ for the operators S, S' constructed from the generators (1.2):

$$S \Psi_{\lambda\mu} = \lambda \Psi_{\lambda\mu}, \qquad S' \Psi_{\lambda\mu} = \mu \Psi_{\lambda\mu}. \quad (2.10)$$

These last expressions enable us to evaluate the integral for $\Psi_{\lambda\mu}$, for they guarantee that $\Psi_{\lambda\mu}$ is a solution of the Helmholtz equation that is separable in the coordinates associated with S, S'. Furthermore, if Ψ is any solution of (1.1) such that $\Psi = I(h)$ for some $h \in L_2(S_2)$, we have the expansion

$$\mathbf{T}(g)\Psi(\mathbf{x}) = \sum_{\lambda, \mu} \langle \mathbf{T}(g)h, f_{\lambda\mu} \rangle \Psi_{\lambda\mu}(\mathbf{x}), \quad (2.11)$$

which converges both pointwise and in the Hilbert space sense.

We now proceed to analyze our model $L_2(S_2)$. Harmonic analysis involving functions on the sphere is itself a topic of considerable interest. Typically, such studies use only spherical coordinates 5 (Table 14) and lead to theorems concerning expansions in spherical harmonics. However, we shall analyze all eleven coordinate systems on S_2 that follow from Table 14. In some cases we shall employ simpler models of our representations than $L_2(S_2)$ to carry forward the analysis.

Since the spherical coordinate system 5 is treated in detail in so many textbooks (e.g., [40, 45, 85, 128]), we shall here list only the most important facts concerning this system, omitting all proofs. The unitary irreducible representations of $SO(3)$ are all finite dimensional. They are denoted by D_l, $l = 0, 1, 2, \ldots$, where $\dim D_l = 2l + 1$. If $\{J_1, J_2, J_3\}$ are the operators on the representation space V_l of D_l which correspond to the Lie algebra generators (1.5), then there is an ON basis $\{f_m^{(l)}: m = l, l-1, \ldots, -l\}$ for V_l such that

$$J^0 f_m^{(l)} = m f_m^{(l)}, \qquad J^{\pm} f_m^{(l)} = \left[(l \pm m + 1)(l \mp m)\right]^{1/2} f_{m \pm 1}^{(l)} \quad (2.12)$$

where $J^{\pm} = \mp J_2 + iJ_1$, $J^0 = iJ_3$. Here, $J^+ f_l^{(l)} \equiv J^- f_{-l}^{(l)} \equiv 0$. If the group is parametrized in terms of Euler angles (1.6), the matrix elements of the

ISBN-0-201-13503-5

operators $\mathbf{D}(A) = \exp(\varphi J_3) \exp(\theta J_1) \exp(\psi J_3)$ with respect to the ON basis $\{f_m^{(l)}\}$,

$$\mathbf{D}(A) f_m^{(l)} = \sum_{n=-l}^{l} D_{nm}^l (A) f_n^{(l)},$$

are given by

$$D_{nm}^l (A) = i^{n-m} \left[\frac{(l+m)!(l-n)!}{(l+n)!(l-m)!} \right]^{1/2} \exp\left[i(n\varphi + m\psi) \right] P_l^{-n,m} (\cos\theta)$$

(2.13)

where

$$P_l^{-n,m} (\cos\theta) = \frac{(\sin\theta)^{m-n} (1+\cos\theta)^{l+n-m} 2^{-l}}{\Gamma(m-n+1)}$$

$$\times \, {}_2F_1 \left(\begin{array}{c} -l-n, m-l \\ m-n+1 \end{array} \middle| \frac{\cos\theta-1}{\cos\theta+1} \right)$$

(2.14)

is a generalized spherical function. (The matrix elements (2.13) are known as the *Wigner D functions* [137].) The D_{nm}^l satisfy the usual group homomorphism and unitary properties

$$D_{nm}^l (AA') = \sum_{j=-l}^{l} D_{nj}^l (A) D_{jm}^l (A'), \qquad A, A' \in SO(3),$$

$$D_{nm}^l (A^{-1}) = \bar{D}_{mn}^l (A).$$

(2.15)

The special matrix element $D_{0m}^l(A)$ is proportional to a spherical harmonic:

$$D_{0m}^l (\varphi, \theta, \psi) = i^m \left(\frac{4\pi}{2l+1} \right)^{1/2} Y_l^m (\theta, \psi),$$

(2.16)

where

$$Y_l^m (\theta, \psi) = \left[\frac{(2l+1)(l-m)!}{4\pi(l+m)!} \right]^{1/2} P_l^m (\cos\theta) e^{im\varphi}$$

(2.17)

and $P_l^m(\cos\theta) = P_l^{0,-m}(\cos\theta)$ is an associated Legendre function.

ISBN-0-201-13503-5

It follows from (2.12) that on V_l

$$\mathbf{J} \cdot \mathbf{J} = J_1^2 + J_2^2 + J_3^2 = -l(l+1)E \qquad (2.18)$$

where E is the identity operator.

Now consider the irreducible representation T of $E(3)$ on $L_2(S_2)$ defined by expression (2.5). The restriction of T to the subgroup $SO(3)$ is no longer irreducible but breaks up into the direct sum

$$T|SO(3) \cong \sum_{l=0}^{\infty} \oplus D_l; \qquad (2.19)$$

that is, $L_2(S_2)$ can be decomposed into a direct sum of mutually orthogonal subspaces V_l,

$$L_2(S_2) \cong \sum_{l=0}^{\infty} \oplus V_l,$$

where $\dim V_l = 2l+1$ and the action of the operators $\mathbf{T}(A)$ on the invariant subspace V_l is unitary equivalent to D_l. The elements h of V_l are characterized as the solutions of the equation $\mathbf{J} \cdot \mathbf{J} h = -l(l+1)h$, or

$$\left(\partial_{\theta\theta} + \cot\theta\, \partial_\theta + \sin^{-2}\theta\, \partial_{\varphi\varphi} \right) h(\theta, \varphi) = -l(l+1)h(\theta, \varphi), \qquad (2.20)$$

in terms of the coordinates (2.3). Here $\mathbf{J} \cdot \mathbf{J}$ is known as the *Laplace operator on the sphere* S_2. It follows from the foregoing results that the self-adjoint extension of this operator (which we also denote $\mathbf{J} \cdot \mathbf{J}$) has discrete spectrum $-l(l+1)$, $l = 0, 1, 2,, \ldots$, each eigenvalue occurring with multiplicity $2l+1$.

There exists a basis for V_l, consisting of eigenfunctions $f_m^{(l)}(\theta, \varphi)$ of the symmetry operator J^0, which satisfy the relations (2.12) where

$$J^\pm = e^{\pm i\varphi} \left(\pm \partial_\theta + i \cot\theta\, \partial_\varphi \right), \qquad J^0 = -i\partial_\varphi. \qquad (2.21)$$

Indeed, from the recurrence relations (2.12) and the differential equation (2.20) we find

$$f_m^{(l)}(\theta, \varphi) = Y_l^m(\theta, \varphi), \qquad \langle Y_l^m, Y_{l'}^{m'} \rangle = \delta_{ll'} \delta_{mm'}. \qquad (2.22)$$

Furthermore, it is straightforward to show that the action of the operators

ISBN-0-201-13503-5

P_j on this basis is given by

$$P^0 f_m^{(l)} = -\omega\left[\frac{(l+m+1)(l-m+1)}{(2l+3)(2l+1)}\right]^{1/2} f_m^{(l+1)} - \omega\left[\frac{(l+m)(l-m)}{(2l+1)(2l-1)}\right]^{1/2} f_m^{(l-1)},$$

$$P^+ f_m^{(l)} = \omega\left[\frac{(l+m+1)(l+m+2)}{(2l+3)(2l+1)}\right]^{1/2} f_{m+1}^{(l+1)} - \omega\left[\frac{(l-m)(l-m-1)}{(2l+1)(2l-1)}\right]^{1/2} f_{m+1}^{(l-1)},$$

$$P^- f_m^{(l)} = -\omega\left[\frac{(l-m+2)(l-m+1)}{(2l+3)(2l+1)}\right]^{1/2} f_{m-1}^{(l+1)} + \omega\left[\frac{(l+m)(l+m-1)}{(2l+1)(2l-1)}\right]^{1/2} f_{m-1}^{(l-1)},$$

$$(2.23)$$

where

$$P^0 = iP_3 = -\omega\cos\theta, \qquad P^\pm = \mp P_2 + iP_1 = -\omega e^{\pm i\varphi}\sin\theta \qquad (2.24)$$

(see [82]).

The matrix elements of the translation operators $T(E,a) = \exp(a_1 P_1 + a_2 P_2 + a_3 P_3)$ are given by

$$T_{lm,l'm'}(\mathbf{a}) = \langle T(E,\mathbf{a}) f_{m'}^{(l')}, f_m^{(l)} \rangle$$

$$= \int_{S_2} e^{i\omega\mathbf{a}\cdot\hat{\mathbf{k}}} Y_{l'}^{m'}(\hat{\mathbf{k}}) \overline{Y}_l^m(\hat{\mathbf{k}}) \, d\Omega(\hat{\mathbf{k}}), \qquad (2.25)$$

or more explicitly,

$$T_{lm,l'm'}(\mathbf{a}) = (4\pi)^{1/2} \sum_{s=0}^{\infty} \left[\frac{(2s+1)(2l+1)}{(2l'+1)}\right]^{1/2} i^s j_s(\omega a)$$

$$\times Y_s^{m'-m}(\alpha,\beta) C(s,0;\,l,0|l',0) C(s,m'-m;\,l,m|l',m') \quad (2.26)$$

where

$$\mathbf{a} = (a\sin\alpha\cos\beta,\, a\sin\alpha\sin\beta,\, a\cos\alpha), \qquad a \geqslant 0,$$

and $C(\cdot)$ is a Clebsch–Gordan coefficient for $SO(3)$ [82, 124, 128]. (In (2.26) the sum is actually finite because the Clebsch–Gordan coefficients vanish except for finitely many values of s. The *spherical Bessel functions* $j_n(z)$ are defined by

$$j_n(z) = (\pi/2z)^{1/2} J_{n+1/2}(z), \qquad n = 0, 1, 2, \ldots. \qquad (2.27)$$

ISBN-0-201-13503-5

Applying the integral transformation I to our ON basis $\{f_m^{(l)}\}$ for $L_2(S_2)$, we obtain an ON basis $\{\Psi_m^{(l)}=I(f_m^{(l)})\}$ of solutions for the Helmholtz equation that satisfy the eigenvalue equations

$$\mathbf{J}\cdot\mathbf{J}\Psi_m^{(l)}=-l(l+1)\Psi_m^{(l)}, \qquad J_3\Psi_m^{(l)}=-im\Psi_m^{(l)}.$$

The eigenfunctions separate in the spherical coordinate system 5 listed in Table 14 and are explicitly given by

$$\Psi_m^{(l)}(r,\theta,\varphi)=4\pi i^l j_l(\omega r)Y_l^m(\theta,\varphi), \qquad l=0,1,2,\ldots,m=l,l-1,\ldots,-l.$$

$$(2.28)$$

These functions are frequently called (*standing*) *spherical waves*. They necessarily satisfy the recurrence relations (2.12) and (2.23) where now the operators are given by (1.2). Furthermore, the matrix elements (2.13) and (2.26) can be used directly to expand the function $\mathbf{T}(g)\Psi_M^{(L)}$ in terms of the spherical basis. In particular, the special case in which $g=(E,\mathbf{a})$ leads to the addition theorem for spherical waves:

$$\Psi_M^{(L)}(R,\Theta,\Phi)=\sum_{l,m}T_{lm,LM}(\mathbf{a})\Psi_m^{(l)}(r,\theta,\varphi) \qquad (2.29)$$

where R,Θ,Φ are spherical coordinates for the three-vector $\mathbf{R}=\mathbf{x}+\mathbf{a}$. Expression (2.29) was first derived in [39].

It is easy to show that the recurrence relations (2.12), (2.23) are also satisfied by the non-Hilbert space solutions

$$\Psi_m'^{(l)}(\rho,\theta,\varphi)=4\pi i^l j_{-l-1}(\omega\rho)Y_l^m(\theta,\varphi), \qquad (2.30)$$

hence by any linear combination $\alpha\Psi_m^{(l)}+\beta\Psi_m'^{(l)}$ [124, p. 229]. As a consequence, the matrix elements (2.13), (2.26) are valid for all of these basis sets, and expansion formulas such as (2.29) hold for the set $\{\Psi_m'^{(l)}\}$ as well as for the Hilbert space basis $\{\Psi_m^{(l)}\}$.

Next we compute the spectral decompositions of the operators corresponding to systems 1–4 in Table 14, via our $L_2(S_2)$ model. These systems are characterized by the fact that P_3 is diagonal. From (2.6) it follows immediately that the bounded self-adjoint operator $iP_3=-\omega\cos\theta$ has continuous spectrum covering the interval $[-\omega,\omega]$ with multiplicity one. Fixing an eigenvalue of iP_3 corresponds to fixing the coordinate θ. The remaining coordinate φ can still vary and sweeps out a circle in S_2 as it goes from $-\pi$ to π. For each of the systems 1–4 the remaining second-order symmetry operator commutes with P_3; hence it leaves the functions

ISBN-0-201-13503-5

on these circles invariant and reduces to one of the four cases studied in Section 1.3. The work of that section carries over immediately to yield the following results:

1. Cartesian System

The eigenvalue equations are

$$iP_3 f^{(1)}_{\alpha,\gamma} = -\omega\cos(\gamma) f^{(1)}_{\alpha,\gamma}, \qquad iP_2 f^{(1)}_{\alpha,\gamma} = -\omega\sin(\gamma)\sin(\alpha) f^{(1)}_{\alpha,\gamma}, \quad (2.31)$$

with basis eigenfunctions

$$f^{(1)}_{\alpha,\gamma}(\theta,\varphi) = \frac{\delta(\varphi-\alpha)\delta(\theta-\gamma)}{(\sin\gamma)^{1/2}}, \qquad -\pi \leqslant \alpha < \pi, \quad 0 \leqslant \gamma \leqslant \pi,$$

$$\langle f^{(1)}_{\alpha,\gamma}, f^{(1)}_{\alpha',\gamma'}\rangle = \delta(\alpha-\alpha')\delta(\gamma-\gamma'). \tag{2.32}$$

The corresponding solutions of the Helmholtz equation are the plane waves

$$\Psi^{(1)}_{\alpha,\gamma}(\mathbf{x}) = I\left(f^{(1)}_{\alpha,\gamma}\right) = (\sin\gamma)^{1/2}\exp\left[i\omega(x_1\sin\gamma\cos\alpha + x_2\sin\gamma\sin\alpha + x_3\cos\gamma)\right].$$

$$\tag{2.33}$$

2. Cylindrical System

The eigenvalue equations are

$$iP_3 f^{(2)}_{n,\gamma} = -\omega\cos(\gamma) f^{(2)}_{n,\gamma}, \qquad iJ_3 f^{(2)}_{n,\gamma} = n f^{(2)}_{n,\gamma}, \tag{2.34}$$

and the basis of eigenfunctions is

$$f^{(2)}_{n,\gamma}(\theta,\varphi) = \frac{e^{in\varphi}\delta(\gamma-\theta)}{(2\pi\sin\gamma)^{1/2}}, \qquad n = 0, \pm 1, \pm 2, \ldots, 0 \leqslant \gamma \leqslant \pi,$$

$$\langle f^{(2)}_{n,\gamma}, f^{(2)}_{n',\gamma'}\rangle = \delta_{nn'}\delta(\gamma-\gamma'). \tag{2.35}$$

Furthermore

$$\Psi^{(2)}_{n,\gamma}(\mathbf{x}) = I(f^{(2)}_{n,\gamma}) = i^n(2\pi\sin\gamma)^{1/2}J_n(\omega\sin(\gamma)r)\exp[i(n\varphi + \omega z\cos\gamma)], \quad (2.36)$$

$$x = r\cos\varphi, \, y = r\sin\varphi, \, z = z.$$

These are *cylindrical wave* solutions of the Helmholtz equation.

ISBN-0-201-13503-5

3. Parabolic Cylindrical System

The eigenvalue equations are

$$iP_3 f^{(3)}_{\mu\pm,\gamma} = -\omega\cos(\gamma) f^{(3)}_{\mu\pm,\gamma}, \qquad \{J_3, P_2\} f^{(3)}_{\mu\pm,\gamma} = 2\mu\omega\sin(\gamma) f^{(3)}_{\mu\pm,\gamma}, \quad (2.37)$$

and the basis of eigenfunctions is

$$f^{(3)}_{\mu+,\gamma}(\theta,\varphi) = \begin{cases} (2\pi\sin\gamma)^{-1/2}(1+\cos\varphi)^{-i\mu/2-\frac{1}{4}}(1-\cos\varphi)^{i\mu/2-\frac{1}{4}}\delta(\theta-\gamma), \\ \qquad 0<\varphi<\pi, \\ 0, \qquad -\pi<\varphi<0, \end{cases}$$

$$f^{(3)}_{\mu-,\gamma}(\theta,\varphi) = f^{(3)}_{\mu+,\gamma}(\theta,-\varphi), \qquad -\infty<\mu<\infty, 0\leqslant\gamma\leqslant\pi, \tag{2.38}$$

$$\langle f^{(3)}_{\mu\pm,\gamma}, f^{(3)}_{\mu'\pm,\gamma'}\rangle = \delta(\mu-\mu')\delta(\gamma-\gamma'), \qquad \langle f^{(3)}_{\mu\pm,\gamma}, f^{(3)}_{\mu'\mp,\gamma'}\rangle = 0.$$

The corresponding solutions of the Helmholtz equation are

$$\Psi^{(3)}_{\mu+,\gamma}(\mathbf{x}) = I(f^{(3)}_{\mu+,\gamma}) = \left(\frac{\sin\gamma}{2}\right)^{1/2}\sec(i\mu\pi)[D_{i\mu-\frac{1}{2}}(\sigma\xi)D_{-i\mu-\frac{1}{2}}(\sigma\eta)$$

$$+ D_{i\mu-\frac{1}{2}}(-\sigma\xi)D_{-i\mu-\frac{1}{2}}(-\sigma\eta)]e^{i\omega z\cos\gamma},$$

$$\Psi^{(3)}_{\mu-,\gamma}(\xi,\eta,z) = \Psi^{(3)}_{\mu+,\gamma}(\xi,-\eta,z), \qquad \sigma = e^{i\pi/4}(2\omega\sin\gamma)^{1/2}, \tag{2.39}$$

$$x = (\xi^2-\eta^2)/2, \ y = \xi\eta, \ z = z.$$

4. Elliptic Cylindrical System

The eigenvalue equations are

$$iP_3 f^{(4)}_{nt,\gamma} = -\omega\cos(\gamma) f^{(4)}_{nt,\gamma}, \qquad (J_3^2+d^2P_1^2) f^{(4)}_{nt,\gamma} = \lambda_{nt} f^{(4)}_{nt,\gamma}, \qquad t=s,c, \quad (2.40)$$

and the basis of eigenfunctions is

$$f^{(4)}_{nc,\gamma}(\theta,\varphi) = (\pi\sin\gamma)^{-1/2}\mathrm{ce}_n(\varphi,q)\delta(\theta-\gamma), \quad n=0,1,2,\ldots,$$

$$f^{(4)}_{ns,\gamma}(\theta,\varphi) = (\pi\sin\gamma)^{-1/2}\mathrm{se}_n(\varphi,q)\delta(\theta-\gamma), \quad n=1,2,\ldots, \tag{2.41}$$

$$q = \frac{d^2\omega^2}{4}\sin^2\gamma, \qquad\qquad 0\leqslant\gamma\leqslant\pi.$$

The eigenvalues $\lambda_{n\pm}$ are discrete, of multiplicity one, and related to the eigenvalues a of the Mathieu equation (B.25) by $a=-\lambda-\frac{1}{2}d^2\omega^2\sin^2\gamma$. The $\{f^{(4)}_{nt,\gamma}\}$ form a basis for $L_2(S_2)$ satisfying

$$\langle f^{(4)}_{nt,\gamma}, f^{(4)}_{n't',\gamma'}\rangle = \delta_{nn'}\delta_{tt'}\delta(\gamma-\gamma'), \qquad t,t'=s,c. \tag{2.42}$$

ISBN-0-201-13503-5

The corresponding solutions of the Helmholtz equation are

$$\Psi^{(4)}_{nc,\gamma}(\mathbf{x}) = C_n(\sin\gamma)^{1/2} \mathrm{Ce}_n(\alpha,q)\, \mathrm{ce}_n(\beta,q)\exp[i\omega z \cos\gamma],$$

$$n = 0,1,2,\ldots,$$

$$\Psi^{(4)}_{ns,\gamma}(\mathbf{x}) = S_n(\sin\gamma)^{1/2} \mathrm{Se}_n(\alpha,q)\, \mathrm{se}_n(\beta,q)\exp[i\omega z \cos\gamma],$$

$$n = 1,2,\ldots,$$

(2.43)

where Ce_n and Se_n are modified Mathieu functions ((3.40), Section 1.3) and C_n, S_n are constants to be determined from the integral equations $\Psi^{(4)}_{nt,\gamma} = I(f^{(4)}_{nt,\gamma})$. The elliptic cylindrical coordinates α, β, z are defined by

$$x = d\cosh\alpha\cos\beta, \qquad y = d\sinh\alpha\sin\beta, \qquad z = z.$$

The spectral decompositions for systems 6–10 were first computed in [22], though 11 was studied earlier in [106]. The results are as follows.

6. Prolate Spheroidal System

The eigenfunction equations are

$$(\mathbf{J}\cdot\mathbf{J} - a^2 P_1^2 - a^2 P_2^2)f^{(6)}_{n,m} = -\lambda^m_n f^{(6)}_{n,m}, \qquad iJ_3 f^{(6)}_{n,m} = m f^{(6)}_{n,m}, \qquad (2.44)$$

and the ON basis of eigenfunctions is

$$f^{(6)}_{n,m}(\theta,\varphi) = \left[\frac{(n-|m|)!(2n+1)}{(n+|m|)!(4\pi)}\right]^{1/2} Ps_n^{|m|}(\cos\theta, a^2\omega^2)e^{im\varphi}. \qquad (2.45)$$

(The first eigenvalue equation (2.44) takes the form of the second equation (1.20).) Here $n = 0,1,2,\ldots$, $m = n, n-1, \ldots, -n$ and the discrete eigenvalues are denoted $\lambda^m_n(a^2\omega^2)$. We have $\langle f^{(6)}_{n,m}, f^{(6)}_{n',m'}\rangle = \delta_{nn'}\delta_{mm'}$ in the normalization adopted by Meixner and Schäfke [79]. The spheroidal wave functions are frequently defined by their expansions in terms of associated Legendre functions:

$$Ps_n^{|m|}(x, a^2\omega^2) = \sum_{2k \geqslant |m|-n} (-1)^k a^{|m|}_{n,2k}(a^2\omega^2)P^{|m|}_{n+2k}(x) \qquad (2.46)$$

(see [7, p. 169]). Indeed, substituting (2.46) into the spheroidal wave equation, one can easily derive a recurrence formula for the coefficients $a^m_{n,2k}$.

The corresponding basis of solutions for the Helmholtz equation is

$$\Psi^{(6)}_{n,m}(\mathbf{x}) = I(f^{(6)}_{n,m}) = C^m_n(a^2\omega^2)Ps_n^{|m|}(\cosh\eta, a^2\omega^2)Ps_n^{|m|}(\cos\alpha, a^2\omega^2)e^{im\varphi}$$

(2.47)

where $C_n^m(a^2\omega^2)$ is a constant to be determined from the integral equation. This result is easily obtained from the fact that $\Psi_{n,m}^{(6)}$ must be separable in the coordinates

$$x = a\sinh\eta\sin\alpha\cos\varphi, \qquad y = a\sinh\eta\sin\alpha\sin\varphi, \qquad z = a\cosh\eta\cos\alpha.$$

(See the corresponding argument for expression (3.38) in Section 1.3.)

7. Oblate Spheroidal System

The eigenvalue equations are

$$\left(\mathbf{J}\cdot\mathbf{J} + a^2 P_1^2 + a^2 P_2^2\right)f_{n,m}^{(7)} = -\lambda_n^{m}f_{n,m}^{(7)}, \qquad iJ_3 f_{n,m}^{(7)} = mf_{n,m}^{(7)}, \qquad (2.48)$$

and the ON basis of eigenfunctions is

$$f_{n,m}^{(7)}(\theta,\varphi) = \left[\frac{(n-|m|)!(2n+1)}{(n+|m|)!4\pi}\right]^{1/2} Ps_n^{|m|}(\cos\theta, -a^2\omega^2)e^{im\varphi}, \quad (2.49)$$

$$n = 0, 1, 2, \ldots, \quad m = n, n-1, \ldots, -n.$$

(Here the first eigenvalue equation (2.48) takes the form of the second equation (1.22).) The discrete eigenvalues are $\lambda_n^{|m|}(-a^2\omega^2)$.

The corresponding solutions of the Helmholtz equation are

$$\Psi_{n,m}^{(7)}(\mathbf{x}) = I\left(f_{n,m}^{(7)}\right)$$

$$= C_n^m(a^2\omega^2)Ps_n^{|m|}(-i\sinh\eta, a^2\omega^2)Ps_n^{|m|}(\cos\alpha, -a^2\omega^2)e^{im\varphi} \quad (2.50)$$

where $C_n^m(a^2\omega^2)$ is a constant to be determined from the integral and

$$x = a\cosh\eta\sin\alpha\cos\varphi, \qquad y = a\cosh\eta\sin\alpha\sin\varphi, \qquad z = a\sinh\eta\cos\alpha.$$

8. Parabolic System

The eigenvalue equations are

$$\left(\{J_1, P_2\} - \{J_2, P_1\}\right)f_{\lambda,m}^{(8)} = 2\lambda\omega f_{\lambda,m}^{(8)}, \qquad iJ_3 f_{\lambda,m}^{(8)} = mf_{\lambda,m}^{(8)}, \quad (2.51)$$

Here $\{J_1, P_2\} - \{J_2, P_1\} = 2i\omega(\cos\theta + \sin\theta\,\partial_\theta)$ is first order and has a unique self-adjoint extension. The eigenfunctions are

$$f_{\lambda,m}^{(8)}(\theta,\varphi) = (2\pi)^{-1}\frac{\left[\tan(\theta/2)\right]^{-i\lambda}}{\sin\theta}e^{im\varphi},$$

$$m = 0, \pm 1, \pm 2, \ldots, \quad -\infty < \lambda < \infty, \quad (2.52)$$

$$\langle f_{\lambda,m}^{(8)}, f_{\lambda',m'}^{(8)}\rangle = \delta(\lambda - \lambda')\delta_{mm'}.$$

ISBN-0-201-13503-5

The corresponding solutions of the Helmholtz equation are

$$\Psi^{(8)}_{\lambda,m}(\mathbf{x}) = I\left(f^{(8)}_{\lambda,m}\right)$$

$$= \frac{i^m\sqrt{2}}{\xi\eta\omega}\Gamma\left(\frac{1-m+i\lambda}{2}\right)\Gamma\left(\frac{1-m-i\lambda}{2}\right)\mathfrak{M}_{i\lambda/2,\,-m/2}$$

$$\times\left[\frac{\exp(-i\pi/2)\omega\xi^2}{\sqrt{2}}\right]\mathfrak{M}_{i\lambda/2,\,-m/2}\left[\frac{\exp(i\pi/2)\omega\eta^2}{\sqrt{2}}\right]\exp(im\varphi).$$

$$(2.53)$$

Here

$$\mathfrak{M}_{\alpha,\mu/2}(z) = \frac{z^{(1+\mu)/2}e^{-z/2}}{\Gamma(1+\mu)}\,{}_1F_1\left(\begin{array}{c}(1+\mu)/2-\alpha\\1+\mu\end{array}\bigg|z\right) \qquad (2.54)$$

is a Whittaker function [26, p. 12], and

$$x = \xi\eta\cos\varphi, \qquad y = \xi\eta\sin\varphi, \qquad z = (\xi^2-\eta^2)/2.$$

9. Paraboloidal System

The eigenvalue equations are

$$\begin{aligned}(J_3^2 - c^2P_3^2 + c\{J_2,P_1\} + c\{J_1,P_2\})f^{(9)}_{nt\lambda} &= -\mu_{n\pm}\,f^{(9)}_{nt\lambda},\\(cP_2^2 - cP_1^2 + \{J_2,P_1\} - \{J_1,P_2\})f^{(9)}_{nt\lambda} &= 2\omega\lambda f^{(9)}_{nt\lambda}\end{aligned} \qquad (2.55)$$

and the basis of eigenfunctions is

$$f^{(9)}_{nt\lambda}(\theta,\varphi) = (2\pi)^{-1/2}\frac{[\tan(\theta/2)]^{i\lambda}}{\sin\theta}\exp\left(-\frac{ic\omega}{2}\cos\theta\cos2\varphi\right)$$

$$\begin{cases}gc_n(\varphi;2c\omega,\lambda)\\gs_n(\varphi;2c\omega,\lambda)\end{cases}, \qquad t=c,s,\ n=0,1,2,\ldots,\ -\infty<\lambda<\infty, \qquad (2.56)$$

where gc_n and gs_n are the even and odd nonpolynomial solutions of the Whittaker–Hill equation. The normalization of these functions is that adopted by Urwin and Arscott [127]. We have

$$\langle f^{(9)}_{nt\lambda},f^{(9)}_{n't'\lambda'}\rangle = \delta_{nn'}\delta_{tt'}\delta(\lambda-\lambda').$$

The corresponding solutions of the Helmholtz equation are

$$\Psi^{(9)}_{nt\lambda}(\mathbf{x}) = K^t_n(\omega c,\lambda)\,gt_n(\beta;2c\omega,\lambda)\,gt_n(i\alpha;2c\omega,\lambda)$$

$$\times gt_n(i\gamma+\pi/2;2c\omega,\lambda), \qquad t=s,c, \qquad (2.57)$$

ISBN-0-201-13503-5

where the constants K_n^t are to be determined from the integral equation $\Psi_{nt\lambda}^{(9)} = I(f_{nt\lambda}^{(9)})$. Here,

$$x = 2c\cosh\alpha\cos\beta\sinh\gamma, \qquad y = 2c\sinh\alpha\sin\beta\cosh\gamma,$$
$$z = c(\cosh 2\alpha + \cos 2\beta - \cosh 2\gamma)/2.$$

10. Ellipsoidal System

We adopt elliptic coordinates on the unit sphere:

$$k_1 = \left[\frac{(s-a)(t-a)}{a(a-1)}\right]^{1/2}, \quad k_2 = \left[\frac{(s-1)(t-1)}{1-a}\right]^{1/2}, \quad k_3 = \left[\frac{st}{a}\right]^{1/2}, \quad (2.58)$$

$$0 < t < 1 < s < a.$$

Then the eigenvalue equations

$$Sf = \lambda f, \qquad S'f = \mu f, \qquad S = P_1^2 + aP_2^2 + (a+1)P_3^2 + \mathbf{J}\cdot\mathbf{J},$$
$$S' = J_2^2 + aJ_1^2 + aP_3^2, \tag{2.59}$$

become

$$\left[\frac{4}{s-t}(\partial_{\alpha\alpha} + \partial_{\beta\beta}) - \omega^2(s+t) - \omega^2(1+a)\right]f = \lambda f,$$

$$\left[\frac{4}{s-t}(t\partial_{\alpha\alpha} + s\partial_{\beta\beta}) - \omega^2 st\right]f = \mu f \tag{2.60}$$

where

$$\partial_\alpha = [(a-s)(s-1)s]^{1/2}\partial_s, \qquad \partial_\beta = [(t-a)(t-1)t]^{1/2}\partial_t.$$

We can find solutions of these equations in the form $f(s,t) = E_1(s)E_2(t)$ where

$$(4\partial_{\alpha\alpha} - \omega^2 s^2 + \lambda's + \mu)E_1(s) = 0,$$
$$(4\partial_{\beta\beta} + \omega^2 t^2 - \lambda't - \mu)E_2(t) = 0, \qquad \lambda' = -\omega^2(1+a) - \lambda. \tag{2.61}$$

These expressions are algebraic forms of the ellipsoidal wave equation (see (1.29)), so the E_j are ellipsoidal functions. Furthermore, if we set $s = \operatorname{sn}^2(\eta,k), t = \operatorname{sn}^2(\psi,k)$ where $k = a^{-1/2}$, then the separated equations take the Jacobian form

$$\left(\partial_{\xi\xi} - k^2\mu - k^2\lambda'\operatorname{sn}^2\xi + k^2\omega^2\operatorname{sn}^4\xi\right)E_j(\xi) = 0, \qquad \xi = \eta,\psi, j = 1,2, \tag{2.62}$$

of the ellipsoidal wave equation (1.33). The new coordinates η,ψ also have

ISBN-0-201-13503-5

the property that they allow parametrization of the entire sphere S_2 rather than just the first octant. Indeed

$$k_1 = k'^{-1} \mathrm{dn}(\eta,k)\,\mathrm{dn}(\psi,k), \qquad k_2 = ikk'^{-1}\mathrm{cn}(\eta,k)\,\mathrm{cn}(\psi,k),$$

$$k_3 = k\,\mathrm{sn}(\eta,k)\,\mathrm{sn}(\psi,k), \qquad k' = (1-k^2)^{1/2} \tag{2.63}$$

and these coordinates cover S_2 exactly once if η varies in the range $-2K < \eta < 2K$ and ψ varies in the range $K \leqslant \psi < K + 2iK'$ where $K = K(k)$ is defined by (C.3) and $K' = K(k')$.

Since k_1, k_2, k_3 remain unchanged when integral multiples of $4K$ and $4iK'$ are added to η or ψ, we are interested only in those single-valued solutions E_j of (2.62) which are also fixed under these substitutions: $E_j(\xi + 4Kn + 4iK'n) = E_j(\xi), n, m$ integers. As we noted in the preceding section, these doubly periodic functions are called the ellipsoidal wave functions. They have been studied in detail by Arscott [7]. The spectrum of S and S' is discrete, each pair of eigenvalues denoted $\lambda_{nm}\mu_{nm}$. The corresponding ellipsoidal wave functions are $\mathrm{el}_n^m(\xi), \xi = \eta, \psi$, and the eigenfunctions of S and S' are denoted

$$f_{nm}^{(10)}(\eta,\psi) = \mathrm{el}p_n^m(\eta,\psi) = \mathrm{el}_n^m(\eta)\,\mathrm{el}_n^m(\psi) \tag{2.64}$$

where $n = 0, 1, \ldots$ and the integer m runs over $2n+1$ values. We assume the basis $\{\mathrm{el}p_n^m\}$ is normalized to be ON:

$$\langle \mathrm{el}p_n^m, \mathrm{el}p_{n'}^{m'} \rangle = \delta_{nn'}\delta_{mm'}.$$

(This determines the solutions (2.64) only to within a factor of absolute value one. An essentially unique normalization is given in [7, p. 240]. Note also that $d\Omega(\hat{\mathbf{k}}) = ik^2(\mathrm{sn}^2\eta - \mathrm{sn}^2\psi)d\eta\,d\psi$.) In general these functions are rather intractable and very little is known about their explicit construction.

The corresponding solutions of the Helmholtz equation $\Psi_{nm}^{(10)}(\mathbf{x}) = I(f_{nm}^{(10)})$ are

$$\Psi_{nm}^{(10)}(\mathbf{x}) = \mathrm{El}_n^m(\alpha,\beta,\gamma) = K_n^m(\omega,k)\,\mathrm{el}_n^m(\alpha)\,\mathrm{el}_n^m(\beta)\,\mathrm{el}_n^m(\gamma) \tag{2.65}$$

where the constant K_n^m is to be evaluated from the integral. Moreover, this integral reads

$$\mathrm{El}_n^m(\alpha,\beta,\gamma) = \iint_{S_2} \exp\left[w\left(-\frac{1}{kk'^2}\,\mathrm{dn}\,\alpha\,\mathrm{dn}\,\beta\,\mathrm{dn}\,\gamma\,\mathrm{dn}\,\eta\,\mathrm{dn}\,\psi \right. \right.$$

$$+ \frac{k^2}{k'^2}\,\mathrm{cn}\,\alpha\,\mathrm{cn}\,\beta\,\mathrm{cn}\,\gamma\,\mathrm{cn}\,\eta\,\mathrm{cn}\,\psi$$

$$\left. \left. + ik^2\,\mathrm{sn}\,\alpha\,\mathrm{sn}\,\beta\,\mathrm{sn}\,\gamma\,\mathrm{sn}\,\eta\,\mathrm{sn}\,\psi \right) \right] \mathrm{el}p(\eta,\psi)\,d\Omega(\hat{\mathbf{k}}), \quad (2.66)$$

ISBN-0-201-13503-5

a nontrivial equation expressing the product of three ellipsoidal wave functions as an integral over a product of two such functions. Here the coordinates α, β, γ are related to x, y, z by expressions (1.32). We were able to evaluate the integral (2.66) to within a constant multiple because we knew in advance that it was separable in α, β, γ.

3.3 Lamé Polynomials and Functions on the Sphere

The eigenvalue problem corresponding to the conical coordinate system 11 (Table 14) is of special interest even though it is relatively intractable. Only for conical and spherical coordinates does the eigenvalue problem become finite dimensional; that is, only in these two cases is the problem reduced to finding the eigenvalues of an $n \times n$ matrix.

For functions f on the sphere S_2 the eigenvalue equations associated with system 11 in Table 14 are

$$\mathbf{J} \cdot \mathbf{J} f = -l(l+1)f, \qquad \left(J_1^2 + bJ_2^2\right)f = \lambda f, \qquad 1 > b > 0. \qquad (3.1)$$

It follows from (2.19) that

$$L_2(S_2) \cong \sum_{l=0}^{\infty} \oplus V_l$$

where $\dim V_l = 2l+1$ and V_l transforms irreducibly under the representation D_l of $SO(3)$. Thus $\mathbf{J} \cdot \mathbf{J}$ has the spectrum $-l(l+1), l=0, 1, 2, \ldots$, each eigenvalue occurring with multiplicity $2l+1$. Since $S = \mathbf{J} \cdot \mathbf{J}$ and $S' = J_1^2 + bJ_2^2$ commute, it follows that the subspaces V_l are invariant under the second operator. Thus we can reduce our search for eigenvalues of S' to the $(2l+1)$-dimensional space V_l. This space has an ON basis $\{f_m^{(l)}\}$, (2.12), and the restriction of S' to V_l can be represented by the $(2l+1) \times (2l+1)$ real symmetric matrix $\mathbb{S}'$ with respect to the basis $\{f_m^{(l)}\}$. The $2l+1$ eigenvalues of $\mathbb{S}'$ are the eigenvalues of S' in V_l.

There is another way to look at this problem. The elements h of V_l are characterized as the solutions of the partial differential equation $\mathbf{J} \cdot \mathbf{J} h = -l(l+1)h$, (2.20). It is straightforward to show that the symmetry algebra $so(3)$ of this equation is three dimensional (neglecting the identity symmetry E) with basis $\{J_1, J_2, J_3\}$, (2.6). The corresponding symmetry group is $SO(3)$. The space $\mathbb{S}^{(2)}/q$ of symmetric second-order symmetries modulo the multiples of $\mathbf{J} \cdot \mathbf{J}$ is five-dimensional with basis $J_2^2, J_3^2, \{J_1, J_2\}$, $\{J_1, J_3\}, \{J_2, J_3\}$. Under the adjoint action of $SO(3)$ this space is decomposed into two oribt types, one orbit with representative J_3^2 and one orbit type with representative $J_1^2 + bJ_2^2$, $1 > b > 0$. Moreover, it is known that the differential equation (2.20) for the Laplace operator on S_2 permits separation in exactly two coordinate systems [106]. One is the spherical coordi-

ISBN-0-201-13503-5

nate system $\{\theta,\varphi\}$ in which we have originally expressed (2.20). It corresponds to the diagonalization of J_3^2. The second is the elliptic coordinate system $\{s,t\}$, (2.58), which corresponds to the diagonalization of $J_1^2 + bJ_2^2$. The elliptic system was first studied from the group-theoretical point of view in [106] (see also [58]).

Whichever point of view is adopted, we need to compute the matrix $\mathcal{S}'$ of the operator

$$S' = J_1^2 + bJ_2^2 = \tfrac{1}{4}(b-1)\left((J^+)^2 + (J^-)^2\right) + \tfrac{1}{2}(b+1)\left((J^0)^2 - l(l+1)\right)$$

with respect to the basis $\{f_m^{(l)}\}$ and compute the $2l+1$ eigenvalues λ of this matrix. As is well known [69, p. 96], this problem is equivalent to computing the roots of the characteristic equation

$$\det(\mathcal{S}' - \lambda\mathcal{E}) = 0 \qquad (3.2)$$

where $\mathcal{E}$ is the $(2l+1)\times(2l+1)$ identity matrix. As shown in [106], for $l \leqslant 7$ one can explicitly find the eigenvalues λ as roots of polynomials of at most fourth order. However, for $l \geqslant 8$ the polynomials are of higher order than four and numerical methods must be used to approximate the roots.

We can use group theory to further aid in the classification of these eigenvalues. Note that both the Helmholtz equation (1.1) and the Laplace equation on the sphere (2.20) are invariant under the full rotation group $O(3)$. (This group is generated by $SO(3)$ and the space inversion operator P: $\mathbf{x} \to -\mathbf{x}$. A matrix realization is the group of all 3×3 real matrices A such that $AA' = E_3$. Here $\det A = \pm 1$ and $\det A = +1$ if and only if $A \in SO(3)$.) The elements of $O(3)$ that do not belong to $SO(3)$ (the rotation-inversions) are bounded away from the identity and are not obtainable by exponentiation of elements from the Lie algebra $so(3)$. The existence of these inversion symmetries must be verified by inspection.

In addition to P we shall be especially interested in the operators Z: $(x,y,z)\to(x,y,-z)$, reflection in the $x-y$ plane; X: $(x,y,z)\to(-x,y,z)$; reflection in the $y-z$ plane; and Y: $(x,y,z)\to(x,-y,z)$, reflection in the $x-z$ plane. Using (2.1) to transfer the action of these operators to the sphere, we find

$$Ph(\hat{\mathbf{k}}) = h(-\hat{\mathbf{k}}), \qquad Zh(\hat{\mathbf{k}}) = h(k_1,k_2,-k_3),$$
$$Xh(\hat{\mathbf{k}}) = h(-k_1,k_2,k_3), \quad Yh(\hat{\mathbf{k}}) = h(k_1,-k_2,k_3), \qquad h \in L_2(S_2). \qquad (3.3)$$

ISBN-0-201-13503-5

Obviously the square of each of the commuting operators P,Z,X,Y is the identity operator E and each operator is self-adjoint. Moreover, these operators each commute with $S' = J_1^2 + bJ_2^2$ and $S = \mathbf{J}\cdot\mathbf{J}$. It follows that there exists an ON basis for V_l consisting of simultaneous eigenvectors of P,Z,X,Y and S'.

The possible eigenvalues of $P,\dots,Y$ are ± 1. To determine the multiplicities of these eigenvalues in V_l, we apply the operators (3.3) to the explicit basis $\{f_m^{(l)}(\theta,\varphi) = Y_l^m(\theta,\varphi)\}$, (2.22). The results are

$$Pf_m^{(l)} = (-1)^l f_m^{(l)}, \qquad Zf_m^{(l)} = (-1)^{l-m} f_m^{(l)},$$

$$Xf_m^{(l)} = f_{-m}^{(l)}, \qquad Yf_m^{(l)} = (-1)^m f_{-m}^{(l)}. \tag{3.4}$$

Note that $P = (-1)^l E$ on V_l. To compute the multiplicities of the other eigenspaces we define eigenspaces

$$\mathcal{C}_l^{pq} = \{h \in V_l : Xh = ph, XYh = qh\}, \qquad p,q = \pm 1, \tag{3.5}$$

and set $n_l^{pq} = \dim \mathcal{C}_l^{pq}$. Since $Y = X(XY)$ and $Z = XYP$, we have

$$Ph = (-1)^l h, \qquad Zh = (-1)^l qh, \qquad Xh = ph, \qquad Yh = pqh, \tag{3.6}$$

for any $h \in \mathcal{C}_l^{pq}$. Furthermore,

$$V_l = \mathcal{C}_l^{++} \oplus \mathcal{C}_l^{+-} \oplus \mathcal{C}_l^{-+} \oplus \mathcal{C}_l^{--}.$$

Using (3.4) we can count the dimensions of these eigenspaces. The results are presented in Table 15.

Since each eigenspace is invariant under S', we can classify the eigenfunctions of S' by their symmetry properties with respect to X and XY. Thus an ON basis for V_l can be denoted $\{f_\lambda^{pq}\}$:

$$\mathbf{J} \cdot \mathbf{J} f_\lambda^{pq} = -l(l+1) f_\lambda^{pq}, \qquad (J_1^2 + bJ_2^2) f_\lambda^{pq} = \lambda f_\lambda^{pq},$$

$$Xf_\lambda^{pq} = p f_\lambda^{pq}, \qquad XY f_\lambda^{pq} = q f_\lambda^{pq}. \tag{3.7}$$

(It can be shown that there is no degeneracy; that is, there do not exist two linearly independent solutions of (3.7) for fixed l, p, q, λ.)

In terms of elliptic coordinates on the sphere, (2.63), the eigenvalue equations (3.1) separate to give the ordinary differential equations

$$E_j''(\xi) + (\lambda - l(l+1)k^2 \mathrm{sn}^2 \xi) E_j(\xi) = 0, \qquad j = 1,2, \; \xi = \eta, \psi, \; k = b^{1/2}, \tag{3.8}$$

where $f(\eta,\psi) = E_1(\eta) E_2(\psi)$. As mentioned in the discussion following expressions (1.35), equation (3.8) is the Lamé equation. It has $2l+1$ linearly

Table 15 Dimensions n_l^{pq} of the Eigenspaces $\mathcal{C}_l^{pq}$

	n_l^{++}	n_l^{+-}	n_l^{-+}	n_l^{--}
l even	$1 + l/2$	$l/2$	$l/2$	$l/2$
l odd	$(1+l)/2$	$(1+l)/2$	$(-1+l)/2$	$(1+l)/2$

ISBN-0-201-13503-5

independent solutions (the Lamé polynomials) that are single valued on S_2, each expressible in the form

$$\text{sn}^s \xi \, \text{cn}^c \xi \, \text{dn}^d \xi \, F_p(\text{sn}^2 \xi), \qquad s,c,d=0,1, \quad s+c+d+2p=l, \quad (3.9)$$

where $F_p(z)$ is a polynomial of order p in z. The eight types of such polynomials correspond to the eight categories listed in Table 15. Since each eigenspace has multiplicity one, the eigenfunctions must take the form $E(\eta)E(\psi)$ where $E(z)$ is a Lamé polynomial.

Rather than continue our analysis of the operator S' on $L_2(S_2)$, we shall study a simpler one-variable model for the spectral resolution of S'. We consider the $(2l+1)$-dimensional space W_l of polynomials $g(z)$ with order $\leqslant 2l$ in the complex variable z. We introduce a scalar product $(\cdot, \cdot)$ on W_l such that

$$(z^{l-m}, z^{l-n}) = (l-m)!(l+m)! \delta_{mn}, \qquad m,n = l, l-1, \ldots, -l, \quad (3.10)$$

or explicitly,

$$(2l+1)!(g_1, g_2) = \pi^{-1} \iint_{-\infty}^{\infty} dx \, dy \, (1+|z|^2)^{-2l-2} g_1(z) \bar{g}_2(z)$$

$$= \pi^{-1} \int_0^{\infty} r \, dr \int_0^{2\pi} d\varphi (1+r^2)^{-2l-2} g_1(re^{i\varphi}) \bar{g}_2(re^{i\varphi}) \quad (3.11)$$

for $g_j \in W_l$. Here $z = x + iy = re^{i\varphi}$ and the integration region is the complex plane. The operators J_1, J_2, J_3 defined by

$$J_1 = -\frac{i}{2}(1-z^2)\frac{d}{dz} - ilz, \qquad J_2 = \frac{1}{2}(1+z^2)\frac{d}{dz} - lz, \qquad J_3 = -iz\frac{d}{dz} + il,$$

$$(3.12)$$

leave W_l invariant and satisfy the commutation relations $[J_j, J_k] = \sum_p \epsilon_{jkp} J_p$ of $so(3)$. Moreover, $\mathbf{J} \cdot \mathbf{J} = -l(l+1)$ in this model. With each function $g \in W_l$ we associate a function $G \in V_l$, defined by

$$G(\hat{\mathbf{k}}) = (g, H(\hat{\mathbf{k}}, \cdot)) = I'(g), \qquad (3.13)$$

$$H(\hat{\mathbf{k}}, z) = (l!)^{-1}[(2l+1)/4\pi]^{1/2}[k_1(1-z^2)/2 + ik_2(1+z^2)/2 + k_3 z]^l.$$

Here, $\hat{\mathbf{k}} \in S_2$. The transformation I' from W_l to V_l is unitary. Indeed it follows from (3.10) that

$$g_m^l(z) = \frac{z^{l+m}}{[(l+m)!(l-m)!]^{1/2}}, \qquad m = l, l-1, \ldots, -l, \quad (3.14)$$

is an ON basis for W_l. Since

$$\bar{H}(\hat{\mathbf{k}},z) = \sum_{m=-l}^{l} Y_l^m(\theta,\psi)\,\bar{g}_m^l(z) \tag{3.15}$$

(see [128, p. 147] for a group-theoretic proof of this fact) for $\hat{\mathbf{k}} = (\sin\theta\cos\psi,\sin\theta\sin\psi,\cos\theta)$ we have

$$I'(g_m^l) = Y_l^m(\theta,\varphi) = f_m^l \tag{3.16}$$

where the spherical harmonics Y_l^m form an ON basis for V_l. From (2.12) and (2.22) we see that the operators (3.12) acting on W_l induce the operators (2.6) on V_l:

$$J_j G(\hat{\mathbf{k}}) = I'(J_j g(z)), \qquad j=1,2,3. \tag{3.17}$$

We will now study the eigenvalue problem for S' on W_l: $(J_1^2 + bJ_2^2)g(z) = \lambda g(z)$. We find

$$S' = \left[(1-k)z^2-(1+k)\right]\left[(1+k)z^2-(1-k)\right]\frac{d^2}{dz^2}$$
$$+2(2l-1)z\left[1+k^2-z^2(1-k^2)\right]\frac{d}{dz}$$
$$+2l\left[1+k^2+(1-k^2)(2l-1)z^2\right], \qquad k=b^{1/2}.$$

If we now write $g(z)=(k')^l[(\alpha-z^2)(1-\alpha z^2)]^{l/2}\mathcal{G}(w)$, where $k'=(1-k^2)^{1/2}$, $\alpha=(1+k)/(1-k)$, and make the change of variable

$$\mathrm{sn}(w,k) = -i(1+\alpha)z\left[(\alpha-z^2)(1-\alpha z^2)\right]^{-1/2}, \tag{3.18}$$

the eigenvalue equation reduces to

$$\left[\frac{d^2}{dw^2}+\lambda-k^2l(l+1)\,\mathrm{sn}^2(w,k)\right]\mathcal{G}(w)=0, \tag{3.19}$$

the Lamé equation.

It follows from (3.9) that the $2l+1$ Lamé polynomial solutions of this equation are exactly the solutions that correspond to elements $g(z)$ of W_l. Let us see how the classification of Lamé polynomials into eight types exhibits itself in our new model. From (3.4), (3.14), and (3.16) it follows that X and XY on W_l take the forms

$$Xg(z)=z^{2l}g(z^{-1}), \qquad XYg(z)=(-1)^l g(-z), \tag{3.20}$$

for $g\in W_l$.

ISBN-0-201-13503-5

Just as in our discussion of the eigenspaces $\mathcal{C}_l^{pq}$ of V_l ((3.5), (3.6)), we can require that the eigenfunctions $g(z) = (k')^l[(\alpha - z^2)(1 - \alpha z^2)]^{l/2}\mathcal{G}(w)$ also satisfy the equations $Xg = pg$, $XYg = qg$, $p, q = \pm 1$. Using these expressions, as well as (3.9) and Table 15, we obtain relations between the exponents a, b, c of (3.9) and the eigenvalues p, q as listed in Table 16. As shown in [7, Chapter 9], the Lamé polynomials in each symmetry class can be labeled by the integer $n = 0, 1, \ldots, n_l^{pq} - 1$ where n is the number of zeros of the polynomial in the interval $0 < w < K(k)$. Recurrence relations for the coefficients in the polynomial $F_\rho(\mathrm{sn}^2 w)$ can be obtained by substituting expression (3.9) into (3.19) and equating coefficients of independent monomials of elliptic functions $\mathrm{sn}^s w\, \mathrm{cn}^c w\, \mathrm{dn}^d w\, \mathrm{sn}^{2j} w$. One obtains polynomial solutions if and only if λ is one of the $2l + 1$ distinct eigenvalues λ_{ln}^{pq}.

Table 16 Symmetry Classes of Lamé Polynomials
$\mathrm{sn}^s w\, \mathrm{cn}^c w\, \mathrm{dn}^d w F\rho(\mathrm{sn}^2 w)$, $s, c, d = 0, 1$, $s + c + d + 2\rho = l$

	(p,q)	s	c	d	Dimension n_l^{pq}
l even	$+,+$	0	0	0	$1 + l/2$
	$+,-$	1	1	0	$l/2$
	$-,+$	0	1	1	$l/2$
	$-,-$	1	0	1	$l/2$
l odd	$+,+$	1	0	0	$(1+l)/2$
	$+,-$	0	1	0	$(1+l)/2$
	$-,+$	1	1	1	$(-1+l)/2$
	$-,-$	0	0	1	$(1+l)/2$

We have shown that these eigenvalues may be obtained in two different ways: either in the traditional manner through the search for polynomial solutions of the Lamé equation as described by Arscott, or by solution of the characteristic equation (3.2). In the second method, the matrix $\mathcal{S}'$ is explicitly determined with respect to the ON basis $\{f_m^{(l)}\}$. Thus, once an eigenvalue is computed, the corresponding eigenvector $f_{\lambda n}^{pq}$ can be directly obtained in terms of its expansion coefficients $a_{n,m}^{pq}$ in the $\{f_m^{(l)}\}$ basis:

$$f_n^{pq} = \sum_m a_{n,m}^{pq} f_m^{(l)}. \tag{3.21}$$

(In practice one obtains three-term recurrence formulas for these coefficients (see [106].) On the other hand, the traditional study of the Lamé equation leads to three-term recurrence formulas for the coefficients in the polynomial $F_\rho(\mathrm{sn}^2 w) = \sum_{j=0}^\rho b_j \mathrm{sn}^{2j} w$. The coefficients $a_{n,m}^{pq}$ are of special interest to us because they define the overlap function between the Lamé basis $\{f_n^{pq}\}$ and the canonical basis $\{f_m^{(l)}\}$. However, it is the coefficients b_k which are tabulated in the literature on Lamé polynomials.

The W_l model can be used to relate these coefficients. Let $\{\Lambda_n^{pql}(z)\}$ be the ON basis for W_l consisting of eigenfunctions of S' classified by symmetry type and number of zeros. Then (3.21) implies

$$\Lambda_n^{pql}(z)=\sum_m a_{n,m}^{pq} g_m^{(l)}=\sum_m a_{n,m}^{pq} z^{l+m}\left[(l+m)!(l-m)!\right]^{-1/2}; \quad (3.22)$$

that is, the overlaps are essentially the coefficients of z^{l+m}, $-l \leqslant m \leqslant l$. On the other hand

$$\Lambda_n^{pql}(z)=(k')^l\left[(\alpha-z^2)(1-\alpha z^2)\right]^{1/2}\operatorname{sn}^s w \operatorname{cn}^c w \operatorname{dn}^d w \sum_{j=0}^{\rho} b_j \operatorname{sn}^{2j}w \quad (3.23)$$

where w is related to z by expression (3.18). Expanding (3.23) as a polynomial in z and equating coefficients of z^{l+m} in (3.22), (3.23), we can express each coefficient $a_{n,m}^{pq}$ as a finite sum of coefficients b_j. Some of the details of the straightforward computation can be found in [58].

The transformation (3.13) can now be used to map our results to V_l. If $\{\Lambda_n^{pql}\}$ is the ON basis of eigenfunctions for S' on W_l, then

$$f_{ln}^{pq}=c_{ln}^{pq}E_{ln}^{p,q}(\eta)E_{ln}^{p,q}(\psi)=I'(\Lambda_n^{pql})=(\Lambda_n^{pql},H(\hat{\mathbf{k}},\cdot)) \quad (3.24)$$

(where η, ψ are elliptic coordinates on S_2, (2.63)) is an ON basis of eigenfunctions for S' on V_l. Here $E_{ln}^{p,q}(\xi)$ is a Lamé polynomial of the same eigenvalue and symmetry type as Λ_n^{pql}. The constant c is to be determined from the double integral once the explicit normalization of $E_{ln}^{p,q}$ and Λ_n^{pql} is fixed. (The integral in (3.24) can be evaluated because we know in advance that it satisfies the Lamé equation in η and ψ, and we can easily check that the integral is periodic in these variables.) Relation (3.21) can now be interpreted as an expansion of products of Lamé polynomials in terms of spherical harmonics.

The totality of all eigenfunctions (3.24) for $l=0,1,2,\ldots$ forms an ON basis for $L_2(S_2)$. Mapping this basis to the Hilbert space of solutions of the Helmholtz equation via the transformation (2.1), we find

$$\Psi_{ln}^{pq}(\mathbf{x})=I(f_{ln}^{pq})=d_n^{pql}j_l(\omega r)E_{ln}^{p,q}(\alpha)E_{ln}^{p,q}(\beta), \quad (3.25)$$

in terms of the conical coordinates (1.34). Here $j_l(z)$ is a spherical Bessel function, (2.27), and d is a constant that can be determined, in principle, from the integral.

Let us note that (3.24) and (3.25) can also be interpreted as nonlinear integral equations satisfied by Lamé polynomials. In this connection we remark that the evaluation of the integral (5.16) in [58] is in error. This integral should be replaced by (3.24).

The W_l model can also be used to study Ince polynomials (see [21]).

ISBN-0-201-13503-5

3.4 Expansion Formulas for Separable Solutions of the Helmholtz Equation

From the discussion in Chapters 1 and 2 it is evident that to expand a solution $T(g)\Psi_\lambda^{(j)}(x)$ of the Helmholtz equation in terms of the eigenfunctions $\{\Psi_\mu^{(i)}\}$ it is sufficient to compute the expansion coefficients $\langle T(g)f_\lambda^{(j)},f_\mu^{(i)}\rangle$ in the $L_2(S_2)$ model:

$$T(g)\Psi_\lambda^{(j)}(x) = \sum_\mu \langle T(g)f_\lambda^{(j)},f_\mu^{(i)}\rangle\Psi_\mu^{(i)}(x). \tag{4.1}$$

Here we list some of the more tractable expansion coefficients in the case where $T(g)$ is the identity operator.

The overlap functions $\langle f_\lambda^{(\ j)},f_{\alpha,\gamma}^{(1)}\rangle$ relating any system $\{f_\lambda^{(\ j)}(\hat{k})\}$ with the Cartesian system (2.32) are trivial:

$$\langle f_\lambda^{(j)},f_{\alpha,\gamma}^{(1)}\rangle = (\sin\gamma)^{1/2} f_\lambda^{(j)}(\sin\gamma\cos\alpha,\,\sin\gamma\sin\alpha,\,\cos\gamma). \tag{4.2}$$

Moreover, the overlaps relating the eigenfunctions for systems 1–4 in Table 14 can easily be obtained from the corresponding overlaps for solutions of the Helmholtz equation $(\Delta_2+\omega^2)\Psi=0$ listed in Section 1.3. Indeed, the overlaps take the form

$$\langle f_{\lambda,\gamma}^{(j)},f_{\mu,\gamma'}^{(i)}\rangle = \delta(\gamma-\gamma')\langle f_\lambda^{(j)'},f_\mu^{(i)'}\rangle, \qquad 1\leqslant i,j\leqslant 4, \tag{4.3}$$

where $\langle f_\lambda^{(\ j)'},f_\mu^{(i)'}\rangle$ is the corresponding overlap computed in Section 1.3 with the $L_2(S_1)$ model.

The overlaps $\langle f_m^{(l)},f_{\lambda,m'}^{(8)}\rangle$ between the spherical and parabolic bases were computed in [95]:

$$\langle f_m^{(l)},f_{\lambda,m'}^{(8)}\rangle = \delta_{mm'}\frac{(-1)^{(m+|m|)/2}}{(|m|!)^2}\left[\frac{(2l+1)(l+|m|)!}{4\pi(l-|m|)!}\right]^{1/2}$$

$$\times\Gamma\left(\frac{i\lambda+|m|+1}{2}\right)\Gamma\left(\frac{-i\lambda+|m|+1}{2}\right)$$

$$\times {}_3F_2\left(\begin{array}{c}|m|-l,\,|m|+l+1,\,(i\lambda+|m|+1)/2\\ |m|+1,\,|m|+1\end{array}\bigg|1\right), \tag{4.4}$$

$$m=0,\pm 1,\ldots,\pm l.$$

The overlaps between the spherical and prolate spheroidal bases are

$$\langle f_m^{(l)}, f_{n,m'}^{(6)} \rangle = \begin{cases} \delta_{mm'} (-1)^{(2m+l-n)/2} \left[\dfrac{(n-m)!(l+m)!(2n+1)}{(n+m)!(l-m)!(2l+1)} \right]^{1/2} a_{n,l-n}^m (a^2\omega^2), \\[4pt] \hspace{6cm} m' \geqslant 0, \\[8pt] \delta_{mm'} (-1)^{(l-n)/2} \left[\dfrac{(n+m)!(2n+1)}{(n-m)!(2l+1)} \right]^{1/2} a_{n,l-n}^{|m|} (a^2\omega^2), \\[4pt] \hspace{6cm} m' < 0, \end{cases}$$

$$(4.5)$$

where the coefficients $a_{n,2k}^{|n|}$ are defined by (2.46).

The overlaps between the cylindrical and prolate spheroidal bases are

$$\langle f_{n,m}^{(6)}, f_{m',\gamma}^{(2)} \rangle = \left[\frac{(n-|m|)!(2n+1)}{(n+|m|)!2} \sin \gamma \right]^{1/2} Ps_n^{|m|} (\cos \gamma, a\omega^2) \delta_{mm'} \quad (4.6)$$

and the overlaps between the parabolic cylindrical and prolate spheroidal bases are

$$\langle f_{n,m}^{(6)}, f_{\mu\pm,\gamma}^{(3)} \rangle = \left[\frac{(n-|m|)!(2n+1)}{(n+|m|)!2} \sin \gamma \right]^{1/2} Ps_n^{|m|} (\cos \gamma, a^2\omega^2) \langle f_m^{(2)}, f_{\mu\pm}^{(3)} \rangle$$

$$(4.7)$$

where the overlap $\langle f_m^{(2)}, f_{\mu\pm}^{(3)} \rangle$ is defined by (3.50), Section 1.3. The overlaps between the elliptic cylindrical and prolate spheroidal bases are

$$\langle f_{n,m}^{(6)}, f_{n'\rho,\gamma}^{(4)} \rangle = \left[\frac{(n-|m|)!(2n+1)}{(n+|m|)!} \sin \gamma \right]^{1/2} Ps_n^{|m|} (\cos \gamma, a^2\omega^2) A_{n'}^m \quad (4.8)$$

where the Fourier coefficient A_n^m is defined in terms of the Mathieu functions $pe_n(\varphi, q)$, $p = s, c$, by

$$pe_n(\varphi, q) = \sum_{m=-\infty}^{\infty} A_n^m e^{im\varphi}. \tag{4.9}$$

The corresponding overlaps for oblate spheroidal coordinates can be obtained from the prolate overlaps (4.5)–(4.8) by making the replacement $a^2\omega^2 \to -a^2\omega^2$ in the spheroidal wave functions.

Arscott [7, p. 247] shows how to compute the overlaps between the conical basis 11 and the ellipsoidal basis, $\langle f_{nm}^{(10)}, f_{lm}^{pq} \rangle$, by exhibiting a three-term recurrence relation obeyed by the overlap function.

ISBN-0-201-13503-5

The remaining overlaps are more complicated than those we have listed.

It is easy to construct a bilinear generating function for all basis sets of solutions of the Helmholtz equation listed here. Let $\{f_{\lambda\mu}(\hat{\mathbf{k}})\}$ be one of the eleven bases for $L_2(S_2)$ constructed earlier and let $\{\Psi_{\lambda\mu}(\mathbf{x})\}$ be the corresponding basis for the solution space of $(\Delta_3 + \omega^2)\Psi(\mathbf{x}) = 0$. Then

$$\Psi_{\lambda\mu}(\mathbf{x}) = I(f_{\lambda\mu}) = \langle f_{\lambda\mu}, H(\mathbf{x}, \cdot)\rangle$$

where $H(\mathbf{x}, \hat{\mathbf{k}}) = \exp[-i\omega\mathbf{x}\cdot\hat{\mathbf{k}}] \in L_2(S_2)$ for each $\mathbf{x} \in R^3$. An explicit computation yields

$$\langle H(\mathbf{x}, \cdot), H(\mathbf{x}', \cdot)\rangle = 4\pi[\sin(\omega R)/\omega R], \qquad R^2 = (\mathbf{x} - \mathbf{x}')\cdot(\mathbf{x} - \mathbf{x}'). \quad (4.10)$$

On the other hand

$$\langle H(\mathbf{x}, \cdot), H(\mathbf{x}', \cdot)\rangle = \sum_{\lambda,\mu} \langle H(\mathbf{x}, \cdot), f_{\lambda\mu}\rangle\langle f_{\lambda\mu}, H(\mathbf{x}', \cdot)\rangle$$

$$= \sum_{\lambda,\mu} \overline{\Psi}_{\lambda\mu}(\mathbf{x})\Psi_{\lambda\mu}(\mathbf{x}'), \quad (4.11)$$

and comparison of (4.10) and (4.11) shows that $4\pi\sin(\omega R)/\omega R$ is a bilinear generating function for each of our bases.

Finally, as shown in [128] and [95], each of our eleven bases $\{\Psi_{\lambda\mu}\}$ considered as functions of ω, $0 < \omega < \infty$, can be used to expand arbitrary functions $f(\mathbf{x})$ on R_3, square integrable with respect to Lebesgue measure.

3.5 Non-Hilbert Space Models for Solutions of the Helmholtz Equation

There are obviously many physically and mathematically interesting solutions of the Helmholtz equation that are not representable in the form $I(h)$, (2.1), for $h \in L_2(S_2)$. We shall investigate a few group-theoretic methods for obtaining such solutions and relating different types of separable non-Hilbert space solutions. These methods are considerably less elegant but more flexible than the techniques discussed earlier. Furthermore, they can be applied to the differential equations treated in Chapters 1 and 2.

We begin by considering transforms $I(h)$, (2.1), where the domain of integration is a complex two-dimensional Riemann surface rather than the real sphere S_2. In particular we set

$$\hat{\mathbf{k}} = (k_1, k_2, k_3) = \left(-\tfrac{1}{2}(t + t^{-1})(1 + \beta^2)^{1/2}, \tfrac{i}{2}(t - t^{-1})(1 + \beta^2)^{1/2}, i\beta\right) \quad (5.1)$$

ISBN-0-201-13503-5

where t and β range over complex values, and write

$$\Psi(\mathbf{x}) = \iint_S d\beta \, \frac{dt}{t} \, h(\beta, t) \exp\left[-\frac{i\omega}{2}(1+\beta^2)^{1/2} \right.$$

$$\left. \times \left\{ x(t+t^{-1}) + iy(t^{-1}-t) \right\} - \omega\beta z \right] = I(h). \qquad (5.2)$$

We assume that the integration surface S and the analytic function h are such that $I(h)$ converges absolutely and arbitrary differentiation with respect to x, y, and z is permitted under the integral sign. Since $k_1^2 + k_2^2 + k_3^2 = 1$ even for arbitrary complex β and t, $t \neq 0$, it follows that $\Psi(\mathbf{x})$ is a solution of the Helmholtz equation

$$(\Delta_3 + \omega^2)\Psi(\mathbf{x}) = 0. \qquad (5.3)$$

Integrating by parts, we find that the operators P_j, J_j, (1.2), acting on the solution space of (5.3) correspond to the operators

$$J^{\pm} = it^{\pm 1}\left(\mp(1+\beta^2)^{1/2}\partial_\beta + \frac{\beta t}{(1+\beta^2)^{1/2}}\partial_t \right), \qquad J^0 = t\partial_t,$$

$$P^{\pm} = \omega(1+\beta^2)^{1/2}t^{\pm 1}, \qquad P^0 = -i\omega\beta, \qquad (5.4)$$

acting on analytic functions $h(\beta, t)$ provided S and h are chosen such that the boundary terms vanish:

$$J^{\pm}\Psi = I(J^{\pm}h), \qquad P^{\pm}\Psi = I(P^{\pm}h),$$

and so on. Here as usual $J^{\pm} = \mp J_2 + iJ_1$, $J^0 = iJ_3$, $P^{\pm} = \mp P_2 + iP_1$, $P^0 = iP_3$.

For our first example we set $h = (2\pi^3)^{-1/2}$ and integrate over the contours C_1 and C_2 in the β and t planes, respectively (Figure 1).

In this case h satisfies the equations $\mathbf{J} \cdot \mathbf{J}h = 0$, $J^0 h = 0$ and it is straightforward to verify that $\Psi(\mathbf{x}) = I(h)$ satisfies the same equations for $z > 0$. Thus, $\Psi(\mathbf{x})$ is independent of the spherical coordinates θ, φ and is a linear combination of the Bessel functions $\rho^{-1/2}J_{1/2}(\omega\rho)$ and $\rho^{-1/2}J_{-1/2}(\omega\rho)$, (1.19). To determine the correct linear combination we evaluate (5.2) in the special case where $x = y = 0$. Then the integral becomes elementary and we find

$$\Psi(0,0,z) = (i/\omega z)(2/\pi)^{1/2}e^{i\omega z}, \qquad z > 0.$$

Thus we obtain

$$\Psi(\mathbf{x}) = -(\omega\rho)^{-1/2}H_{1/2}^{(1)}(\omega\rho) \qquad (5.5)$$

ISBN-0-201-13503-5

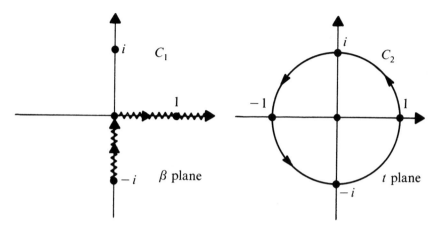

Figure 1

where $H_{n+\frac{1}{2}}^{(j)}(z)$ are Hankel functions of the first $(j=1)$ and the second $(j=2)$ kind:

$$H_{\nu}^{(1)}(z)=(i\sin\pi\nu)^{-1}\left[J_{-\nu}(z)-J_{\nu}(z)e^{-i\pi\nu}\right],$$

$$H_{\nu}^{(2)}(z)=(i\sin\pi\nu)^{-1}\left[J_{\nu}(z)e^{i\pi\nu}-J_{-\nu}(z)\right], \qquad (5.6)$$

$$H_{n+\frac{1}{2}}^{(1,2)}(z)=\mp i(-1)^{n}\left(\frac{\pi z}{2}\right)^{-1/2}z^{n+1}\left(\frac{d}{z\,dz}\right)^{n}\frac{e^{\pm iz}}{z}, \qquad n=0,1,2,\ldots.$$

The solution (5.5) is a *(traveling) spherical wave.*
More generally, we set

$$h=f_{m}^{(l)}(\beta,t)=\left[\frac{(2l+1)(l-m)!}{4\pi(l+m)!}\right]^{1/2}P_{l}^{m}(i\beta)(-t)^{m},$$

$$l=0,1,\ldots,m=l,l-1,\ldots,-l, \qquad (5.7)$$

where $P_{l}^{m}(z)$ is an associated Legendre function. (This expression makes sense for all $\beta\in C_{1}$ since, from (B.6iv), for $m\geqslant 0$ $P_{l}^{-m}(z)$ is a polynomial in $z=i\beta$ times the factor $[(i\beta-1)/(i\beta+1)]^{m/2}$ which remains bounded on C_{1} and vanishes at $\beta=-i$. Moreover, $P_{l}^{-m}(z)=(-1)^{m}(l-m)!P_{l}^{m}(z)/(l+m)!$. It follows from (2.17), (2.22), and (2.24) that the operators (5.4) acting on the functions $\{f_{m}^{(l)}(\beta,t)\}$ satisfy the recurrence relations (2.12) and (2.23). Thus, the solutions $\Psi_{m}^{(l)}(\mathbf{x})=I(f_{m}^{(l)})$ of the Helmholtz equation also satisfy these relations.

We have already computed the spherical wave $\Psi_{0}^{(0)}(\mathbf{x})$, (5.5). Using the fact that both the functions (2.28) and (2.30), hence a fixed linear combination of these functions, satisfy recurrence relations (2.12), (2.23), we can conclude from (5.5) and (5.6) that

$$\Psi_m^{(l)}(\rho,\theta,\varphi) = -i^l(\omega\rho)^{-1/2}H_{l+\frac{1}{2}}^{(1)}(\omega\rho)Y_l^m(\theta,\varphi). \qquad (5.8)$$

Next we consider the cylindrical system corresponding to the operators (5.4),

$$P^0 f_{m,\gamma}^{(2)} = -i\omega\gamma f_{m,\gamma}^{(2)}, \qquad J^0 f_{m,\gamma}^{(2)} = m f_{m,\gamma}^{(2)}, \qquad f_{m,\gamma}^{(2)}(\beta,t) = t^m\delta(\beta-\gamma). \quad (5.9)$$

Using the integration contours C_1, C_2, we easily find

$$\Psi_{m,\gamma}^{(2)}(r,\theta,z) = i^{m+1}(-1)^m(2\pi)J_m\left(\omega(1+\gamma^2)^{1/2}r\right)e^{im\theta-\omega\gamma z} \qquad (5.10)$$

for $\gamma \in C_1$. Here $\{r,\theta,z\}$ are cylindrical coordinates (2.36).

From (5.7), (5.9) and the corresponding integral representations $\Psi = I(f)$ there follows easily the expansion

$$\Psi_m^{(l)}(\mathbf{x}) = \left[\frac{(2l+1)(l-m)!}{4\pi(l+m)!}\right]^{1/2}(-1)^m\int_{C_1} P_l^m(i\beta)\Psi_{m,\beta}^{(2)}(\mathbf{x})\,d\beta, \qquad (5.11)$$

$$z > 0.$$

More generally, if $\Psi_m^{(l)}$ is subjected to a translation $\mathbf{T}(g) = \exp(a_1 P_1 + a_2 P_2 + a_3 P_3)$, we obtain the expansion formula

$$\mathbf{T}(g)\Psi_m^{(l)}(\mathbf{x}) = \left[\frac{(2l+1)(l-m)!}{4\pi(l+m)!}\right]^{1/2}\sum_{n=-\infty}^{\infty}\int_{C_1}(-1)^{n+m}(ie^{-i\alpha})^n$$

$$\times P_l^m(i\beta)J_n[\omega a(1+\beta^2)^{1/2}]\exp(-a_3\omega\beta)\Psi_{m+n,\beta}^{(2)}(\mathbf{x})\,d\beta, \qquad (5.12)$$

$$z + a_3 > 0, \ a_1 + ia_2 = ae^{i\alpha}, \ a > 0.$$

Similar techniques can be used to expand traveling spherical waves in other bases. In each case one derives the expansion for the complex sphere model and then attempts to map the results to the solution space of the Helmholtz equation via the transformation (5.2). The procedure is no longer so straightforward as for our Hilbert space models, and special techniques may have to be developed for each example. Some important cases are worked out (by another method) in [26, Section 16].

We can obtain other expansions by varying the integration contours in (5.2). For example, consider the contour C_1' in the β plane as drawn in Figure 2. We retain the contour C_2 in the t plane as drawn in Figure 1. It is easily verified that the J and P operators on β–t space and on the solution space of the Helmholtz equation correspond, under the mapping (5.2) induced by this choice of contours.

ISBN-0-201-13503-5

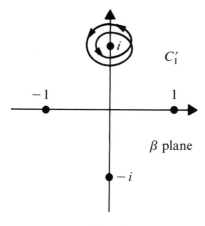

Figure 2

Now consider the eigenvalue equations for the parabolic system 8 (Table 14) in β–t space:

$$(\{J_1, P_2\} - \{J_2, P_1\})f^{(8)}_{\lambda,m} = -2i\lambda\omega f^{(8)}_{\lambda,m}, \qquad iJ_3 f^{(8)}_{\lambda,m} = mf^{(8)}_{\lambda,m}.$$

It is straightforward to show that the eigenfunctions are

$$f^{(8)}_{\lambda,m}(\beta, t) = (1+\beta^2)^{-1/2}[(1+i\beta)/(1-i\beta)]^{\lambda/2}t^m, \qquad \lambda, m \in \mathbb{C}. \quad (5.13)$$

For convenience we restrict ourselves to the case where λ and m are integers. Then, substituting (5.13) into (5.2) for the contours C_1', C_2 and integrating, we find

$$\Psi^{(8)}_{\lambda m}(\mathbf{x}) = I(f^{(8)}_{\lambda,m})$$

$$= \frac{8\pi^2(i)^{|m|}(-1)^k k!}{(|m|+k)!}(i\omega\xi^2)^{|m|/2}(-i\omega\eta^2)^{|m|/2}$$

$$\times \exp\left[i\omega\frac{(\eta^2-\xi^2)}{2}\right]L_k^{(|m|)}(i\omega\xi^2)L_k^{(|m|)}(-i\omega\eta^2)e^{im\varphi}$$

if

$$\lambda = -|m| - 2k - 1, \ k = 0, 1, 2, \ldots, \ m = 0, \pm 1, \pm 2, \ldots,$$

$$\Psi^{(8)}_{\lambda m}(\mathbf{x}) = 0 \qquad \text{otherwise.} \qquad (5.14)$$

Here ξ, η, φ are parabolic coordinates

$$x = \xi\eta\cos\varphi, \qquad y = \xi\eta\sin\varphi, \qquad z = (\xi^2 - \eta^2)/2.$$

(See [95] for the details of this computation.) Note that some nonzero

ISBN-0-201-13503-5

functions $f^{(8)}_{\lambda,m}$ are mapped to zero by the transformation I. Here the $L^{(\alpha)}_n(z)$ are generalized Laguerre polynomials.

We can use our model to compute the matrix elements of the operators $T(g)$ with respect to this basis. For example, the operator $T(a)=\exp(aP_3)$ acts on the $\{f^{(8)}_{\lambda,m}\}$ basis to yield

$$T(a)f^{(8)}_{\lambda,m}(\beta,t)=e^{-a\omega\beta}(1+\beta^2)^{-1/2}\left[(1+i\beta)/(1-i\beta)\right]^{\lambda/2}t^m$$

$$=\sum_{s=0}^{\infty}e^{-i\omega a}(-1)^sL_s^{(-1)}(2ia\omega)f^{(8)}_{\lambda+2s,m}(\beta,t).$$

This result is obtained from the generating function (4.11), Section 2.4. It is not hard to show that this identity is mapped by the transformation I to the identity

$$T(a)\Psi^{(8)}_{\lambda m}(\mathbf{x})=\sum_{s=0}^{\infty}e^{-i\omega a}(-1)^sL_s^{(-1)}(2ia\omega)\Psi^{(8)}_{\lambda+2s,m}(\mathbf{x}).\qquad(5.15)$$

(Note that the sum is actually finite.) Details of the computation as well as general $E(3)$ matrix elements with respect to the parabolic basis can be found in [95]. The first (nongroup-theoretic) proof of these expansion formulas was given by Hochstadt [50].

Next we consider identities for solutions of the Helmholtz equation which are derivable by Weisner's method. The natural setting for application of this method is the complex Helmholtz equation obtained by allowing all variables in equation (1.1) to assume complex values. To treat this equation systematically we should determine all complex analytic coordinate systems in which variables separate. Here, however, we will consider only a few separable systems that are of particular importance.

Of greatest practical importance is the spherical system

$$\mathbf{J}\cdot\mathbf{J}\Psi=-l(l+1)\Psi,\qquad J^0\Psi=m\Psi.\qquad(5.16)$$

We will now study solutions Ψ of the complex Helmholtz equation that satisfy (5.16) in those cases where l and m are complex numbers, not necessarily integers. To treat the group-theoretic properties of these solutions it is convenient first to analyze the corresponding eigenfunctions in our complex sphere model. Thus we begin with the operators (5.4). In terms of the new complex variables τ, ρ where

$$\tau=t(1+\beta^2)^{1/2},\qquad\rho=-i\beta,\qquad(5.17)$$

these operators assume the form

$$J^+=-\tau\partial_\rho,\quad J^-=\tau^{-1}((1-\rho^2)\partial_\rho-2\rho\tau\partial_\tau),\quad J^0=\tau\partial_\tau,$$
$$P^+=\omega\tau,\qquad P^-=\omega(1-\rho^2)\tau^{-1},\qquad\qquad P^0=\omega\rho.\qquad(5.18)$$

ISBN-0-201-13503-5

It follows easily from these expressions that the solution $f_l^{(l)}$ of the equations

$$J^0 f_l^{(l)} = l f_l^{(l)}, \qquad J^+ f_l^{(l)} = 0 \qquad (5.19)$$

is

$$f_l^{(l)}(\rho,\tau) = \Gamma\left(l + \tfrac{1}{2}\right)(2\tau)^l,$$

unique to within a multiplicative constant. (The factor $\Gamma(l + \tfrac{1}{2})2^l$ is inserted for convenience in the computations to follow. Here l is an arbitrary complex constant except that we assume $l + \tfrac{1}{2}$ is not an integer. It follows from (5.19) that $\Psi = f_l^{(l)}$ satisfies (5.16) for $m = l$. To obtain more solutions we consider the expansion

$$\exp(\alpha J^-) f_l^{(l)} = \sum_{n=0}^{\infty} (\alpha^n / n!)(J^-)^n f_l^{(l)}. \qquad (5.20)$$

Setting $f_{l-n}^{(l)} = [(-1)^n \Gamma(2l - n + 1)/\Gamma(2l + 1)](J^-)^n f_l^{(l)}$, $n = 0, 1, 2, \ldots$, we see that the commutation relations for the J operators imply

$$J^0 f_m^{(l)} = m f_m^{(l)}, \qquad J^+ f_m^{(l)} = (m - l) f_{m+1}^{(l)}, \qquad J^- f_m^{(l)} = -(m + l) f_{m-1}^{(l)},$$

$$\mathbf{J} \cdot \mathbf{J} f_m^{(l)} = -l(l + 1) f_m^{(l)}, \qquad m = l, l - 1, l - 2, \ldots . \qquad (5.21)$$

Lie theory arguments applied to the left-hand side of (5.20) yield the generating function

$$\Gamma\left(l + \tfrac{1}{2}\right)\left[2\tau - 4\alpha\rho - 2\alpha^2(1 - \rho^2)/\tau\right]^l = \sum_{n=0}^{\infty} (-\alpha)^n \binom{2l}{n} j_n^{(l)}(\rho)\tau^{l-n},$$

$$f_m^{(l)}(\rho,\tau) = j_n^{(l)}(\rho)\tau^{l-n}, \qquad m = l - n, \qquad (5.22)$$

valid for $\tau \neq 0$ and α in a sufficiently small neighborhood of 0. Comparing coefficients of α^n on both sides of this equation, we find

$$f_m^{(l)}(\rho,\tau) = \Gamma(l - m + 1)\Gamma\left(m + \tfrac{1}{2}\right)C_{l-m}^{m+\frac{1}{2}}(\rho)(2\tau)^m \qquad (5.23)$$

where $C_n^\nu(x)$ is a Gegenbauer (ultraspherical) polynomial (B.6ii). This polynomial is commonly defined by the generating function

$$(1 - 2\alpha x + \alpha^2)^{-\nu} = \sum_{n=0}^{\infty} C_n^\nu(x)\alpha^n, \qquad (5.24)$$

ISBN-0-201-13503-5

whose group-theoretic significance will be explained in Section 3.7. For l a positive integer and $m = l, l-1, \ldots, -l$, the functions (5.23) are proportional to complexifications of the spherical harmonics Y_l^m. However, we shall be primarily interested in the case where $2l$ is not an integer.

From the recurrence relation for the Gegenbauer polynomials,

$$x C_n^\nu(x) = \frac{n+1}{2(\nu+n)} C_{n+1}^\nu(x) + \frac{(2\nu+n-1)}{2(\nu+n)} C_{n-1}^\nu(x),$$

which can be verified directly from either (5.24) or (B.6ii), it follows that

$$P^0 f_m^{(l)} = \frac{\omega}{2l+1} f_m^{(l+1)} + \frac{\omega(l+m)(l-m)}{2l+1} f_m^{(l-1)}. \tag{5.25}$$

Furthermore, from the commutation relations $[P^0, J^\pm] = \pm P^\pm$ it follows that

$$P^+ f_m^{(l)} = \frac{\omega}{2l+1} f_{m+1}^{(l+1)} - \frac{\omega(l-m)(l-m-1)}{2l+1} f_{m+1}^{(l-1)},$$

$$P^- f_m^{(l)} = \frac{-\omega}{2l+1} f_{m-1}^{(l+1)} + \frac{\omega(l+m)(l+m-1)}{2l+1} f_{m-1}^{(l-1)}. \tag{5.26}$$

Relations (5.21), (5.25), (5.26) determine the action of $\mathcal{E}(3)$ on the basis $\{f_m^{(l)}\}$ where $l = l_0, l_0 \pm 1, l_0 \pm 2, \ldots, m = l, l-1, \ldots,$ and $2l_0$ is not an integer. (As is well known, the simple form of (5.25) is related to the fact that the Gegenbauer polynomials are orthogonal with respect to a suitable measure [37, Chapter X]. This property of these polynomials, along with many others, is related to the wave equation and will be studied in the next chapter.)

It is well known that any entire function of x can be expanded uniquely in a series of Gegenbauer polynomials $C_n^\nu(x)$, $n = 0, 1, 2, \ldots$ ($2\nu \neq$ integer), uniformly convergent in compact subsets of the complex plane (e.g., [116, p. 238]). Thus we can exponentiate the P and J operators, and compute the matrix elements of these operators in an $\{f_m^{(l)}\}$ basis. The rather complicated results are presented in [83]. Except for (5.20), (5.22), we present here only one of these results: An induction argument based on the operator relation (5.25) shows that

$$e^{\alpha \rho} = (2/\alpha)^\nu \Gamma(\nu) \sum_{n=0}^\infty (\nu+n) I_{\nu+n}(\alpha) C_n^\nu(\rho), \qquad \nu, \alpha \in \mathcal{C}; \tag{5.27}$$

ISBN-0-201-13503-5

that is,

$$\exp(\alpha P^0)f_l^{(l)} = \left(\frac{2}{\alpha}\right)^{l+\frac{1}{2}}\Gamma\left(l+\frac{1}{2}\right)\sum_{n=0}^{\infty}\frac{\left(l+n+\frac{1}{2}\right)}{n!}I_{l+n+\frac{1}{2}}(\alpha)f_l^{(l+n)}. \quad (5.27')$$

Here, $I_\nu(\alpha) = \exp(-i\nu\pi/2)J_\nu[\alpha\exp(i\pi/2)]$ is a modified Bessel function [37].

Now we consider the relationship between these results and solutions of the complex Helmholtz equation in the spherical basis. Instead of the complex spherical coordinates r, θ, φ (5 in Table 14), it is more convenient to use the equivalent separable coordinates

$$\rho = -\cos\theta, \qquad \tau = -e^{i\varphi}\sin\theta, \qquad s = ir. \quad (5.28)$$

In terms of these coordinates the symmetry operators for the Helmholtz equation are

$$J^+ = -\tau\partial_\rho, \qquad J^- = \tau^{-1}\left((1-\rho^2)\partial_\rho - 2\rho\tau\partial_\tau\right), \qquad J^0 = \tau\partial_\tau,$$

$$P^+ = \tau\partial_s - \frac{\rho\tau}{s}\partial_\rho - \frac{\tau^2}{s}\partial_\tau,$$

$$P^- = \frac{(1-\rho^2)}{\tau}\partial_s - \frac{\rho(1-\rho^2)}{s\tau}\partial_\rho + \frac{(\rho^2+1)}{s}\partial_\tau,$$

$$P^0 = \rho\partial_s + \frac{(1-\rho^2)}{s}\partial_\rho - \frac{\rho\tau}{s}\partial_\tau. \quad (5.29)$$

We search for a set of solutions $\{\Psi_m^{(l)}(\mathbf{x})\}$ of the Helmholtz equation which satisfy the recurrence relations (5.21), (5.25), (5.26) when acted on by the symmetry operators (5.29). Since the J operators in (5.18) and (5.29) are identical, it follows that

$$\Psi_m^{(l)}(\mathbf{x}) = S^{(l)}(s)f_m^{(l)}(\rho, \tau).$$

Substituting this expression into (5.25) and (5.26), we find that the $S^{(l)}$ must satisfy the recurrence formulas

$$\left(\frac{d}{ds} - \frac{l}{s}\right)S^{(l)}(s) = \omega S^{(l+1)}(s), \qquad \left(\frac{d}{ds} + \frac{l+1}{s}\right)S^{(l)}(s) = \omega S^{(l-1)}(s). \quad (5.30)$$

It follows that $s^{1/2}S^{(l)}(s)$ is a solution of the modified Bessel equation and

ISBN-0-201-13503-5

that the choices

$$S^{(l)}(s) = (\omega s)^{-1/2} I_{l+\frac{1}{2}}(\omega s) \quad \text{or} \quad (\omega s)^{-1/2} I_{-l-\frac{1}{2}}(\omega s) \qquad (5.31)$$

separately satisfy the recurrence formulas. Adopting the first of these choices, we conclude that the functions

$$\Psi_m^{(l)}(s,\rho,\tau) = (l-m)! \Gamma\left(m+\tfrac{1}{2}\right)(\omega s)^{-1/2} I_{l+\frac{1}{2}}(\omega s) C_{l-m}^{m+\frac{1}{2}}(\rho)(2\tau)^m \qquad (5.32)$$

and the operators (5.29) satisfy the recurrence formulas (5.21), (5.25), (5.26). Thus the matrix elements giving the $E(3)$ group action that were computed for the $\{f_m^{(l)}\}$ basis are also valid for the $\{\Psi_m^{(l)}\}$ basis. For example, (5.27) leads to the addition theorem of Gegenbauer:

$$I_{l+\frac{1}{2}}(sS)(2S)^{-l-\frac{1}{2}} = \Gamma\left(l+\tfrac{1}{2}\right) \sum_{n=0}^{\infty} \left(l+n+\tfrac{1}{2}\right) I_{l+n+\frac{1}{2}}(s) I_{l+n+\frac{1}{2}}(\gamma) C_n^{l+\frac{1}{2}}(\rho),$$

$$S = (1+2\gamma\rho/s + \gamma^2/s^2)^{1/2}, \qquad |2\gamma\rho/s + \gamma^2/s^2| < 1.$$

$$(5.33)$$

We can also use the complex sphere model to prove operational identities relating solutions of the Helmholtz equation. For example, from (5.18), (5.23) we obtain the virtually trivial identity

$$(l-m)! C_{l-m}^{m+\frac{1}{2}}(\omega^{-1} P^0) f_m^{(m)} = f_m^{(l)}, \qquad l-m = 0,1,2,\dots. \qquad (5.34)$$

However, for the model (5.29), (5.32) this identity assumes the nontrivial form

$$C_{l-m}^{m+\frac{1}{2}}\left(\rho\partial_s + \frac{(1-\rho^2)}{s}\partial_\rho - \frac{\rho m}{s}\right) I_{m+\frac{1}{2}}(\omega) s^{-1/2} = I_{l+\frac{1}{2}}(s) C_{l-m}^{m+\frac{1}{2}}(\rho) s^{-1/2}.$$

$$(5.35)$$

Many other operational identities and addition theorems can be found in [83].

Weisner's method in its general form can also be applied to derive identities for spherical waves. For example, consider the (cylindrical wave) solution of the simultaneous equations

$$(\mathbf{P}\cdot\mathbf{P}+\omega^2)\Psi = 0, \qquad P^0\Psi = \lambda\Psi, \qquad J^0\Psi = m\Psi, \qquad \lambda, m \in \mathcal{C},$$

$$\Psi(s,\rho,\tau) = \left[\tau(\rho^2-1)^{1/2}(\lambda^2-1)^{1/2}\right]^m e^{\omega\lambda s\rho} I_{\pm m}\left(\omega s(\rho^2-1)^{1/2}(\lambda^2-1)^{1/2}\right).$$

$$(5.36)$$

ISBN-0-201-13503-5

Choosing the I_m solution, we note the validity of the expansion

$$\Psi(s,\rho,\tau)=(\omega s)^{-1/2}\sum_{n=0}^{\infty}a_n(\lambda)I_{m+n+\frac{1}{2}}(\omega s)C_n^{m+\frac{1}{2}}(\rho)\tau^m$$

expressing Ψ as a sum of spherical wave solutions of the Helmholtz equation. It remains only to compute $a_n(\lambda)$. Since Ψ is symmetric in ρ and λ, we have $a_n(\lambda)=b_nC_n^{m+\frac{1}{2}}(\lambda)$. Furthermore, if $\lambda=1$, then

$$\Psi(s,\rho,\tau)=\frac{(\omega s\tau/2)^m e^{\omega s\rho}}{\Gamma(m+1)}$$

and the identity (5.27) permits computation of the coefficients $a_n(\lambda)$ with the final result ($\omega=1$)

$$\left[(\rho^2-1)(\lambda^2-1)\right]^{-m/2}e^{s\lambda\rho}I_m\left(s\left[(\rho^2-1)(\lambda^2-1)\right]^{1/2}\right)$$

$$=\frac{2^{2m+1}}{(2\pi s)^{1/2}}\Gamma\left(m+\tfrac{1}{2}\right)^2\sum_{n=0}^{\infty}\frac{n!\left(m+n+\tfrac{1}{2}\right)}{\Gamma(2m+n+1)}I_{m+n+\frac{1}{2}}(s)C_n^{m+\frac{1}{2}}(\rho)C_n^{m+\frac{1}{2}}(\lambda),$$

$$(5.37)$$

convergent for all $\rho,\lambda\in\mathcal{C}$ (see [37, p. 102]).

Another example is provided by the solutions (5.14) corresponding to the parabolic system 8 in Table 14. Expressing these solutions in terms of coordinates (5.28) and expanding in the spherical basis, we obtain

$$e^{s\rho}L_k^{(m)}(-s(1+\rho))L_k^{(m)}(s(1-\rho))=\sum_{n=0}^{\infty}a_n s^{-m-\frac{1}{2}}I_{m+n+\frac{1}{2}}(s)C_n^{m+\frac{1}{2}}(\rho).$$

$$(5.38)$$

The coefficients a_n can be determined by setting $\rho=\alpha/s$ and letting $s\to0$ to obtain

$$2^{m+\frac{1}{2}}\Gamma\left(m+\tfrac{1}{2}\right)e^{\alpha}\left[L_k^{(m)}(-\alpha)\right]^2=\sum_{n=0}^{\infty}\frac{a_n\alpha^n}{n!\left(m+n+\tfrac{1}{2}\right)}.$$

Use of the transformation formula for a $_1F_1$ (Appendix B, Section 3) allows us to explicitly compute the coefficient of α^n on the left-hand side of this equation, with the result

$$a_n=\frac{2^{m+\frac{1}{2}}\left(m+n+\tfrac{1}{2}\right)\Gamma\left(m+\tfrac{1}{2}\right)\Gamma(m+k+1)\Gamma(m+k+n+1)}{(k!)^2\Gamma(m+1)\Gamma(m+n+1)}$$

$$\times{}_3F_2\left(\begin{matrix}-k,-m-n,-n\\m+1,-m-k-n\end{matrix}\Big|1\right).$$

$$(5.39)$$

For $k=0$ this expression reduces to (5.27).

ISBN-0-201-13503-5

3.6 The Laplace Equation $\Delta_3 \Psi = 0$

The known coordinate systems that permit R-separation of variables in the real Laplace equation

$$\Delta_3 \Psi(\mathbf{x}) = 0, \qquad \mathbf{x} = (x_1, x_2, x_3) = (x, y, z), \tag{6.1}$$

are derived and studied in the classic book of Bôcher [17]. However, the explicit relationship between these systems and the symmetry group of (6.1) has been discussed only very recently [22]. Apart from the trivial symmetry E, the symmetry algebra of this equation is ten dimensional with basis

$$P_j = \partial_j = \partial_{x_j}, \qquad j = 1, 2, 3; \qquad J_3 = x_2 \partial_1 - x_1 \partial_2,$$

$$J_2 = x_1 \partial_3 - x_3 \partial_1, \quad J_1 = x_3 \partial_2 - x_2 \partial_3, \quad D = -(\tfrac{1}{2} + x_1 \partial_1 + x_2 \partial_2 + x_3 \partial_3),$$

$$K_1 = x_1 + (x_1^2 - x_2^2 - x_3^2) \partial_1 + 2 x_1 x_3 \partial_3 + 2 x_1 x_2 \partial_2,$$

$$K_2 = x_2 + (x_2^2 - x_1^2 - x_3^2) \partial_2 + 2 x_2 x_3 \partial_3 + 2 x_2 x_1 \partial_1,$$

$$K_3 = x_3 + (x_3^2 - x_1^2 - x_2^2) \partial_3 + 2 x_3 x_1 \partial_1 + 2 x_3 x_2 \partial_2. \tag{6.2}$$

The P_l and J_l operators generate a subalgebra isomorphic to $\mathcal{E}(3)$ and D is the generator of dilatations. The operators K_j are generators of *special conformal transformations* and will be discussed later. Only the elements of the $\mathcal{E}(3)$ subalgebra actually commute with the Laplace operator Δ_3. The remaining elements of the Lie algebra merely leave the solution space of (6.1) invariant.

The symmetry algebra of the Laplace equation is isomorphic to $so(4, 1)$, the Lie algebra of all real 5×5 matrices $\mathcal{C}$ such that $\mathcal{C} G^{4,1} + G^{4,1} \mathcal{C}^t = 0$ where

$$G^{4,1} = \begin{pmatrix} 1 & & & & \\ & 1 & & & \\ & & 1 & & \\ & & & 1 & \\ & & & & -1 \end{pmatrix} = \sum_{j=1}^{4} \mathcal{E}_{jj} - \mathcal{E}_{55}. \tag{6.3}$$

Here $\mathcal{E}_{ij}$ is the 5×5 matrix with a one in row i, column j, and zeros everywhere else.

$$\mathcal{E}_{ij} = \begin{pmatrix} & & \overset{\displaystyle j}{} & \\ & & & \\ & & 1 & \end{pmatrix} i \;. \tag{6.4}$$

ISBN-0-201-13503-5

A basis for $so(4,1)$ is provided by the ten elements

$$
\begin{aligned}
\Gamma_{ab} &= \mathscr{E}_{ab} - \mathscr{E}_{ba} = -\Gamma_{ba}, && 1 \leqslant a,b \leqslant 4, \\
\Gamma_{a5} &= \mathscr{E}_{a5} + \mathscr{E}_{5a} = \Gamma_{5a}
\end{aligned}
\tag{6.5}
$$

with commutation relations

$$
\begin{aligned}
\left[\Gamma_{ab}, \Gamma_{cd}\right] &= \delta_{bc}\Gamma_{ad} + \delta_{ad}\Gamma_{bc} + \delta_{ca}\Gamma_{db} + \delta_{db}\Gamma_{ca}, \\
\left[\Gamma_{a5}, \Gamma_{cd}\right] &= -\delta_{ad}\Gamma_{c5} + \delta_{ac}\Gamma_{d5}, && \left[\Gamma_{a5}, \Gamma_{b5}\right] = \Gamma_{ab}.
\end{aligned}
\tag{6.6}
$$

One can verify that the correct commutation relations for the operators (6.2) result if the following identifications are made.

$$
\begin{aligned}
J_3 &= \Gamma_{32}, & J_2 &= \Gamma_{24}, & J_1 &= \Gamma_{43}, & D &= \Gamma_{15}, \\
P_1 &= \Gamma_{12} + \Gamma_{25}, & P_2 &= \Gamma_{13} + \Gamma_{35}, & P_3 &= \Gamma_{14} + \Gamma_{45}, \\
K_1 &= \Gamma_{12} - \Gamma_{25}, & K_2 &= \Gamma_{13} - \Gamma_{35}, & K_3 &= \Gamma_{14} - \Gamma_{45}.
\end{aligned}
\tag{6.7}
$$

The symmetry group of (6.1), the *conformal group*, is thus locally isomorphic to $SO(4,1)$, the group of all real 5×5 matrices A such that

$$
AG^{4,1}A' = G^{4,1}.
\tag{6.8}
$$

The identity component of this group consists of those matrices satisfying (6.8), $\det A = 1$, and $A_{55} \geqslant 1$. The Lie algebra of $SO(4,1)$ is $so(4,1)$ [46].

Exponentiating the operators (6.2), we can obtain the local action of $SO(4,1)$ as a transformation group of symmetry operators. In particular, the linear momentum and angular momentum operators generate the subgroup of symmetries (1.12) isomorphic to $E(3)$; the dilatation operator generates

$$
\exp(\lambda D)\Psi(\mathbf{x}) = \exp(-\lambda/2)\Psi[\exp(-\lambda)\mathbf{x}], \qquad \lambda \in R;
\tag{6.9}
$$

and the K_j generate the special conformal transformations

$$
\begin{aligned}
&\exp(a_1 K_1 + a_2 K_2 + a_3 K_3)\Psi(\mathbf{x}) \\
&= \left[1 - 2\mathbf{x} \cdot \mathbf{a} + (\mathbf{a} \cdot \mathbf{a})(\mathbf{x} \cdot \mathbf{x})\right]^{-1/2} \Psi\!\left(\frac{\mathbf{x} - \mathbf{a}(\mathbf{x} \cdot \mathbf{x})}{1 - 2\mathbf{a} \cdot \mathbf{x} + (\mathbf{a} \cdot \mathbf{a})(\mathbf{x} \cdot \mathbf{x})}\right).
\end{aligned}
\tag{6.10}
$$

In addition, we shall consider the inversion and space reflection symmetries of the Laplace equation:

$$
\begin{aligned}
I\Psi(\mathbf{x}) &= (\mathbf{x} \cdot \mathbf{x})^{-1/2}\Psi(\mathbf{x}/\mathbf{x} \cdot \mathbf{x}), & I &= I^{-1}, \\
R\Psi(\mathbf{x}) &= \Psi(-x_1, x_2, x_3), & R &= R^{-1}.
\end{aligned}
\tag{6.11}
$$

These are well-known symmetries of (6.1) that are not generated by the infinitesimal operators (6.2) [12, p. 31]. It follows from the definitions of these operators that

$$IP_jI^{-1}=-K_j, \qquad IDI^{-1}=-D, \qquad IJ_jI^{-1}=J_j. \qquad (6.12)$$

By a tedious computation we can verify that the Laplace equation is class I. Furthermore, although the space of symmetric second-order operators in the enveloping algebra of $so(4,1)$ is 35 dimensional, on the solution space of (6.1) there are 20 linearly independent relations between these operators. Thus, only 15 operators can be regarded as linearly independent on the solution space. For example, we have the relations

     (i) $\mathbf{P}\cdot\mathbf{P}=\mathbf{K}\cdot\mathbf{K}=0,$

     (ii) $\mathbf{J}\cdot\mathbf{J}=\frac{1}{4}-D^2,$

     (iii) $\Gamma_{45}^2+\Gamma_{41}^2-\Gamma_{51}^2=\frac{1}{4}+\Gamma_{23}^2,$

     (iv) $\{P_1,K_1\}+\{P_2,K_2\}+\{P_3,K_3\}=2+4D^2. \qquad (6.13)$

(Note that these relations are valid only on the solution space of (6.1), not in general. We are considering the $\Gamma_{\alpha\beta}$ as differential operators on this space via the definitions (6.7).)

The reader may be wondering why we have not applied a similar analysis to the Laplace equation $\Delta_2\Psi(\mathbf{x})=0$. The reason is that the symmetry algebra of this equation is infinite dimensional. In fact, every transformation $\Psi(x,y)\rightarrow\Psi(u(x,y),v(x,y))$, where $u+iv=f(z),z=x+iy$, and $f(z)$ is an analytic function, defines a symmetry of the Laplace equation. The group of all analytic transformations $z\rightarrow f(z)$ is the symmetry group of this equation, but it is not a Lie group. (Indeed each group transformation is determined by an infinite number of parameters $\{a_n\}$ where $f(z)=\sum_{n=0}^{\infty}a_nz^n$.) Thus, Lie theory methods are not particularly useful for this Laplace equation. It can be shown that infinite-dimensional symmetry algebras can occur for second-order partial differential equations in n variables only in the case where $n=2$ [105].

We now return to the separation of variables problem for equation (6.1). We will see that each R-separable coordinate system is characterized by a pair of commuting second-order symmetry operators in the enveloping algebra of $so(4,1)$. As usual, two coordinate systems will be regarded as equivalent if one can be obtained from the other by a transformation from the connected component of the identity of the conformal group, augmented by the discrete symmetries (6.11).

Note first that the eleven separable coordinate systems for the Helmholtz equation, listed in Table 14, are also separable for the Laplace

ISBN-0-201-13503-5

equation. The separation equations can be obtained from the corresponding Helmholtz results by setting $\omega = 0$ in Section 3.1. We briefly indicate the form of the separated solutions Ψ of the eigenvalue equations $S_j \Psi = \lambda_j \Psi, \Delta_3 \Psi = 0$.

For the Cartesian system 1 the solutions take the form

$$\exp(\alpha x + \beta y + \gamma z), \qquad \alpha^2 + \beta^2 + \gamma^2 = 0, \tag{6.14}$$

whereas for cylindrical coordinates 2 they are

$$\Psi^{(2)}_{\lambda,n}(r,\varphi,z) = J_{\pm n}(\lambda r)\exp(\lambda z + in\varphi),$$
$$iJ_3 \Psi_{\lambda,n} = n\Psi_{\lambda,n}, \qquad P_3 \Psi_{\lambda,n} = \lambda \Psi_{\lambda,n}. \tag{6.15}$$

The results for parabolic cylinder coordinates 3 are

$$\Psi^{(3)}_{\lambda,\mu}(\xi,\eta,z) = D_{i\mu - \frac{1}{2}}(\pm \sigma\xi)D_{-i\mu - \frac{1}{2}}(\pm \sigma\eta)e^{\lambda z},$$
$$\sigma = \exp(i\pi/4)(2\lambda)^{1/2}, \tag{6.16}$$
$$P_3 \Psi_{\lambda,\mu} = \lambda \Psi_{\lambda,\mu}, \qquad \{J_3, P_2\}\Psi_{\lambda,\mu} = 2\mu\lambda\Psi_{\lambda,\mu},$$

and for elliptic cylinder coordinates 4 they are

$$\Psi^{(4)}_{\lambda,n}(\alpha,\beta,z) = \begin{cases} Ce_n(\alpha,q)ce_n(\beta,q)e^{\lambda z}, \\ Se_n(\alpha,q)se_n(\beta,q)e^{\lambda z}, \end{cases} \quad q = d^2\lambda^2/4,$$
$$(J_3^2 + d^2 P_1^2)\Psi_{\lambda,n} = \mu_n \Psi_{\lambda,n}, \qquad P_3 \Psi_{\lambda,n} = \lambda \Psi_{\lambda,n}. \tag{6.17}$$

Corresponding to spherical coordinates 5 we have solutions

$$\Psi^{(5)}_{l,m}(\rho,\theta,\varphi) = \begin{cases} \rho^l \\ \rho^{-l-1} \end{cases} P_l^m(\theta)e^{im\varphi}, \qquad iJ_3 \Psi^{(5)}_{l,m} = m\Psi^{(5)}_{l,m},$$
$$\mathbf{J} \cdot \mathbf{J} \Psi^{(5)}_{l,m} = -l(l+1)\Psi^{(5)}_{l,m}. \tag{6.18}$$

For prolate spheroidal coordinates 6 the separated equations take the form (1.20) with $\omega = 0$ and typical solutions are

$$P_n^m(\cosh\eta)P_n^m(\cos\alpha)e^{im\varphi}; \tag{6.19}$$

that is, (1.21) with $\omega = 0$. Similarly, for oblate spheroidal coordinates 7 the separated equations are (1.22) with $\omega = 0$ and the eigenfunctions are of the form .

$$P_n^m(-i\sinh\eta)P_n^m(\cos\alpha)e^{im\varphi}. \tag{6.20}$$

ISBN-0-201-13503-5

For parabolic coordinates 8 the separated equations are (1.24) with $\omega=0$ and the separated solutions are

$$J_{\pm m}\left(i\sqrt{\lambda}\,\xi\right)J_{\pm m}\left(\sqrt{\lambda}\,\eta\right)e^{im\varphi}. \tag{6.21}$$

For paraboloidal coordinates 9 the separated equations are (1.26) with $\omega=0$ and the separated solutions are Mathieu functions of the form

$$\begin{aligned}
&Ce_n(\alpha,-\lambda c/2)\,ce_n(\beta,-\lambda c/2)\,Ce_n(\gamma+i\pi/2,-\lambda c/2),\\
&Se_n(\alpha,-\lambda c/2)\,se_n(\beta,-\lambda c/2)\,Se_n(\gamma+i\pi/2,-\lambda c/2).
\end{aligned} \tag{6.22}$$

For ellipsoidal coordinates 10 the separation equations take the form (1.29) or (1.33) with $\omega=0$. Thus the three separation equations reduce to the Lamé equation and the single-valued solutions in R^3 are products of three Lamé polynomials (see [7, p. 228]).

Finally, for conical coordinates 11 the separation equations are (1.35) with $\omega=0$. The single-valued solutions in R^3 take the form

$$\left\{\begin{matrix} r^l \\ r^{-l-1} \end{matrix}\right\} E_{ln}^{p,q}(\alpha)E_{ln}^{p,q}(\beta), \qquad l=0,1,2,\ldots, \tag{6.23}$$

where the E functions are Lamé polynomials; see (3.24). Such products of Lamé polynomials are called *ellipsoidal harmonics* in analogy with the spherical harmonics $Y_l^m(\theta,\varphi)$, (6.18) [136a]. The overlap functions relating spherical and ellipsoidal harmonics have already been computed in Section 3.3.

The remaining separable coordinate systems for the Laplace equation are purely R-separable and do not lead to separation for the Helmholtz equation. The coordinate surfaces for these systems are orthogonal families of confocal cyclides. A cyclide is a surface with equation

$$a\left(x^2+y^2+z^2\right)^2+P(x,y,z)=0 \tag{6.24}$$

where a is a constant and P is a polynomial of order two. If $a=0$, the cyclide reduces to a quadric surface. Now it is well known that the coordinate surfaces of the eleven separable systems listed in Table 14 are confocal families of quadrics

$$\frac{x^2}{a_1+\lambda}+\frac{y^2}{a_2+\lambda}+\frac{z^2}{a_3+\lambda}=1, \qquad a_j\text{ constant}, \tag{6.25}$$

and their limiting cases, (see [13, 97, 98, 136a]). In particular, all these coordinates are limiting cases of the ellipsoidal coordinates and the coordinate surfaces are ellipsoids, hyperboloids, and their various limits, such as paraboloids, spheres, and planes.

ISBN-0-201-13503-5

We know that under any conformal symmetry of the Laplace equation an R-separable system is mapped to an R-separable system. However, the inversion operator I, (6.11), maps a quadric surface to a cyclide with $a\neq0$, as the reader can easily verify. Thus one cannot avoid the appearance of cyclides in the study of R-separable coordinate systems for the Laplace equation.

It is straightforward to check that the family of all cyclides is invariant under the action of the conformal group and that this group maps orthogonal surfaces to orthogonal surfaces. Instead of using families of confocal quadrics to construct orthogonal coordinate systems, one can more generally use families of confocal cyclides. By direct computation it can be shown that such families define orthogonal, R-separable coordinate systems for the Laplace equation. Moreover, all separable systems for the Laplace equation can be obtained in this manner.

Since we regard coordinate systems related by a transformation from the conformal group as equivalent, to obtain all distinct cyclidic systems it is obviously necessary to decompose the family of cyclides (6.24) into conformal equivalence classes. Among the equivalence classes of cyclides are some which contain cyclides (6.24) with $a=0$. These correspond to the eleven separable systems listed in Table 14. The remaining classes contain only cyclides with $a\neq0$ and lead to new R-separable systems. The details of this construction can be found in the classic book of Bôcher [17]. Our primary aim is to provide a group-theoretic characterization of the coordinate systems listed by Bôcher. This characterization was first given in [22] and is contained in Table 17.

For each coordinate system $\{\mu,\nu,\rho\}$ the R-separable solutions of (6.1) take the form $\Psi(\mathbf{x})=\mathcal{R}^{1/2}(\mu,\nu,\rho)A(\mu)B(\nu)C(\rho)$ and these solutions are characterized by the eigenvalue equations $S_j\Psi=\lambda_j\Psi$ where λ_1,λ_2 are the separation constants.

More specifically, for system 12 the parameters vary over the range

$$0<\rho<1<\nu<b<\mu<a$$

and each factor in the separated solution satisfies the equation

$$\left[(f(\xi))^{1/2}\frac{d}{d\xi}(f(\xi))^{1/2}\frac{d}{d\xi}-\left(\frac{3}{16}\xi^2+\frac{\lambda_1}{4}\xi+\frac{\lambda_2}{4}\right)\right]A(\xi)=0,\qquad(6.26)$$

$$f(\xi)=(\xi-a)(\xi-b)(\xi-1)\xi,\xi=\mu,\nu,\rho.$$

Here (6.26) is the standard form of an equation with five elementary singularities [51, p. 500]. Very little is known about the solutions. For

ISBN-0-201-13503-5

Table 17 Additional R-Separable Systems for the Laplace Equation

Commuting operators S_1, S_2	Separable coordinates
12 $S_1 = \dfrac{a+1}{4}(P_2+K_2)^2 + \dfrac{b+1}{4}(P_1+K_1)^2$	$x = \mathcal{R}^{-1}\left[\dfrac{(\mu-a)(\nu-a)(\rho-a)}{(b-a)(a-1)a}\right]^{1/2}$,
$\qquad + \dfrac{a+b}{4}(P_3+K_3)^2 + J_3^2 + bJ_2^2 + aJ_1^2,$	$y = \mathcal{R}^{-1}\left[\dfrac{(\mu-b)(\nu-b)(\rho-b)}{(a-b)(b-1)b}\right]^{1/2}$,
$S_2 = \dfrac{a}{4}(P_2+K_2)^2 + \dfrac{b}{4}(P_1+K_1)^2$	$z = \mathcal{R}^{-1}\left[\dfrac{(\mu-1)(\nu-1)(\rho-1)}{(a-1)(b-1)}\right]^{1/2}$,
$\qquad + \dfrac{ab}{4}(P_3+K_3)^2$	$\mathcal{R} = 1 + \left[\dfrac{\mu\nu\rho}{ab}\right]^{1/2}$
13 $S_1 = 2\alpha J_3^2 + \dfrac{\alpha+1}{2}\{P_2, K_2\} + \dfrac{\beta}{2}(P_2^2-K_2^2)$	$x = \mathcal{R}^{-1}\left[\dfrac{(\mu-1)(\nu-1)(\rho-1)}{(a-1)(b-1)}\right]^{1/2}$,
$\qquad + \dfrac{\alpha}{2}\{P_1, K_1\} + \dfrac{\beta}{2}(K_1^2-P_1^2),$	$y = \mathcal{R}^{-1}\left[-\dfrac{\mu\nu\rho}{ab}\right]^{1/2},$
$S_2 = \dfrac{\alpha}{2}\{P_2, K_2\} + \dfrac{\beta}{2}(P_2^2-K_2^2)$	$z = \mathcal{R}^{-1}$
$\qquad + (\alpha^2+\beta^2)J_3^2$	$\mathcal{R} = 2\,\mathrm{Re}\left[-\dfrac{i(\mu-a)(\nu-a)(\rho-a)}{(a-b)(a-1)a}\right]^{1/2}$,
	$a = \bar{b} = \alpha + i\beta, \alpha, \beta\ \text{real}$
14 $S_1 = J_3^2$	$x = \mathcal{R}^{-1}\cos\varphi,$
$\quad 4S_2 = (P_3+K_3)^2 - a(P_3-K_3)^2$	$y = \mathcal{R}^{-1}\sin\varphi,$
	$z = \mathcal{R}^{-1}\left[-\dfrac{\mu\rho}{a}\right]^{1/2},$
	$\mathcal{R} = \left[\dfrac{(\mu-a)(a-\rho)}{a(a-1)}\right]^{1/2} - \left[\dfrac{(\mu-1)(1-\rho)}{a-1}\right]^{1/2}$
15 $S_1 = J_3^2$	$x = \mathcal{R}^{-1}\cos\varphi,$
$\quad 4S_2 = -4aD^2 - (P_3-K_3)^2$	$y = \mathcal{R}^{-1}\sin\varphi,$
	$z = \mathcal{R}^{-1}\left[\dfrac{(\mu-a)(a-\rho)}{a(a-1)}\right]^{1/2}$,
	$\mathcal{R} = \left[\dfrac{\mu\rho}{a}\right]^{1/2} + \left[\dfrac{(\mu-1)(\rho-1)}{(a-1)}\right]^{1/2}$
16 $S_1 = J_3^2$	$x = \mathcal{R}^{-1}\cos\varphi,$
$\quad 2S_2 = \alpha\{P_3, K_3\} + \beta(K_3^2 - P_3^2)$	$y = \mathcal{R}^{-1}\sin\varphi,$
	$z = \mathcal{R}^{-1}\left[-\dfrac{\mu\rho}{ab}\right]^{1/2},$
	$\mathcal{R} = 2\,\mathrm{Re}\left[\dfrac{i(\rho-a)(\mu-a)}{a(a-b)}\right]^{1/2}$,
	$a = \bar{b} = \alpha + i\beta$
17 $S_1 = J_3^2$	$x = \mathcal{R}^{-1}\sinh\xi\cos\varphi,$
$\quad 4S_2 = (P_3+K_3)^2$	$y = \mathcal{R}^{-1}\sinh\xi\sin\varphi,$
	$z = \mathcal{R}^{-1}\cos\psi,$
	$\mathcal{R} = \cosh\xi + \sin\psi$

ISBN-0-201-13503-5

system 13 the parameters vary in the range

$$-\infty<\rho<0<\mu<1<\nu<\infty.$$

The separated equations are (6.26) with $a=\bar{b}=\alpha+i\beta$. For system 14 the parameters vary in the range $\mu>a>1,\rho<0,0\leqslant\varphi<2\pi$ and the solutions of Laplace's equation have the form $\Psi=\mathcal{R}^{1/2}E_1(\mu)E_2(\rho)e^{im\varphi}$ where

$$\left[4(\mathcal{P}(\xi))^{1/2}\frac{d}{d\xi}(\mathcal{P}(\xi))^{1/2}\frac{d}{d\xi}+\left(\frac{1}{4}-m^2\right)\xi-\lambda\right]E_j(\xi)=0,$$

$$j=1,2,\quad \xi=\mu,\rho,\quad \mathcal{P}(\xi)=(\xi-a)(\xi-1)\xi, \tag{6.27}$$

$$iJ_3\Psi=m\Psi,\quad S_2\Psi=\lambda\Psi.$$

If we set $\mu=\operatorname{sn}^2(\alpha,k),\rho=\operatorname{sn}^2(\beta,k)$ where $a=k^{-2}$, then we find

$$x=\mathcal{R}^{-1}\cos\varphi,\quad y=\mathcal{R}^{-1}\sin\varphi,\quad z=ik\mathcal{R}^{-1}\operatorname{sn}\alpha\operatorname{sn}\beta,$$

$$\mathcal{R}=i(k')^{-1}\operatorname{dn}\alpha\operatorname{dn}\beta-i(kk')^{-1}\operatorname{cn}\alpha\operatorname{cn}\beta$$

$$\Psi=\mathcal{R}^{1/2}\Lambda_{m-\frac{1}{2}}^p(\alpha,k)\Lambda_{m-\frac{1}{2}}^p(\beta,k)e^{im\varphi} \tag{6.28}$$

where $\Lambda_n^p(z,k)$ is a solution of the Lamé equation

$$\frac{d^2\Lambda}{dz^2}+\left(h_n^p-n(n+1)k^2\operatorname{sn}^2(z,k)\right)\Lambda=0. \tag{6.29}$$

The parameters α,β range over the intervals $\alpha\in[iK',iK'+2K],\beta\in[2K-iK',2K+iK']$ in the complex plane.

For system 15 the parameters vary in the range

$$1<\rho<a<\mu<\infty,\quad 0\leqslant\varphi<2\pi$$

and the separation equations are (6.27). Making the same elliptic function substitutions as in the previous case, we find

$$x=\mathcal{R}^{-1}\cos\varphi,\quad y=\mathcal{R}^{-1}\sin\varphi,\quad z=i(k'\mathcal{R})^{-1}\operatorname{dn}\alpha\operatorname{dn}\beta,$$

$$\mathcal{R}=k(\operatorname{sn}\alpha\operatorname{sn}\beta+\operatorname{cn}\alpha\operatorname{cn}\beta/k'), \tag{6.30}$$

$$\Psi=\mathcal{R}^{1/2}\Lambda_{m-\frac{1}{2}}^p(\alpha,k)\Lambda_{m-\frac{1}{2}}^p(\beta,k)e^{im\varphi}$$

where α,β range over the intervals $\alpha\in[iK',iK'+2K],\beta\in[K,K+2iK']$ in the complex plane.

For system 16 the parameters satisfy $\mu>0,\rho<0,0\leqslant\varphi<2\pi$, and the separation equations are (6.27) with

$$\mathcal{P}(\xi)=(\xi-a)(\xi-b)\xi,\quad a=\bar{b}=\alpha+i\beta.$$

ISBN-0-201-13503-5

Setting $\mu = \text{sn}^2(\gamma, t), \rho = \text{sn}^2(\theta, t)$ where $t = (s + is')(s - is')^{-1}, s^2 = (|a| - \text{Re}\,a)/2|a|$, we obtain solutions

$$\Psi = \mathcal{R}^{1/2}\Lambda^p_{m-\frac{1}{2}}(\gamma, t)\Lambda^p_{m-\frac{1}{2}}(\theta, t)e^{im\varphi} \qquad (6.31)$$

where $\gamma \in [-iK', iK'], \theta \in [2K - iK', 2K + iK']$.

Finally, for system 17, toroidal coordinates, the eigenfunctions have the form

$$\Psi = (\cosh\xi + \sin\psi)^{1/2}E(\xi)\exp[i(l\psi + m\varphi)]$$

$$iJ_3\Psi = m\Psi, \qquad (P_3 + K_3)\Psi = -2il\Psi,$$

$$\left[(\sinh\xi)^{-1}\frac{d}{d\xi}\sinh\xi\frac{d}{d\xi} + \left(1/4 - l^2 - \frac{m^2}{\sinh^2\xi}\right)\right]E(\xi) = 0. \qquad (6.32)$$

The associated Legendre functions $P^m_{l-\frac{1}{2}}(\cosh\xi), Q^m_{l-\frac{1}{2}}(\cosh\xi)$ provide a basis of solutions for this last equation.

We can check explicitly that the coordinate surfaces are cyclides in all these cases. For systems 14–17 some of the surfaces are cyclides of revolution. Systems 12–16 are relatively intractable and only the toroidal system 17 has been widely used in studies of the Laplace equation. The toroidal and spherical coordinate systems have much in common. (Indeed, for the complex Laplace equation these two systems become equivalent under the complex conformal group.) Bipolar coordinates [12, p. 108] are frequently used in connection with separation of variables for the Laplace equation but these coordinates are conformally equivalent to spherical coordinates. They are, however, inequivalent to spherical coordinates with respect to the more physical scale Euclidean group, generated by $E(3)$ and dilatations $\exp(\alpha D)$.

Nine of the seventeen R-separable systems for the Laplace equation correspond to diagonalization of the operator J_3: systems 2, 5–8, 14–17. These special systems have the property that their eigenfunctions take the form $\Psi(\mathbf{x}) = \Phi e^{im\varphi}, iJ_3\Psi = m\Psi$, where Φ is a function of the remaining two variables. If we substitute this Ψ into the Laplace equation and factor out $e^{im\varphi}$, we obtain a differential equation for Φ which in cylindrical coordinates is

$$\left(\partial_{rr} + r^{-1}\partial_r - r^{-2}m^2 + \partial_{zz}\right)\Phi(r, z) = 0. \qquad (6.33)$$

Expression (6.33) for fixed $m \geqslant 0$ is the equation of generalized axial-symmetric potential theory. The real symmetry algebra of this equation is isomorphic to $sl(2, R)$. Indeed, a basis is provided by the operators

ISBN-0-201-13503-5

K_3, P_3, D, (6.2), with commutation relations

$$[D, P_3] = P_3, \qquad [D, K_3] = -K_3, \qquad [P_3, K_3] = -2D, \qquad (6.34)$$

and, from the identity (6.13iii) it follows that (6.33) can be written in the equivalent operator form

$$\left(\tfrac{1}{2} P_3^2 + \tfrac{1}{2} K_3^2 - D^2\right)\Phi = \left(\tfrac{1}{4} + m^2\right)\Phi. \qquad (6.35)$$

It is shown in [139] (see also [63]) that the space of symmetric second-order symmetry operators in the enveloping algebra of $sl(2, R)$ modulo the subspace generated by the Casimir operator $\tfrac{1}{2} P_3^2 + \tfrac{1}{2} K_3^2 - D^2$ decomposes into nine orbit types under the action of the symmetry group $SL(2, R)$. The nine coordinate systems listed above are exactly those which permit separation of variables in (6.33) and it is straightforward to check that these systems correspond one to one with the nine orbit types. That is, there is perfect correspondence between the list of operators S_2 where J_3^2, S_2 defines each system and a list of representatives of the orbit types.

3.7 Identities Relating Separable Solutions of the Laplace Equation

It is not possible to find a Hilbert space model for the solutions of the Laplace equation such that the action of the conformal group is given by a unitary representation. Indeed, if such a model existed, the momentum operators iP_j, $j = 1, 2, 3$, would be self-adjoint on this Hilbert space. However, the identity $P_1^2 + P_2^2 + P_3^2 = 0$ and the spectral theorem for self-adjoint operators imply $P_j = 0$, which is a contradiction.

Nevertheless we can use Weisner's method to relate separable solutions of the Laplace equation and we can construct non-Hilbert space models of this equation in a manner analogous to that of Section 3.5. Consider the expression

$$\Psi(x, y, z) = \int_{C_1} d\beta \int_{C_2} \frac{dt}{t} h(\beta, t) \exp\left[\frac{ix\beta}{2}(t + t^{-1}) \right.$$

$$\left. + \frac{y\beta}{2}(t - t^{-1}) - \beta z \right] = I(h), \qquad (7.1)$$

where h is analytic on a domain in $\mathcal{C} \times \mathcal{C}$ that contains the integration contours $C_1 \times C_2$ and is chosen such that $I(h)$ converges absolutely and arbitrary differentiation with respect to x, y, z is permitted under the integral sign. It is easy to verify that for each such h, $\Psi = I(h)$ is a solution of the Laplace equation (6.1). Moreover, integrating by parts, we find that the operators P_j, J_j, K_j, D, (6.2), acting on the solution space of (6.1)

ISBN-0-201-13503-5

correspond to the operators

$$P^+ = -\beta t, \qquad P^- = -\beta t^{-1}, \qquad P^0 = -i\beta, \qquad D = \beta\partial_\beta + \tfrac{1}{2},$$

$$J^+ = it\beta\partial_\beta - it^2\partial_t, \qquad J^- = -i\beta t^{-1}\partial_\beta - i\partial_t, \qquad J^0 = t\partial_t,$$

$$K^+ = t\beta^{-1}(\beta\partial_\beta - t\partial_t)(\beta\partial_\beta - t\partial_t - 1),$$

$$K^- = t^{-1}\beta^{-1}(\beta\partial_\beta + t\partial_t)(\beta\partial_\beta + t\partial_t - 1),$$

$$K^0 = i\beta^{-1}((t\partial_t)^2 - (\beta\partial_\beta)^2), \qquad\qquad\qquad (7.2)$$

where

$$J^\pm = \mp J_2 + iJ_1, \qquad J^0 = iJ_3$$

with similar expressions for $P^\pm$, $K^\pm$, and so on. Here we are assuming C_1, C_2, and h are chosen such that the boundary terms vanish for each integration by parts:

$$P^\pm\Psi = I(P^\pm h), \qquad J^\pm\Psi = I(J^\pm h),$$

and so on.

For our first example we choose C_1, C_2 as unit circles in the β and t planes, respectively, with centers at the origin and oriented in the counterclockwise direction. Then for

$$h(\beta,t) = \beta^{-l-1}j(t), \qquad j(t) = \sum_{m=-l}^{l} a_m t^m, \qquad l = 0,1,2,\ldots, \quad (7.3)$$

we can evaluate the β integral by residues to obtain

$$\Psi(x,y,z) = I(h) = -\frac{2\pi}{l!}\int_0^{2\pi}\left[ix\cos\alpha + iy\sin\alpha - z \right]^l j(e^{i\alpha})\,d\alpha. \quad (7.4)$$

From (7.3), h is an eigenfunction of D with eigenvalue $-l-\tfrac{1}{2}$, so by (6.13ii)

$$\mathbf{J}\cdot\mathbf{J}\Psi = -l(l+1)\Psi.$$

Furthermore, Ψ is a solution of the Laplace equation which is a homogeneous polynomial in x,y,z of order l. In particular, for $j(t) = t^m$, $-l \leq m \leq l$, we have $J^0\Psi = m\Psi$, so Ψ must be multiple of the solid harmonic $\rho^l Y_l^m(\theta,\varphi)$, expressed in spherical coordinates (5 in Table 14). Evaluating

ISBN-0-201-13503-5

the integral in the special case where $\theta=0$, we find

$$I(\beta^{-l-1}t^m) = -\frac{2\pi\rho^l}{l!}\int_0^{2\pi}\left[i\sin\theta\cos(\varphi-\alpha)-\cos\theta\right]^l e^{i\alpha m}\,d\alpha$$

$$= 16\pi^3(-1)^{l+1}i^m\rho^l\left[4\pi(2l+1)(l-m)!(l+m)!\right]^{-1/2}Y_l^m(\theta,\varphi).$$

(7.5)

Another example is provided by the contour C_2 in the t plane, the contour C_1', which goes from $\beta=0$ to $+\infty$ along the positive real axis in the β plane, and the analytic function $h(\beta,t)=\beta^l t^m, l=0,1,2,\ldots, m=l,$ $l-1,\ldots,-l$. Here $\Psi=I(h)$ satisfies $D\Psi=(l+\frac{1}{2})\Psi, \mathbf{J}\cdot\mathbf{J}\Psi=-l(l+1)\Psi, J^0\Psi=m\Psi$ and it is easy to verify that

$$I(\beta^l t^m) = il!\rho^{-l-1}\int_0^{2\pi}\left[-i\sin\theta\cos(\varphi-\alpha)+\cos\theta\right]^{-l-1}e^{im\alpha}\,d\alpha$$

$$= i^{1-m}\rho^{-l-1}\left[16\pi^3(l-m)!(l+m)!/(2l+1)\right]^{1/2}Y_l^m(\theta,\varphi), \quad (7.6)$$

where ρ,θ,φ are spherical coordinates and $0\leqslant\theta<\pi/2$.

Now consider the equations

$$(\{J_1,P_2\}-\{P_1,J_2\})f=-\lambda f, \qquad J^0 f=mf, \tag{7.7}$$

for eigenfunctions corresponding to the parabolic system. In terms of the model (7.2) these eigenfunctions are

$$f_{\lambda,m}^{(8)}(\beta,t)=\exp(-\lambda/2\beta)\beta^{-1}t^m. \tag{7.8}$$

Setting $h=f_{\lambda,m}^{(8)}$ in (7.1) and choosing the contours C_1,C_2, we find

$$\Psi_{\lambda,m}^{(8)}=I\left(f_{\lambda,m}^{(8)}\right)=-2\pi\int_0^{2\pi}J_0\left[i(2\lambda)^{1/2}(z-ix\cos\alpha-iy\sin\alpha)^{1/2}\right]e^{im\alpha}\,d\alpha$$

$$= -4\pi^2 J_m\left(-i\sqrt{\lambda}\,\xi\right)J_m\left(\sqrt{\lambda}\,\eta\right)e^{im\varphi}, \tag{7.9}$$

$$x=\xi\eta\cos\varphi, y=\xi\eta\sin\varphi, z=(\xi^2-\eta^2)/2.$$

As usual, the fact that variables separate enables us to compute the integral. For $h=f_{\lambda,m}^{(8)}$ in (7.1) and the contours C_1', C_2 we obtain

$$\Psi_{\lambda,m}^{(8)'}=I\left(f_{\lambda,m}^{(8)}\right)=2\pi i^{m+1}e^{im\varphi}\int_0^\infty J_m(\beta r)\exp(-\beta z-\lambda/2\beta)\,d\beta/\beta$$

$$= 2i\int_0^{2\pi}K_0\left[(2\lambda)^{1/2}(z-ix\cos\alpha-iy\sin\alpha)^{1/2}\right]e^{im\alpha}\,d\alpha$$

$$= 4\pi i K_m\left(\sqrt{\lambda}\,\xi\right)I_m\left(i\sqrt{\lambda}\,\eta\right)e^{im\varphi}, \qquad \lambda>0,\xi>|\eta|. \tag{7.10}$$

The second and third equalities are obtained by performing only one of the integrations. Note that the second equality yields the expansion of our solution in terms of cylindrical waves.

Similarly, performing the t integration in (7.6) first, we find the expansion

$$I(\beta^l t^m) = 2\pi i^{m+1} e^{im\varphi} \int_0^\infty J_m(\beta r) e^{-\beta z}\beta^l \, d\beta, \qquad z > 0, \qquad (7.6')$$

of a solid spherical harmonic in terms of cylindrical waves.

Applying the transformation I (with contours C_1, C_2) to both sides of the identity

$$f_{\lambda,m}^{(8)}(\beta,t) = t^m \sum_{l=0}^\infty (-\lambda/2)^l \beta^{-l-1}/l!,$$

we find the expansion

$$\Psi_{\lambda,m}^{(8)}(\mathbf{x}) = - \sum_{l=|m|}^\infty 16\pi^3 i^m \left[4\pi(2l+1)(l-m)!(l+m)! \right]^{-1/2}$$

$$\times (\lambda\rho/2)^l (l!)^{-1} Y_l^m(\theta,\varphi) \qquad (7.9')$$

of products of Bessel functions in terms of spherical harmonics.

Corresponding to the oblate spheroidal system 7, the eigenvalue equations

$$\left(\mathbf{J}\cdot\mathbf{J} + a^2 P_1^2 + a^2 P_2^2\right) f = -\lambda f, \qquad J^0 f = mf,$$

in the model (7.2) yield the eigenfunctions

$$f_{\lambda,m}^{(7)}(\beta,t) = \beta^{-1/2} J_\nu(a\beta) t^m, \qquad \nu^2 = \lambda + \tfrac{1}{4}. \qquad (7.11)$$

Choosing the case where m is a positive integer and $\nu = l + \tfrac{1}{2}$ (where $l > -1$) and applying the transformation I (contours C_1', C_2), we find

$$\Psi_{\lambda,m}^{(7)}(\mathbf{x}) = I(f_{\lambda,m}^{(7)}) = 2\pi i^{m+1} e^{im\varphi} \int_0^\infty J_m(\beta r) J_\nu(a\beta) e^{-\beta z} \, d\beta/\beta^{1/2}$$

$$= 2\pi i^{m+1}(a\cosh\eta)^{-1/2}\Gamma(m+l+1) e^{im\varphi} \qquad (7.12)$$

$$\times P_l^{-m}(\cos\alpha) P_{m-\frac{1}{2}}^{-l-\frac{1}{2}}(\tanh\eta), \qquad 0 < \alpha < \frac{\pi}{2}, 0 < \eta,$$

where α, η, φ are oblate spheroidal coordinates (7 in Table 14). Note that the second equality gives the expansion of our solution in terms of

ISBN-0-201-13503-5

cylindrical waves. Again the integrals are rather easy to evaluate because we know in advance that variables separate in the solution. To determine the remaining four constants we need only examine the behavior of the integral near the values $\eta=0$ and $\alpha=0$, $\pi/2$.

In the case where $\nu=l+\frac{1}{2}$, $l=0,1,2,\ldots$, we can expand (7.11) as a power series in β and apply the transformation I term by term to obtain

$$\Psi^{(7)}_{\lambda,m}(\mathbf{x})= \sum_{n=\max\left(\frac{|m|-l}{2},0\right)}^{\infty} \left(\frac{a}{2}\right)^{l+2n+\frac{1}{2}} (i)^{2n-m+1} \frac{\rho^{-l-2n-1}}{n!\Gamma(l+n+3/2)}$$

$$\times \left[16\pi^3 \frac{(l+2n-m)!(l+2n+m)!}{(2l+4n+1)}\right]^{1/2} Y^m_{l+2n}(\theta,\varphi), \quad (7.13)$$

which is an expansion of a spheroidal solution in solid spherical harmonics.

For the toroidal system 17 the eigenvalue equations

$$(P^0+K^0)f=2lf, \qquad J^0f=mf,$$

in the model (7.2) yield the eigenfunctions

$$f^{(17)}_{n,m}(\beta,t)=e^{-i\beta}(\beta t)^m {}_1F_1\left(\frac{-n}{2m+1}\bigg|2i\beta\right), \qquad n=-l-m-\tfrac{1}{2}. \quad (7.14)$$

We choose n, $m=0$, 1, 2,... and apply I (contours C'_1, C_2) to obtain

$$\Psi^{(17)}_{n,m}(\mathbf{x})= I\left(f^{(17)}_{n,m}\right)$$

$$=2\pi i^{m+1}e^{im\varphi} \int_0^\infty e^{-\beta z-i\beta} J_m(r\beta)\beta^m {}_1F_1\left(\frac{-n}{2m+1}\bigg|2i\beta\right)d\beta$$

$$=\sqrt{2}\,\pi\left(-\tfrac{1}{2}\right)^m(-i)^n(2m)!(\cosh\xi+\sin\psi)^{1/2}$$

$$\times \exp\left[i(m\varphi+l\psi+\pi/4)\right]P^{-m}_{l-\frac{1}{2}}(\cosh\xi). \quad (7.15)$$

An explicit computation yields

$$\exp(\alpha P_3)f^{(17)}_{n,m}= \sum_{s=m}^{\infty} a^s_{n,m}\beta^s t^m$$

$$\qquad\qquad (7.16)$$

$$a^s_{n,m}= \frac{(-\alpha-i)^{s-m}}{(s-m)!} {}_2F_1\left(\frac{-n,m-s}{2m+1}\bigg|\frac{2i}{\alpha+i}\right),$$

so

$$\exp(\alpha P_3)\Psi_{n,m}^{(17)}(\mathbf{x}) = \sum_{s=m}^{\infty} a_{n,m}^s i^{1-m}\rho^{-s-1}\left[16\pi^3\frac{(s-m)!(s+m)!}{(2s+1)}\right]^{1/2} Y_s^m(\theta,\varphi)$$

$$(7.17)$$

is the expansion of this toroidal system solution in solid spherical harmonics. (The term-by-term integration used to derive (7.13) and (7.17) can be justified with the Lebesgue-dominated convergence theorem [69].)

As the foregoing examples indicate, the non-Hilbert space model permits us to derive integral representations and expansion formulas for the Laplace separable systems. (In some cases, however, the models yield third- and fourth-order differential operators.) The analysis for systems related to the Lamé and Whittaker–Hill equations proceeds in analogy with Section 3.3. The number of examples can be greatly multiplied by choosing other contours in the β and t planes. In addition, the Hilbert space expansions for solutions of the wave equation (Section 3.9) can be reinterpreted as Laplace equation expansions by replacing t with iz for $z>0$.

The most useful functions for application of Weisner's method are those associated with the spherical system. These functions are characterized as common eigenfunctions of the commuting operators D and J^0. We shall now study the eigenfunctions in greater generality than earlier by first considering the model (7.2). In this model the solutions of the equations

$$J^0 g = mg, \qquad Dg = \left(l+\tfrac{1}{2}\right)g, \qquad m,l\in\mathcal{C},$$

are multiples of $\beta^l t^m$. If the eigenfunctions are normalized so that

$$g_m^{(l)} = i^{l-m}\beta^l t^m, \qquad (7.18)$$

it follows easily that the action of the operators (7.2) on this basis is

$$J^{\pm}g_m^{(l)} = (-l\pm m)\,g_{m\pm1}^{(l)}, \qquad J^0 g_m^{(l)} = mg_m^{(l)},$$

$$P^0 g_m^{(l)} = -g_m^{(l+1)}, \qquad P^{\pm}g_m^{(l)} = \mp g_{m\pm1}^{(l+1)},$$

$$Dg_m^{(l)} = (l+\tfrac{1}{2})\,g_m^{(l)}, \qquad K^0 g_m^{(l)} = (l^2-m^2)\,g_m^{(l-1)},$$

$$K^{\pm}g_m^{(l)} = \mp(l\mp m)(l\mp m-1)\,g_{m\pm1}^{(l-1)}. \qquad (7.19)$$

We shall study our model in the case where $l_0\in\mathcal{C}$ is fixed with $l_0+\tfrac{1}{2}$ not an integer, $l=l_0,l_0\pm1,l_0\pm2,\dots$, and $m=l,l-1,l-2,\dots$. Note that the corresponding set of basis functions $\{g_m^{(l)}\}$ is invariant under the action of $so(4,1)$. In particular, the eigenfunction $g_l^{(l)}$ is mapped to zero by each of the operators J^+, K^0, K^+.

Due to the simplicity of the recurrence relations (7.19), we can easily exponentiate the Lie algebra operators L to obtain the local action $\exp(\alpha L)$ of the conformal group with respect to the basis (7.18). (Indeed one can use local Lie theory to exponentiate all operators (7.2) except the second-order K operators. However, the K operators can be formally exponentiated in the $\{g_m^{(l)}\}$ basis by using the recurrence relations (7.19) and the results will be valid for the Laplace equation model (6.2).) The matrix elements of the group action have been worked out in some detail in [84] and the results applied to derive identities for the Gegenbauer polynomials.

To see how these functions arise, we consider a complex coordinate system $\{w,t,\rho\}$ that is complex equivalent to the complex spherical coordinates $\{\theta,\varphi,\rho\}$ (5 in Table 14). (Since we are interested in analytic expansions, it is now useful to consider solutions of the complex Laplace equation.)

$$w = \cos\theta = z/\rho, \qquad t = e^{i\varphi}(1-w^2)^{1/2} = (x+iy)/\rho,$$
$$\rho = (x^2 + y^2 + z^2)^{1/2}. \tag{7.20}$$

In terms of these coordinates the operators (6.2) become

$$J^0 = t\,\partial_t, \qquad J^+ = -t\,\partial_w, \qquad J^- = t^{-1}((1-w^2)\partial_w - 2wt\,\partial_t),$$

$$D = -(\tfrac{1}{2} + \rho\,\partial_\rho), \qquad -iP^0 = w\,\partial_\rho + \rho^{-1}(1-w^2)\partial_w - \rho^{-1}wt\,\partial_t,$$

$$-iP^+ = t\,\partial_\rho - \rho^{-1}tw\,\partial_w - \rho^{-1}t^2\,\partial_t,$$

$$-iP^- = t^{-1}(1-w^2)\partial_\rho - \rho^{-1}t^{-1}w(1-w^2)\partial_w + \rho^{-1}(1+w^2)\partial_t,$$

$$-iK^0 = \rho w + \rho^2 w\,\partial_\rho + \rho(w^2-1)\partial_w + \rho tw\,\partial_t,$$

$$-iK^+ = \rho t + \rho^2 t\,\partial_\rho + \rho tw\,\partial_w + \rho t^2\,\partial_t,$$

$$-iK^- = \rho t^{-1}(1-w^2) + \rho^2 t^{-1}(1-w^2)\partial_\rho - \rho(1+w^2)\partial_t + \rho t^{-1}w(1-w^2)\partial_w. \tag{7.21}$$

Now we search for functions $\Psi_m^{(l)}(w,t,\rho)$ that satisfy the recurrence relations (7.19) when acted on by operators (7.21). (Since $\mathbf{P}\cdot\mathbf{P}g_m^{(l)}=0$ in model (7.19), the $\Psi_m^{(l)}$ will automatically be solutions of the Laplace equation corresponding to system 5.)

The relations

$$J^0\Psi_l^{(l)} = l\Psi_l^{(l)}, \qquad D\Psi_l^{(l)} = \left(l + \tfrac{1}{2}\right)\Psi_l^{(l)}, \qquad K^0\Psi_l^{(l)} = 0$$

imply $\Psi_l^{(l)} = \Gamma(l+\tfrac{1}{2})(2t)^l(\rho/i)^{-l-1}$ to within a constant multiple. From

(7.19) we have

$$\exp(-i\alpha P^0)\Psi_m^{(l)} = \sum_{n=0}^{\infty} \frac{(i\alpha)^n}{n!} \Psi_m^{(l+n)} \qquad (7.22)$$

and from (7.21),

$$\exp(-i\alpha P^0)\Psi_m^{(l)}(w,t,\rho) = \Psi_m^{(l)}\left[(w+\alpha/\rho)(1+\alpha^2/\rho^2+2\alpha w/\rho)^{-1/2},\right.$$
$$\left. t(1+\alpha^2/\rho^2+2\alpha w/\rho)^{-1/2}, \rho(1+\alpha^2/\rho^2+2\alpha w/\rho)^{1/2}\right]. \qquad (7.23)$$

Substituting (7.23) into (7.22), setting $m = l$, and using our explicit expression for $\Psi_l^{(l)}$, we obtain a simple generating function for the eigenfunctions $\Psi_l^{(l+n)}$. Comparing this expression with (5.24), we find

$$\Psi_m^{(l)}(w,t,\rho) = (l-m)!\,\Gamma\left(m+\tfrac{1}{2}\right)C_{l-m}^{m+\frac{1}{2}}(w)(2t)^m(\rho/i)^{-l-1}. \qquad (7.24)$$

Indeed, we can check directly that all of the recurrence relations (7.19) are satisfied by these functions. (The relations coincide exactly with the known differential recurrence relations obeyed by the Gegenbauer polynomials.) The general identity for Gegenbauer polynomials obtained by substituting (7.23) into (7.22) is

$$\left[1-2w+\alpha^2\right]^{-\nu-k/2}C_k^{\nu}\left[(w-\alpha)(1-2\alpha w+\alpha^2)^{-1/2}\right]$$
$$= \sum_{n=0}^{\infty} \alpha^n \binom{k+n}{n} C_{n+k}^{\nu}(w), \qquad |\alpha^2-2\alpha w| < 1, \qquad (7.25)$$

which reduces to (5.24) when $k = 0$.

Similarly, consideration of the expression

$$\exp(-\alpha P^+)\Psi_m^{(l)} = \sum_{n=0}^{\infty} \frac{\alpha^n}{n!} \Psi_{m+n}^{(l+n)}$$

leads to the identity

$$(1-\alpha)^{-\nu-k/2}C_k^{\nu}\left[w(1-\alpha)^{-1/2}\right] = \sum_{n=0}^{\infty} \frac{\alpha^n}{n!}\frac{\Gamma(\nu+n)}{\Gamma(\nu)}C_k^{\nu+n}(w), \qquad |\alpha| < 1; \qquad (7.26)$$

ISBN-0-201-13503-5

consideration of $\exp(\alpha K^0)\Psi_m^{(l)}$ leads to

$$(1+2\alpha w+\alpha^2)^{k/2}C_k^\nu\left[(w+\alpha)(1+2\alpha w+\alpha^2)^{-1/2}\right]$$

$$=\sum_{n=0}^{k}\alpha^l\binom{2\nu+k-1}{n}C_{k-n}^\nu(w),\qquad(7.27)$$

and so on. For a more complete list of such expansions see [84].

Another type of identity obtainable from (7.19) is closely related to the Maxwell theory of poles. The identity $(P^0)^n g_l^{(l)}=(-1)^n g_l^{(l+n)}$, obvious from (7.19), leads to

$$n!\rho^{-\nu-n-\frac{1}{2}}C_n^\nu(w)=(w\,\partial_\rho+\rho^{-1}(1-w^2)\,\partial_w-\rho^{-1}w(\nu-\tfrac{1}{2}))^n\rho^{-\nu-\frac{1}{2}}$$

$$n=0,1,2,\ldots.$$

More generally we can use Weisner's method to derive expansions of the form

$$\mathbf{T}(g)\Psi(w,t,\rho)=\sum_{m,l}a_{m,l}C_{l-m}^{m+\frac{1}{2}}(w)t^m\begin{Bmatrix}\rho^l\\\rho^{-l-1}\end{Bmatrix}\qquad(7.28)$$

even when g is bounded away from the identity element in the conformal group or Ψ is a solution of the Laplace equation not on the spherical orbit. We give one simple example related to the cylindrical orbit. A solution of the equations

$$\mathbf{P}\cdot\mathbf{P}\Psi=0,\qquad -iP^0\Psi=\lambda\Psi,\qquad J^0\Psi=m\Psi,\qquad m,\lambda\in\mathbb{C},$$

is

$$\Psi(w,t,\rho)=\left[t/\left(\lambda(w^2-1)^{1/2}\right)\right]^m e^{\lambda w\rho}I_m\left[\lambda\rho(w^2-1)^{1/2}\right]$$

where $I_m(z)$ is a modified Bessel function. In this case (7.28) yields

$$\Psi(w,t,\rho)=\sum_{n=0}^{\infty}a_n(\lambda)\rho^{m+n}t^m C_n^{m+\frac{1}{2}}(w).$$

The constants $a_n(\lambda)$ can be evaluated by setting $w=1$ on both sides of the equation, yielding the final result

$$\Gamma(m+1)\left[\rho(w^2-1)^{1/2}\right]^{-m}e^{\rho w}I_m\left[\rho(w^2-1)^{1/2}\right]$$

$$=\sum_{n=0}^{\infty}\frac{\Gamma(2m+1)}{\Gamma(2m+n+1)}C_n^{m+\frac{1}{2}}(w)\rho^n.\qquad(7.29)$$

Every analytic Ψ obtainable from separation of variables in the complex Laplace equation will lead to an expansion (7.28). Such functions can be obtained by analytic continuation of the separable solutions of the real Laplace equation and by continuation of separable solutions of the wave equation

$$(\partial_{tt} - \Delta_2)\Phi(x,y,t) = 0$$

to be studied in the following sections. (Set $t = iz$.) Thus there are an enormous number of generating functions for Gegenbauer polynomials that are obtainable in this way. In general (7.28) is a double sum but if $\mathbf{T}(g)\Psi$ is an eigenfunction of J^0, then m is fixed and only l is summed. These functions are just the solutions of (6.33) and can be obtained by choosing Ψ as one of the separable solutions corresponding to this equation and g as an element in the complex group $SL(2,\mathcal{C})$ generated by P^0, K^0, D. In [129], Viswanathan has given a detailed derivation of the generating functions that can be obtained in this manner, with the exception of the difficult Lamé systems. Equation (7.28) also reduces to a single sum when $\mathbf{T}(g)\Psi$ is an eigenfunction of D. Then l is fixed and only m is summed. Coordinate systems in which D is diagonal are discussed in Section 4.3.

Finally, we remark that quadratic transformation formulas for the hypergeometric function $_2F_1$ can be obtained from the conformal symmetry of the complex Laplace equation [93].

Exercises

1. Show that $\mathcal{E}(3)$ is decomposed into three orbits under the adjoint action of $E(3)$.

2. Verify that the Helmholtz equation separates in parabolic cylindrical coordinates $x = (\xi^2 - \eta^2)/2$, $y = \xi\eta$, $z = z$, and that the corresponding defining operators are $\{J_3, P_2\}$ and P_3^2.

3. Use expressions (4.10), (4.11) to compute the bilinear expansions of the function $\sin(\omega R)/\omega R$ in terms of separable solutions of the Helmholtz equation in spherical and prolate spheroidal coordinates.

4. Compute the symmetry algebra of the Laplace equation $\Delta_3\psi = 0$.

5. Show that the change of variables $x = u$, $y - iz = s$, $y + iz = 2t$ and the substitution $\Psi = e^{\lambda s}\Phi(t,u)$ reduce the complex Laplace equation $(\partial_{xx} + \partial_{yy} + \partial_{zz})\Psi = 0$ to the heat equation for Φ.

ISBN-0-201-13503-5

CHAPTER 4 _____

The Wave Equation

4.1 The Equation $\Psi_{tt} - \Delta_2 \Psi = 0$

Here we are concerned with the real wave equation

$$(\partial_{00} - \partial_{11} - \partial_{22})\Psi(x) = 0, \qquad x = (x_0, x_1, x_2). \tag{1.1}$$

It is well known that the symmetry algebra of (1.1) is ten dimensional with basis the momentum and energy operators

$$P_\alpha = \partial_\alpha, \qquad \alpha = 0, 1, 2, \tag{1.2}$$

the generators of homogeneous Lorentz transformations

$$M_{12} = x_1 \partial_2 - x_2 \partial_1, \qquad M_{01} = x_0 \partial_1 + x_1 \partial_0, \qquad M_{02} = x_0 \partial_2 + x_2 \partial_0, \tag{1.3}$$

the generator of dilatations

$$D = -\left(\tfrac{1}{2} + x_0 \partial_0 + x_1 \partial_1 + x_2 \partial_2\right), \tag{1.4}$$

and the generators of special conformal transformations

$$
\begin{aligned}
K_0 &= -x_0 + \left(x \cdot x - 2x_0^2\right)\partial_0 - 2x_0 x_1 \partial_1 - 2x_0 x_2 \partial_2, \\
K_1 &= x_1 + \left(x \cdot x + 2x_1^2\right)\partial_1 + 2x_1 x_0 \partial_0 + 2x_1 x_2 \partial_2, \\
K_2 &= x_2 + \left(x \cdot x + 2x_2^2\right)\partial_2 + 2x_2 x_0 \partial_0 + 2x_2 x_1 \partial_1
\end{aligned}
\tag{1.5}
$$

where

$$x \cdot y = x_0 y_0 - x_1 y_1 - x_2 y_2 = x_0 y_0 - \mathbf{x} \cdot \mathbf{y}.$$

(We are ignoring the trivial symmetry E.)

———————

ENCYCLOPEDIA OF MATHEMATICS and Its Applications, Gian-Carlo Rota (ed.).
Vol. 4: Willard Miller, Jr., Symmetry and Separation of Variables

ISBN-0-201-13503-5

It is convenient to introduce another basis for the symmetry algebra which clearly displays the isomorphism between this algebra and $so(3,2)$. We define $so(3,2)$ as the ten-dimensional Lie algebra of 5×5 real matrices $\mathcal{C}$ such that $\mathcal{C} G^{3,2} + G^{3,2} \mathcal{C}^t = 0$ where

$$
G^{3,2} = \begin{pmatrix} 1 & & & & \\ & 1 & & & \\ & & 1 & & \\ & & & -1 & \\ & & & & -1 \end{pmatrix} = \sum_{j=1}^{3} \mathcal{E}_{jj} - \sum_{k=4}^{5} \mathcal{E}_{kk}
$$

and $\mathcal{E}_{ij}$ is defined by (6.4), Section 3.6. It is straightforward to check that the matrices

$$
\begin{aligned}
\Gamma_{ab} &= \mathcal{E}_{ab} - \mathcal{E}_{ba} = -\Gamma_{ba}, & a \neq b, \\
\Gamma_{aB} &= \mathcal{E}_{aB} + \mathcal{E}_{Ba} = \Gamma_{Ba}, & 1 \leqslant a,b \leqslant 3, \quad 4 \leqslant A,B \leqslant 5, \quad (1.6) \\
\Gamma_{AB} &= -\mathcal{E}_{AB} + \mathcal{E}_{BA} = -\Gamma_{BA},
\end{aligned}
$$

form a basis for $so(3,2)$ with commutation relations

$$
\begin{aligned}
[\Gamma_{ab}, \Gamma_{cd}] &= \delta_{bc}\Gamma_{ad} + \delta_{ad}\Gamma_{bc} + \delta_{ca}\Gamma_{db} + \delta_{db}\Gamma_{ca}, \\
[\Gamma_{aB}, \Gamma_{cd}] &= -\delta_{ad}\Gamma_{cB} + \delta_{ac}\Gamma_{dB}, \quad [\Gamma_{Ab}, \Gamma_{45}] = \delta_{A5}\Gamma_{4b} - \delta_{A4}\Gamma_{5b}, \quad (1.7) \\
[\Gamma_{aB}, \Gamma_{cD}] &= \delta_{BD}\Gamma_{ac} - \delta_{ac}\Gamma_{BD}, \quad [\Gamma_{ab}, \Gamma_{45}] = 0.
\end{aligned}
$$

This Γ basis is related to our other basis via the identifications

$$
\begin{aligned}
P_0 &= \Gamma_{14} + \Gamma_{45}, & P_1 &= \Gamma_{12} + \Gamma_{25}, & P_2 &= \Gamma_{13} + \Gamma_{35}, \\
K_0 &= \Gamma_{14} - \Gamma_{45}, & K_1 &= \Gamma_{12} - \Gamma_{25}, & K_2 &= \Gamma_{13} - \Gamma_{35}, \quad (1.8) \\
M_{12} &= \Gamma_{23}, & M_{01} &= \Gamma_{42}, & M_{02} &= \Gamma_{43}, & D &= \Gamma_{15}.
\end{aligned}
$$

The symmetry operators can be exponentiated to obtain a local Lie transformation group of symmetries of (1.1). In particular, the momentum and Lorentz operators generate the Poincaré group of symmetries

$$
\Psi(x) \rightarrow \Psi(x\Lambda + a), \qquad a = (a_0, a_1, a_2), \quad \Lambda \in SO(1,2); \qquad (1.9)
$$

the dilatation operator generates

$$
\exp(\lambda D)\Psi(x) = \exp(-\lambda/2)\Psi[\exp(-\lambda)x]; \qquad (1.10)
$$

ISBN-0-201-13503-5

and the K_α generate the special conformal transformations

$$\exp(a \cdot K)\Psi(x) = \left[1 + 2x \cdot a + (a \cdot a)(x \cdot x)\right]^{-1/2} \Psi\left(\frac{x + a(x \cdot x)}{1 + 2x \cdot a + (a \cdot a)(x \cdot x)}\right).$$

$$(1.11)$$

In addition, we shall consider the inversion, space reflection, and time reflection symmetries,

$$I\Psi(x) = \left[-x \cdot x\right]^{-1/2}\Psi(-x/(x \cdot x)), \qquad S\Psi(x) = \Psi(x_0, -x_1, x_2),$$
$$T\Psi(x) = \Psi(-x_0, x_1, x_2), \qquad I = I^{-1}, \qquad S = S^{-1}, \qquad T = T^{-1},$$

$$(1.12)$$

which are not generated by the local symmetry operators. It follows from the expression for the inversion I that

$$IK_\alpha I^{-1} = -P_\alpha, \qquad IDI^{-1} = -D,$$
$$IM_{\alpha\beta}I^{-1} = M_{\alpha\beta}.$$

$$(1.13)$$

In analogy with the treatment of the Laplace equation in Section 3.6, we can verify that the wave equation is class I. Furthermore, although the space of symmetric second-order elements in the enveloping algebra of $so(3,2)$ is 35 dimensional, there are 20 linearly independent relations between these operators on the solution space of (1.1). For example, we have

$$\text{(i)} \quad P_0^2 - P_1^2 - P_2^2 = K_0^2 - K_1^2 - K_2^2 = 0,$$

$$\text{(ii)} \quad \Gamma_{12}^2 + \Gamma_{13}^2 + \Gamma_{23}^2 = \tfrac{1}{4} + \Gamma_{45}^2,$$

$$\text{(iii)} \quad M_{12}^2 - M_{01}^2 - M_{02}^2 = \tfrac{1}{4} - D^2,$$

$$\text{(iv)} \quad \Gamma_{45}^2 - \Gamma_{41}^2 - \Gamma_{51}^2 = \tfrac{1}{4} + \Gamma_{23}^2,$$

$$(1.14)$$

valid when applied to solutions of (1.1).

As is well known [44, 66, 118], by formally taking the Fourier transform in the variables x_α we can express a solution $\Psi(x)$ of (1.1) in the form

$$\Psi(x) = (4\pi)^{-1} \iint_{-\infty}^{\infty} \left[\exp(ik \cdot x)f(\mathbf{k}) + \exp(i\tilde{k} \cdot x)\tilde{f}(\mathbf{k})\right] d\mu(\mathbf{k}) \quad (1.15)$$

where $k_0 = (k_1^2 + k_2^2)^{1/2}$, $\tilde{k} = (-k_0, k_1, k_2)$ and $d\mu(\mathbf{k}) = dk_1 \, dk_2 / k_0$. Let $\mathcal{H} = \mathcal{H}_+ \oplus \mathcal{H}_-$ be the space of all ordered pairs of complex-valued functions

ISBN-0-201-13503-5

$\mathbf{F(k)} = \{f(\mathbf{k}), \tilde{f}(\mathbf{k})\}$ defined on R^2 such that

$$\int\int (|f|^2 + |\tilde{f}|^2)\, d\mu(\mathbf{k}) < \infty$$

(Lebesgue integral), and consider the indefinite inner product on $\mathcal{H}$ given by

$$\langle \mathbf{F}, \mathbf{G} \rangle = \int\int \left(f\bar{g} - \tilde{f}\bar{\tilde{g}} \right) d\mu(\mathbf{k}). \qquad (1.16)$$

Then [44, 66] the functions Ψ, Φ related to $\mathbf{F}$, $\mathbf{G}$ by (1.15) satisfy

$$\langle \Psi, \Phi \rangle \equiv \langle \mathbf{F}, \mathbf{G} \rangle = 2i \int\int_{x_0 = t} \left(\Psi(x)\partial_0\bar{\Phi}(x) - [\partial_0\Psi(x)]\bar{\Phi}(x) \right) dx_1\, dx_2, \quad (1.17)$$

independent of t. (More precisely, (1.17) can be derived from (1.16) by first considering the dense subspace of $\mathcal{H}$ consisting of C^∞ functions with compact support bounded away from $(0,0)$ and then passing to the limit. For $\mathbf{F} \in \mathcal{H}$ the corresponding $\Psi(x)$ is a solution of (1.1) in the weak sense of distribution theory; it may not be true that Ψ is two times continuously differentiable in each variable.)

The operators (1.2)–(1.5) acting on solutions of (1.1) induce corresponding operators on $\mathcal{H}$ under which $\mathcal{H}_+$ and $\mathcal{H}_-$ are separately invariant. Indeed, with repeated integrations by parts we can establish that the action of these operators on $\mathcal{H}_+$ is

$$P_0 = ik_0, \qquad P_j = -ik_j, \quad j = 1, 2, \quad M_{12} = k_1\partial_2 - k_2\partial_1,$$

$$M_{01} = k_0\partial_1, \quad M_{02} = k_0\partial_2, \qquad\qquad D = \tfrac{1}{2} + k_1\partial_1 + k_2\partial_2,$$

$$K_0 = ik_0(\partial_{11} + \partial_{22}), \qquad\qquad K_1 = i(k_1\partial_{11} - k_1\partial_{22} + 2k_2\partial_{12} + \partial_1),$$

$$K_2 = i(-k_2\partial_{11} + k_2\partial_{22} + 2k_1\partial_{12} + \partial_2). \qquad (1.18)$$

The action on $\mathcal{H}_-$ is the same except that k_0 is replaced by $-k_0$ in each of expressions (1.18). Moreover, it is straightforward to verify that these operators are skew-Hermitian on $\mathcal{H}_+$ and $\mathcal{H}_-$ separately.

The induced operators S, T on $\mathcal{H}$ are

$$S\mathbf{F}(k_1, k_2) = \mathbf{F}(-k_1, k_2), \qquad T\mathbf{F}(\mathbf{k}) = (\tilde{f}(\mathbf{k}), f(\mathbf{k})). \qquad (1.19)$$

Thus, $\mathcal{H}_+$ and $\mathcal{H}_-$ are invariant under S but are interchanged by T. In view of this interchange property of T, we will henceforth limit ourselves to consideration of elements in the Hilbert space $\mathcal{H}_+$, that is, the positive

ISBN-0-201-13503-5

energy solutions

$$\Psi(x)=(4\pi)^{-1}\iint_{-\infty}^{\infty}\exp(ik\cdot x)f(\mathbf{k})\,d\mu(\mathbf{k}).\qquad(1.20)$$

The inner product on $\mathcal{H}_+$ is

$$\langle f,g\rangle=\iint_{-\infty}^{\infty}f(\mathbf{k})\bar{g}(\mathbf{k})\,d\mu(\mathbf{k})\qquad(1.21)$$

and

$$\langle\Psi,\Phi\rangle\equiv\langle f,g\rangle=4i\iint_{x_0=t}\Psi(x)\partial_0\bar{\Phi}(x)\,dx_1\,dx_2$$
$$=-4i\iint_{x_0=t}\bar{\Phi}(x)\partial_0\Psi(x)\,dx_1\,dx_2.\qquad(1.22)$$

Furthermore, if Ψ is given by (1.20), we have

$$f(\mathbf{k})=k_0\pi^{-1}\iint_{-\infty}^{\infty}\Psi(x)\exp(-ik\cdot x)\,dx_1\,dx_2.\qquad(1.23)$$

By employing arguments analogous to those in [66], we can show that $\mathcal{H}_+$ is invariant under I and

$$If(\mathbf{k})=(2\pi)^{-1}\iint_{-\infty}^{\infty}\cos\left[(2l\cdot k)^{1/2}\right]f(\mathbf{l})\,d\mu(\mathbf{l}),\qquad f\in\mathcal{H}_+,\ I^2=E,\quad(1.24)$$

where E is the identity operator on $\mathcal{H}_+$. Clearly, I extends to a unitary self-adjoint operator on $\mathcal{H}_+$ with eigenvalues ±1.

If $\{\Psi_\alpha(x)\}$ is an ON basis for $\mathcal{H}_+$, then (in the sense of distributions)

$$\sum_\alpha\bar{\Psi}_\alpha(x)\Psi_\alpha(x')=\Delta_+(x-x')=(16\pi^2)^{-1}\int\int\exp\left[ik(x'-x)\right]d\mu(\mathbf{k})\quad(1.25)$$

where the distribution Δ_+ has the explicit expression

$$\Delta_+(x)=\begin{cases}2\pi i(t^2-r^2)^{-1/2}, & t>r\\-2\pi i(t^2-r^2)^{-1/2}, & t<-r,\\2\pi(r^2-t^2)^{-1/2}, & -r<t<r,\end{cases}\quad r=(x_1^2+x_2^2)^{1/2},\quad(1.26)\quad t=x_0.$$

The computation of (1.26) is carried out in analogy with the corresponding result for four-dimensional space-time [32]. It follows immediately that

$$\Psi(x)=\langle\Psi,\Delta_+(x'-x)\rangle \tag{1.27}$$

where the integration is carried out over x'.

It is well known that the representation of $so(3,2)$ on $\mathcal{K}_+$ induced by the operators (1.18) exponentiates to a global irreducible unitary representation of a covering group $\widetilde{SO}(3,2)$ of the identity component of $SO(3,2)$ [44]. The maximal connected compact subgroup of $\widetilde{SO}(3,2)$ is $SO(3)\times SO(2)$ where $SO(3)$ is generated by $\Gamma_{12}, \Gamma_{13}, \Gamma_{23}$ and $SO(2)$ by Γ_{45}. We will determine the explicit action of this subgroup on $\mathcal{K}_+$ as well as the action of several other interesting subgroups of $\widetilde{SO}(3,2)$.

The operators M_{01}, M_{02}, M_{12} generate a subgroup of $\widetilde{SO}(3,2)$ isomorphic to $SO(2,1)$ (see section 4.3). The action of this subgroup on $\mathcal{K}_+$ is determined by

$$\exp(\theta M_{12})f(\mathbf{k})=f(k_1\cos\theta-k_2\sin\theta,\,k_1\sin\theta+k_2\cos\theta),$$

$$\exp(aM_{01})f(\mathbf{k})=f(k_1(a),k_2), \tag{1.28}$$

$$k_1(a)=\left[e^a(k_1+k_0)^2-e^{-a}k_2^2\right]/2(k_1+k_0), \qquad f\in\mathcal{K}_+.$$

(The result for M_{02} follows easily from that for M_{01}.)

The P_α generate a translation subgroup of $\widetilde{SO}(3,2)$:

$$\exp\left(\sum a_\alpha P_\alpha\right)f(\mathbf{k})=\exp(ia\cdot k)f(\mathbf{k}). \tag{1.29}$$

Unitary operators of the form $\exp(\sum a_\alpha K_\alpha)$ are more difficult to compute. In [60] it is shown that

$$\exp(aK_0)f(\mathbf{s})=-i(2\pi a)^{-1}\iint_{-\infty}^{\infty}\exp\left[i(s_0+l_0)/a\right]$$

$$\times\cos\left\{a^{-1}\left[2(s_0l_0+s_1l_1+s_2l_2)\right]^{1/2}\right\}f(\mathbf{l})\,d\mu(\mathbf{l}), \quad (1.30)$$

$$\exp(aK_1)f(\mathbf{s})=(8\pi|a|)^{-1}\iint_{-\infty}^{\infty}\exp\left[i\frac{(s_1+l_1)}{a}\right]$$

$$\times\cos\left[\frac{s_1(l_2+l_0)-l_1(s_2+s_0)}{a(s_2+s_0)^{1/2}(l_2+l_0)^{1/2}}\right]f(\mathbf{l})\,d\mu(\mathbf{l}) \quad (1.31)$$

ISBN-0-201-13503-5

for $f \in \mathcal{K}_+$, and

$$\exp\left[a(K_0 + K_1)\right]f(\mathbf{s}) = \left[4\pi i a(s_0 - s_1)\right]^{-1/2} \int_{-\infty}^{\infty}$$

$$\times \exp\left[\frac{-(s_2 - w)^2}{4ia(s_0 - s_1)}\right] f\left(\frac{w^2 - (s_0 - s_1)^2}{2(s_0 - s_1)}, w\right) dw. \qquad (1.32)$$

The dilatation operator D generates the subgroup

$$\exp(aD)f(\mathbf{k}) = \exp(a/2)f(e^a\mathbf{k}). \qquad (1.33)$$

We can now easily exponentiate the compact generator $\Gamma_{45} = (P_0 - K_0)/2$. Indeed, the operators P_0, D, and K_0 generate a $SL(2,R)$ subgroup of $\widetilde{SO}(3,2)$. From (1.17) in Chapter 2 it is easy to verify the relation

$$\exp(2\theta\Gamma_{45}) = \exp(\tan(\theta)P_0)\exp(-K_0\sin\theta\cos\theta)\exp(-2D\ln\cos\theta),$$

and evaluating the right-hand side we find

$$\exp(2\theta\Gamma_{45})f(\mathbf{k}) = i(2\pi)^{-1}\csc(\theta)\int\int\exp[-i(k_0 + l_0)\cot\theta]$$

$$\times\cos\{\csc(\theta)[2(k_0l_0 + k_1l_1 + k_2l_2)]^{1/2}\}f(\mathbf{l})\,d\mu(\mathbf{l}), \qquad (1.34)$$

$$\theta \neq n\pi.$$

Similarly, the operators P_1, D, K_1 generate an $SL(2,R)$ subgroup of $\widetilde{SO}(3,2)$ and we can verify the relation

$$\exp(2\theta\Gamma_{12}) = \exp(\tan(\theta)P_1)\exp(K_1\sin\theta\cos\theta)\exp(-2D\ln\cos\theta),$$

$$2\Gamma_{12} = K_1 + P_1,$$

or

$$\exp(2\theta\Gamma_{12})f(\mathbf{k}) = (8\pi|\sin\theta|)^{-1}\exp(ik_1\cot\theta)\int\int\exp(il_1\cot\theta)$$

$$\times\cos\left[\frac{k_1(l_2 + l_0) - l_1(k_2 + k_0)}{\sin\theta(k_2 + k_0)^{1/2}(l_2 + l_0)^{1/2}}\right]f(\mathbf{l})\,d\mu(\mathbf{l}), \qquad (1.35)$$

$$\theta \neq n\pi.$$

The operators (1.35) together with the operators $\exp(\theta M_{12})$, (1.28), determine the action of the $SO(3)$ subgroup.

The known R-separable coordinate systems for (1.1) each correspond to a two-dimensional (commuting) subspace of the space of second-order symmetric elements in the enveloping algebra of $so(3,2)$. If the commuting operators form a basis for such a subspace, then the corresponding separated solutions of (1.1) are characterized by the eigenvalue equations

$$S_j\Psi = \lambda_j\Psi, \qquad j = 1,2,$$

see [60–62]. Coordinate systems are considered equivalent if they can be mapped into one another by transformations generated by $\widetilde{SO}(3,2)$, S, T, and I. If a separable system corresponds to a subspace with basis $S_j = L_j^2$, $j = 1,2$, such that $[L_1, L_2] = 0$ and $L_j \in so(3,2)$, we call these coordinates *split*. In this case one can diagonalize the first-order operators L_j. Such systems are the best known and most tractable. More generally, if a system corresponds to a subspace in which there exists a basis $S_1 = L^2$, S_2, $L \in so(3,2)$, we call these coordinates *semisplit*. Here, we can diagonalize the first-order operator L. If there exists no basis S_1, S_2 such that S_1 is the square of some $L \in so(3,2)$, we call the coordinates *nonsplit*. Nonsplit coordinates are the most intractable of all separable coordinates and appear the least frequently in applications.

A detailed (but still incomplete) study of R-separable solutions of (1.1) was carried out in [60–62]. Here we will be content with an examination of some of the most important semisplit systems. A given $L \in so(3,2)$ may correspond to several (or to no) semisplit systems. Indeed, if Ψ satisfies (1.1) and $L\Psi = i\lambda\Psi$, then since L is a symmetry of (1.1) we can introduce new variables y_0, y_1, y_2 such that $L = \partial_{y_0} + f(y)$ (where f may be zero) and $\Psi(y) = r(y)\exp(i\lambda y_0)\Phi_\lambda(y_1, y_2)$ where r is a fixed function satisfying $\partial_{y_0} r + fr = 0$. Then (1.1) reduces to a second-order partial differential equation for Φ_λ in the two variables y_1, y_2. The semisplit systems we will study each correspond to systems such that the reduced equation separates. In particular, $S_1 = L^2$ and S_2 corresponds to a second-order symmetry of the reduced equation.

In the next few sections we shall examine various possibilities for L that lead to semisplit systems.

4.2 The Laplace Operator on the Sphere

The first systems we shall study correspond to diagonalization of the operator Γ_{45}, (1.8). On restriction of our unitary irreducible representation of $\widetilde{SO}(3,2)$ on $\mathcal{H}_+$ to the compact subgroup $SO(3)$, this representation decomposes into a direct sum of irreducible representations D_l of $SO(3)$, $\dim D_l = 2l + 1$. We will determine a convenient basis for $\mathcal{H}_+$ which exhibits the decomposition. This is a basis of eigenfunctions of the com-

ISBN-0-201-13503-5

muting operators Γ_{45} and $\Gamma_{23} = M_{12}$:

$$\Gamma_{45}f = i\lambda f, \qquad \Gamma_{23}f = imf, \qquad -i\Gamma_{45} = (k_0/2)(-\partial_{11} - \partial_{22} + 1). \quad (2.1)$$

With $k_1 = k\cos\theta$, $k_2 = k\sin\theta$, $k_0 = k$ it is easy to show that the ON basis of eigenvectors is

$$f_m^{(l)}(\mathbf{k}) = [(l-m)!/\pi(l+m)!]^{1/2}(2k)^m e^{-k} L_{l-m}^{(2m)}(2k) e^{im\theta},$$

$$\lambda = l + \tfrac{1}{2}, \, l = 0, 1, \ldots, m = l, l-1, \ldots, -l. \quad (2.2)$$

From this result and (1.14ii), we see that the $\{f_m^{(l)}\}$ for fixed l form an ON basis for the representation D_l of $SO(3)$. Furthermore, the restriction of our representation of $\widetilde{SO}(3,2)$ to $SO(3)$ decomposes as $\sum_{l=0}^{\infty} \oplus D_l$. The known recurrence formulas for Laguerre polynomials imply

$$\Gamma_{15}f_m^{(l)} = \tfrac{1}{2}\big[(l-m+1)(l+m+1)\big]^{1/2} f_m^{(l+1)} - \tfrac{1}{2}\big[(l-m)(l+m)\big]^{1/2} f_m^{(l-1)},$$

$$\Gamma_{42}f_m^{(l)} = -\tfrac{1}{4}\big[(l+m+2)(l+m+1)\big]^{1/2} f_{m+1}^{(l+1)} + \tfrac{1}{4}\big[(l-m)(l-m-1)\big]^{1/2} f_{m+1}^{(l-1)}$$

$$+ \tfrac{1}{4}\big[(l+m)(l+m-1)\big]^{1/2} f_{m-1}^{(l-1)} - \tfrac{1}{4}\big[(l-m+1)(l-m+2)\big]^{1/2} f_{m-1}^{(l+1)}.$$

$$(2.3)$$

Using (2.1), (2.3) and taking commutators, we can compute the action of $\Gamma_{\alpha\beta}$ on this basis.

Note the close connection between the eigenvalue equation $\Gamma_{45}f = i\lambda f$ and the quantum Kepler problem in two-dimensional space:

$$Hg = \mu g, \qquad H = -\partial_{xx} - \partial_{yy} + e/r,$$

$$r = (x^2 + y^2)^{1/2}, \quad \iint_{R^2} |g|^2 \, dx\, dy < \infty. \quad (2.4)$$

The two eigenvalue equations can be identified provided $k_1 = x(-\mu)^{1/2}$, $k_2 = y(-\mu)^{1/2}$, $\mu = -e^2/4\lambda^2$. The eigenvalue problems are defined on Hilbert spaces with different inner products, but from the Virial theorem [31, p. 51] we see that if the energy eigenvalue μ belongs to the point spectrum of H and g is a corresponding eigenvector, then g also has finite norm in $\mathcal{H}_+$. Conversely, if f is an eigenfunction of Γ_{45}, then $\iint |f|^2 \, dx\, dy < \infty$ and f corresponds to an energy eigenvalue μ in the point spectrum of H. Since the eigenvalues λ of Γ_{45} are $\lambda = l + \tfrac{1}{2}$, $l = 0, 1, \ldots$, it follows that the point eigenvalues of H are $\mu_l = -e^2/4(l + \tfrac{1}{2})^2$. Although this is a satisfying explanation of the point spectrum of H, it sheds no light on the continuous spectrum of H, since Γ_{45} has only a point spectrum.

ISBN-0-201-13503-5

Using the mapping (1.20) we can compute the corresponding ON basis of positive energy solutions of (1.1):

$$\Psi_m^{(l)}(x) = \left[\frac{(l-m)!}{4\pi(l+m)!} \right]^{1/2} \exp\left[im\left(\alpha - \frac{\pi}{2}\right) \right]$$

$$\times \int_0^\infty \exp[(ix_0-1)k](2k)^m J_m(kr) L_{l-m}^{(2m)}(2k)dk, \qquad (2.5)$$

$$x_1 = r\cos\alpha, \; x_2 = r\sin\alpha.$$

In terms of the coordinates

$$x_0 = \sin\psi/(\cos\sigma - \cos\psi), \qquad x_1 = \sin\sigma\cos\alpha/(\cos\sigma - \cos\psi),$$
$$\qquad \qquad (2.6)$$
$$x_2 = \sin\sigma\sin\alpha/(\cos\sigma - \cos\psi),$$

variables R-separate in (1.1), (2.5) to give

$$\Psi_m^{(l)}(x) = (-i)^{m-1}\left[(\cos\sigma - \cos\psi)/(4l+2)\right]^{1/2}\exp\left[-i\psi\left(l+\tfrac{1}{2}\right)\right]Y_l^m(\sigma,\alpha),$$
$$\qquad \qquad (2.7)$$

where Y_l^m is a spherical harmonic. (We can always parametrize so $\cos\sigma - \cos\psi > 0$, see [63].) Indeed, on the solution space of (1.1) we find

$$\Gamma_{45} = -\partial_\psi + \tfrac{1}{2}\frac{\sin\psi}{\cos\sigma - \cos\psi}, \qquad \Gamma_{23} = \partial_\alpha, \qquad (2.8)$$

so

$$\Psi_m^{(l)} = (\cos\sigma - \cos\psi)^{1/2}\exp\left[-i\psi\left(l+\tfrac{1}{2}\right)\right]\exp(im\alpha)g(\sigma),$$

and substituting into (1.1), we see that variables R-separate and $g(\sigma)$ is a linear combination of $P_l^m(\cos\sigma)$, $Q_l^m(\cos\sigma)$. Evaluating the integral (2.5) for special values of the parameters (e.g., $\sigma=0,\pi$), we establish (2.7).

There is another model of our irreducible representation of $\widetilde{SO}(3,2)$ in which the eigenfunctions of Γ_{45} and Γ_{23} take an especially simple form. The representation space is the Bargmann–Segal Hilbert space $\mathcal{F}_2$ consisting of all entire functions $h(z_1,z_2)$ such that [11]

$$\int_{\mathbb{C}\times\mathbb{C}} |h|^2 d\xi(\mathbf{z}) < \infty, \quad d\xi(\mathbf{z}) = \pi^{-2}\exp[-(|z_1|^2+|z_2|^2)]dx_1\,dx_2\,dy_1\,dy_2, \quad (2.9)$$

$$z_j = x_j + iy_j.$$

ISBN-0-201-13503-5

The inner product is

$$\langle f, h \rangle = \int_{\mathcal{C} \times \mathcal{C}} f\bar{h}\, d\xi(\mathbf{z}).$$

The carrier space for our representation is not $\mathcal{F}_2$ but the subspace $\mathcal{F}_2^+$ consisting of all $h \in \mathcal{F}_2$ such that $h(-z_1, -z_2) = h(z_1, z_2)$. The functions

$$f_m^{(l)}(z) = z_1^{l+m} z_2^{l-m} / \left[(l+m)!(l-m)! \right]^{1/2},$$
$$l = 0, 1, 2, \ldots, m = l, \ldots, -l, \tag{2.10}$$

form an ON basis for $\mathcal{F}_2^+$. Setting

$$\Gamma_{45} = \frac{i}{2}(z_1 \partial_{z_1} + z_2 \partial_{z_2} + 1), \quad \Gamma_{15} = \frac{1}{2}(z_1 z_2 - \partial_{z_1 z_2}),$$
$$\Gamma_{23} = \frac{i}{2}(z_1 \partial_{z_1} - z_2 \partial_{z_2}), \quad \Gamma_{42} = \frac{1}{4}\left(\partial_{z_1 z_1} + \partial_{z_2 z_2} - z_1^2 - z_2^2 \right), \tag{2.11}$$

and comparing with expressions (2.3), we see that there is a new model of our representation of $\widetilde{SO}(3,2)$ in which the functions $f_m^{(l)}(\mathbf{k})$ can be identified with the functions (2.10). The explicit unitary mapping U from $\mathcal{K}^+$ to $\mathcal{F}_2^+$ that commutes with the group action is

$$Uf(\mathbf{z}) = \iint_{R^2} U(\mathbf{k}, \mathbf{z}) f(\mathbf{k})\, d\mu(\mathbf{k}), \quad f \in \mathcal{K}^+, \tag{2.12}$$

where

$$U(\mathbf{k}, \mathbf{z}) = \sum_{l,m} \bar{f}_m^{(l)}(\mathbf{k}) f_m^{(l)}(\mathbf{z}) = \pi^{-1/2} \exp(-k + z_1 z_2)$$
$$\times \cosh\left\{ \sqrt{2k}\, \left[z_1 \exp(i\theta/2) - z_2 \exp(i\theta/2) \right] \right\}, \tag{2.13}$$
$$k_1 = k \cos\theta, \quad k_2 = k \sin\theta.$$

(Note that $f_m^{(l)}(\mathbf{k}) \in \mathcal{K}^+$ and $f_m^{(l)}(\mathbf{z}) \in \mathcal{F}_2^+$.)

To understand more clearly the significance of the coordinates (2.6), note that if Ψ is a solution of (1.1) such that $\Gamma_{45}\Psi = i(l + \frac{1}{2})\Psi$, then $\Psi(\sigma, \alpha, \psi) = (\cos\alpha - \cos\psi)^{1/2} \exp[-i\psi(l + \frac{1}{2})]\Phi(\sigma, \alpha)$ where Φ is an eigenfunction of the Laplace operator on the sphere ((2.20), Section 3.2) $(\sigma = \theta, \alpha = \varphi)$. Equation (2.20) separates in two coordinate systems, as we saw in Section 3.3. The first system (spherical coordinates $\{\sigma, \alpha\}$) leads to the R-separable solutions (2.7) of (1.1) that are characterized by diagonalization of the operators

1. Γ_{45}^2, Γ_{23}^2.

However, there is also a Lamé-type system which leads to R-separable

ISBN-0-201-13503-5

solutions of (1.1) characterized by diagonalization of

2. $\Gamma_{45}^2,\ \Gamma_{12}^2+a^2\Gamma_{13}^2,\qquad a>0.$

The overlaps between these bases are just those computed in Section 3.3.

4.3 Diagonalization of P_0, P_2, and D

We next look for those coordinate systems permitting separation of variables in (1.1) such that the corresponding basis functions Ψ are eigenfunctions of P_0: $P_0\Psi=i\omega\Psi$. For such systems we have $\Psi(x)=\exp(i\omega x_0)\Phi(x_1,x_2)$ where

$$\left(\partial_{11}+\partial_{22}+\omega^2\right)\Phi=0. \tag{3.1}$$

Thus the equation for the eigenfunctions reduces to the Helmholtz equation. Now P_0 commutes with every element in the Euclidean Lie algebra $\mathcal{E}(2)$ generated by P_1, P_2, M_{12} and, as we know from Chapter 1, $\mathcal{E}(2)$ is the symmetry algebra of (3.1). Furthermore, (3.1) separates in four coordinate systems, each system corresponding to a symmetric second-order element in the enveloping algebra of $\mathcal{E}(2)$ (see Table 1). The four associated separable systems for (1.1) are characterized by diagonalization of the operators in Table 18.

On $\mathcal{H}_+$ the requirement $P_0 f=i\omega f$ implies $f(\mathbf{k})=\delta(k-\omega)g_\omega(\theta)$ where $\omega>0$, $k_1=k\cos\theta$, $k_2=k\sin\theta$. The search for the functions g_ω reduces to a study of the Hilbert space $L_2(S_2)$ on which $E(2)$ acts via

$$P_1=-i\omega\cos\theta,\qquad P_2=-i\omega\sin\theta,\qquad M_{12}=\partial_\theta.$$

These operators determine a unitary irreducible representation of $E(2)$ on $L_2(S_2)$. Once the eigenfunctions $g_{\omega\mu}(\theta)$ of the second operator in 3–6 in Table 18 have been determined, the corresponding separated solutions $\Psi_{\omega\mu}$ of (1.1) can be obtained from the relation

$$\Psi_{\omega\mu}(x)=(4\pi)^{-1}\exp(i\omega x_0)\int_{-\pi}^{\pi}\exp\left[-i\omega(x_1\cos\theta+x_2\sin\theta)\right]g_{\omega\mu}(\theta)\,d\theta. \tag{3.2}$$

Table 18

3	P_0^2, P_1^2	Cartesian
4	P_0^2, M_{12}^2	Polar
5	$P_0^2, \{M_{12}, P_2\}$	Parabolic cylinder
6	$P_0^2, M_{12}^2+d^2P_2^2$	Elliptic

ISBN-0-201-13503-5

Note that this model is essentially identical to the circle model studied in Chapter 1. Thus, the spectral resolutions and overlaps computed there can be immediately carried over to the wave equation.

Now we search for coordinate systems allowing separation of variables in (1.1) such that the basis functions Ψ are eigenfunctions of P_2: $P_2\Psi = -i\omega\Psi$. Here we have $\Psi(x) = \exp(-i\omega x_2)\Phi(x_0, x_1)$ where

$$\left(\partial_{00} - \partial_{11} + \omega^2\right)\Phi = 0. \tag{3.3}$$

The operator P_2 commutes with the subalgebra $\mathcal{E}(1,1)$ generated by P_0, P_1, M_{01} and, indeed, $\mathcal{E}(1,1)$ is the symmetry algebra of (3.3). This equation separates in ten coordinate systems associated with ten symmetric second-order operators in the enveloping algebra of $\mathcal{E}(1,1)$ (see Table 2). The pairs of commuting operators associated with the corresponding separable solutions of (1.1) are listed in Table 19. The case 3' is equivalent to 3 in Table 18.

On $\mathcal{H}_+$ the requirement $P_2 f = -i\omega f$ implies $f(\mathbf{k}) = \delta(k_2 - \omega)g_\omega(\zeta)$ where $-\infty < \omega < \infty, k_1 = |k_2|\sinh\zeta, k_0 = |k_2|\cosh\zeta$. The search for eigenfunctions reduces to a study of the Hilbert space $L_2(R)$ on which $E(1,1)$ acts via

$$P_0 = i|\omega|\cosh\xi, \qquad P_1 = -i|\omega|\sinh\xi, \qquad M_{01} = \partial_\xi. \tag{3.4}$$

These operators define a unitary irreducible representation of $E(1,1)$ on $L_2(R)$. After the eigenfunctions $g_{\omega\mu}(\xi)$ of the second operator in 7–15 (Table 19) have been determined, the corresponding separable solutions $\Psi_{\omega\mu}$ of (1.1) follow from

$$\Psi_{\omega\mu}(x) = (4\pi)^{-1}\exp(-i\omega x_2)\int_{-\infty}^{\infty}\exp\left[i|\omega|(x_0\cosh\xi - x_1\sinh\xi)\right]g_{\omega\mu}|(\xi)d\xi. \tag{3.5}$$

This is virtually identical to the $L_2(R)$ model discussed in Chapter 1, and the spectral resolutions and overlaps derived there can be carried over to the wave equation.

Next we look for coordinate systems yielding separation of variables in (1.1) such that the basis functions Ψ are eigenfunctions of D: $D\Psi = -i\nu\Psi$.

Table 19

3'	P_2^2, P_0, P_1	11 $P_2^2, M_{01}^2 - P_0 P_1$
7	P_2^2, M_{01}^2	12 $P_2^2, M_{01}^2 + (P_0 + P_1)^2$
8	$P_2^2, \{M_{01}, P_1\}$	13 $P_2^2, M_{01}^2 - (P_0 + P_1)^2$
9	$P_2^2, \{M_{01}, P_0\}$	14 $P_2^2, M_{01}^2 + P_1^2$
10	$P_2^2, \{M_{01}, P_0 - P_1\} + (P_0 + P_1)^2$	15 $P_2^2, M_{01}^2 - P_1^2$

ISBN-0-201-13503-5

In this case we have $\Psi(x) = \rho^{i\nu - \frac{1}{2}}\Phi(s)$ where

$$x_\alpha = \rho s_\alpha \qquad (\rho \geqslant 0), \qquad s_0^2 - s_1^2 - s_2^2 = \varepsilon,$$

and $\varepsilon = \pm 1$ or 0 depending on whether $x \cdot x > 0$, < 0, or $= 0$. It follows from (1.14iii) that

$$\left(M_{12}^2 - M_{01}^2 - M_{02}^2\right)\Phi(s) = \left(\nu^2 + \tfrac{1}{4}\right)\Phi(s). \tag{3.6}$$

The operators $M_{\alpha\beta}$, (1.3), satisfy the commutation relations

$$\left[M_{12}, M_{01}\right] = -M_{02}, \qquad \left[M_{12}, M_{02}\right] = M_{01}, \qquad \left[M_{01}, M_{02}\right] = M_{12}, \tag{3.7}$$

so they form a basis for the subalgebra $sl(2, R) \cong so(2, 1)$ (see Section 2.1). Now D commutes with this subalgebra and in fact $SO(2, 1)$ is the symmetry group of (3.6). The Casimir operator $M_{12}^2 - M_{01}^2 - M_{02}^2$ commutes with all elements of $so(2, 1)$. As shown in [139], the space of second-order symmetry operators in the enveloping algebra of $so(2, 1)$, modulo the Casimir operator, is decomposed into nine orbit types under the adjoint action of $SO(2, 1)$. (The groups $SO(2, 1)$ and $SL(2, R)$ are locally isomorphic.) Moreover, the reduced equation (3.6) separates in nine coordinate systems, each system associated with a unique operator orbit. The coordinate systems for $\varepsilon = 1$ can be found in [58, 104], in which case (3.6) is the eigenvalue equation for the *Laplace operator on the hyperboloid*. Coordinates for all cases $\varepsilon = \pm 1, 0$ are derived in [61]. Referring to the papers just cited for details, we give here (Table 20) only the functional forms of the separated solutions of (3.6), the names of the coordinate systems, and the pairs of commuting operators associated with the corresponding separable solutions of (1.1). System 7' is equivalent to 7.

On $\mathcal{H}_+$ the requirement $Df = -i\nu f$ implies $f(\mathbf{k}) = k^{-i\nu - \frac{1}{2}} h_\nu(\theta)$ where $-\infty < \nu < \infty$, $k_1 = k\cos\theta$, $k_2 = k\sin\theta$. The eigenfunction problem thus reduces to a study of the Hilbert space $L_2(S_2)$ on which $SO(2, 1)$ acts via

$$M_{12} = \partial_\theta, \qquad M_{01} = -\sin\theta\,\partial_\theta - \left(i\nu + \tfrac{1}{2}\right)\cos\theta,$$
$$M_{02} = \cos\theta\,\partial_\theta - \left(i\nu + \tfrac{1}{2}\right)\sin\theta. \tag{3.8}$$

These operators define a unitary irreducible representation of $SO(2, 1)$ that is single valued and belongs to the principal series: $l = -\tfrac{1}{2} + i|\nu|$ (see [10, 115]). Once the eigenfunctions $h_{\nu\alpha}(\theta)$ of the second operator in 16–23 (Table 20) have been determined, the corresponding separable solutions

ISBN-0-201-13503-5

Table 20

Operators	Coordinates	Separated functions
16 D^2, M_{12}^2	Spherical	Exponential Associated Legendre
17 D^2, M_{01}^2	Equidistant	Exponential Associated Legendre
7' $D^2, (M_{12} - M_{02})^2$	Horocyclic	Exponential Macdonald
18 $D^2, M_{12}^2 + a^2 M_{01}^2$	Elliptic	Periodic Lamé Periodic Lamé
19 $D^2, M_{01}^2 - a^2 M_{12}^2,$ $0 < a < 1$	Hyperbolic	Lamé–Wangerin Lamé–Wangerin
20 $D^2, aM_{01}^2 - \{M_{12}, M_{02}\},$ $0 < a$	Semihyperbolic	Lamé–Wangerin Lamé–Wangerin
21 $D^2, aM_{01}^2 + M_{02}^2$ $+ M_{12}^2 - \{M_{12}, M_{02}\},$ $0 < a$	Elliptic-parabolic	Associated Legendre Associated Legendre
22 $D^2, -aM_{01}^2 + M_{02}^2 + M_{12}^2$ $- \{M_{12}, M_{02}\},$ $0 < a$	Hyperbolic-parabolic	Associated Legendre Associated Legendre
23 $D^2, \{M_{01}, M_{02}\}$ $- \{M_{12}, M_{01}\}$	Semicircular-parabolic	Bessel Macdonald

$\Psi_{\nu\alpha}$ of (1.1) can be obtained from

$$\Psi_{\nu\alpha}(x) = \rho^{i\nu - \frac{1}{2}} (4\pi)^{-1} \Gamma\left(\tfrac{1}{2} - i\nu\right) \int_0^{2\pi} \exp\left[\pm i\pi \left(\tfrac{1}{2} - i\nu\right)/2 \right]$$

$$\times |s_0 - s_1 \cos\theta - s_2 \sin\theta|^{i\nu - \frac{1}{2}} h_{\nu\alpha}(\theta) \, d\theta \qquad (3.9)$$

where the plus sign occurs when $s_0 - s_1\cos\theta - s_2\sin\theta > 0$ and the minus sign occurs when this expression is < 0. The spectral resolutions of the operators 16–23 and various overlaps computed in the $L_2(S_2)$ model can be found in [58]. (See also [54] for mixed-basis matrix elements corresponding to subgroup systems.)

4.4 The Schrödinger and EPD Equations

Of special interest are the coordinate systems permitting separation of variables in (1.1) such that the basis functions Ψ are eigenfunctions of $P_0 + P_1 : (P_0 + P_1)\Psi = i\beta\Psi$. For this case we have $\Psi(x) = e^{is\beta}\Phi(t, x_2)$ where $2s = x_0 + x_1, 2t = x_1 - x_0$. The reduced equation for Φ is the free-particle Schrödinger equation

$$\left(i\beta \partial_t + \partial_{x_2 x_2}\right)\Phi(t, x_2) = 0. \qquad (4.1)$$

This equation admits as symmetries the operators

$$\mathcal{K}_{-1}=P_2, \quad \mathcal{K}_{-2}=P_1-P_0, \quad \mathcal{K}_0=P_0+P_1, \quad \mathcal{K}_1=\tfrac{1}{2}(M_{02}-M_{12}),$$

$$\mathcal{K}^0=-D-M_{01}, \quad \mathcal{K}_2=-\tfrac{1}{2}(K_0+K_1), \tag{4.2}$$

which all commute with $P_0+P_1=\mathcal{K}_0$. As we showed in Section 2.1, these operators form a basis for the six-dimensional Schrödinger algebra $\mathcal{G}_2$, the symmetry algebra of (4.1). (Note that the constant β can be set equal to 1 in (4.1) by a renormalization of t and x_2.) The pairs of commuting operators associated with separable systems for (4.1) are listed in Table 21. (Coordinates $3''$ are essentially equivalent to 3 in Table 18.) These results follow from Table 6. Note that here the defining operators are first order, rather than second order, in the enveloping algebra. This is because they appear as first order in the explicit separation equations. All of the earlier listed semisubgroup coordinate systems have been orthogonal with respect to the Minkowski metric. However, the four systems in Table 21 are nonorthogonal.

On $\mathcal{K}_+$ the requirement $(P_0+P_1)f=i\beta f$ implies $f(\mathbf{k})=u\delta(u-\beta)l_\beta(v)$ where $\beta>0, u=k_0-k_1, v=k_2$. Thus, the search for the l_β reduces to a study of the Hilbert space $L_2(R)$ on which the Schrödinger group acts via

$$\mathcal{K}_0=i\beta, \quad \mathcal{K}_{-1}=-iv, \quad \mathcal{K}_1=\frac{\beta}{2}\partial_v, \quad \mathcal{K}^0=-\tfrac{1}{2}-v\partial_v,$$

$$\mathcal{K}_{-2}=-\frac{iv^2}{\beta}, \quad \mathcal{K}_2=-i\frac{\beta}{2}\partial_{vv}. \tag{4.3}$$

As shown in Section 2.1, these operators determine an irreducible unitary representation of the Schrödinger group on $L_2(R)$. (Indeed, for $\beta=1$ the operators (1.24), Section 2.1, are unitary equivalent via the Fourier transform to the operators given here, and for $\beta\neq 1$ our earlier results can easily be modified to yield the global group action.) Once the eigenfunctions $l_{\beta\mu}(v)$ of the second operators in $3'', 24$–26 (Table 21) have been determined, the corresponding separable solutions $\Psi_{\beta\mu}$ of (1.1) can be computed from

$$\Psi_{\beta\mu}(x)=(4\pi)^{-1}\exp(i\beta s)\int_{-\infty}^{\infty}\exp\left[-i(v^2t/\beta+vx_2)\right]l_{\beta\mu}(v)\,dv. \tag{4.4}$$

Table 21

$3''$	P_0+P_1, P_2	Free particle
24	$P_0+P_1, P_0-P_1-\tfrac{1}{4}K_0-\tfrac{1}{4}K_1$	Oscillator
25	$P_0+P_1, P_0-P_1+aM_{12}-aM_{02},$	Linear potential
	$a\neq 0$	
26	$P_0+P_1, D+M_{01}$	Repulsive oscillator

ISBN-0-201-13503-5

Next we look for coordinate systems yielding separation of variables for (1.1) such that the basis functions Ψ are eigenfunctions of M_{12}: $M_{12}\Psi = im\Psi$. We have $\Psi(x) = e^{im\varphi}\Phi(x_0, r)$ where $x_1 = r\cos\varphi$, $x_2 = r\sin\varphi$, and Φ satisfies the Euler–Poisson–Darboux (EPD) equation

$$\left(\partial_{00} - \partial_{rr} - r^{-1}\partial_r + m^2 r^{-2}\right)\Phi = 0 \tag{4.5}$$

or

$$\left(\Gamma_{45}^2 - \Gamma_{41}^2 - \Gamma_{51}^2\right)\Phi = \left(\Gamma_{23}^2 + \tfrac{1}{4}\right)\Phi = -\left(m + \tfrac{1}{2}\right)\left(m - \tfrac{1}{2}\right)\Phi \tag{4.6}$$

from (1.14iv). The symmetry algebra of (4.5) is $sl(2, R)$, generated by the operators $\Gamma_{45}, \Gamma_{41}, \Gamma_{51}$, and the symmetry group (for m integer) is $SL(2, R)$:

$$\left[\Gamma_{41}, \Gamma_{51}\right] = -\Gamma_{45}, \qquad \left[\Gamma_{41}, \Gamma_{45}\right] = -\Gamma_{51}, \qquad \left[\Gamma_{51}, \Gamma_{45}\right] = \Gamma_{41}. \tag{4.7}$$

In [63] it is shown that the EPD equation R-separates in exactly nine coordinate systems corresponding to the nine $SL(2, R)$ orbit types of symmetric second-order operators in the enveloping algebra of $sl(2, R)$, modulo the Casimir operator $\Gamma_{45}^2 - \Gamma_{41}^2 - \Gamma_{51}^2$. We list in Table 22 only the operator characterizations of the R-separable solutions of (1.1) together with the functional forms of the associated solutions of (4.5) ($\Gamma_{23} = M_{12}, \Gamma_{51} = D$, $\Gamma_{45} = (P_0 - K_0)/2$, $\Gamma_{41} = (P_0 + K_0)/2$). The truly R-separable systems are $1'$ and 29–31.

Table 22

Operators		Separated functions
$1'$	$\Gamma_{23}^2, \Gamma_{45}^2$	Exponential Gegenbauer
$4'$	$\Gamma_{23}^2, (\Gamma_{45} + \Gamma_{41})^2$	Exponential Bessel
$16'$	$\Gamma_{23}^2, \Gamma_{51}^2$	Exponential Associated Legendre
27	$\Gamma_{23}^2, 2\Gamma_{41}^2 + \{\Gamma_{45}, \Gamma_{41}\}$	Associated Legendre Associated Legendre
28	$\Gamma_{23}^2, 2\Gamma_{45}^2 + \{\Gamma_{45}, \Gamma_{41}\}$	Associated Legendre Associated Legendre
29	$\Gamma_{23}^2, \Gamma_{41}^2 + a\{\Gamma_{45}, \Gamma_{51}\}$	Lamé–Wangerin Lamé–Wangerin
30	$\Gamma_{23}^2, \Gamma_{45}^2 + a\Gamma_{51}^2,$ $a > 0$	Lamé–Wangerin Lamé–Wangerin
31	$\Gamma_{23}^2, a\Gamma_{41}^2 + \Gamma_{51}^2,$ $a > 1$	Lamé–Wangerin Lamé–Wangerin
32	$\Gamma_{23}^2, \{\Gamma_{51}, \Gamma_{41} + \Gamma_{45}\}$	Bessel Bessel

ISBN-0-201-13503-5

On $\mathcal{H}_+$ the requirement $M_{12}f = imf$ implies $f(\mathbf{k}) = e^{im\theta}j_m(k)$ where $m = 0, \pm 1, \ldots, k_1 = k\cos\theta, k_2 = k\sin\theta$. The eigenfunction problem reduces to a study of the Hilbert space $L_2[0, \infty]$ on which $SL(2,R)$ acts via

$$\Gamma_{45} = \frac{ik}{2}(-\partial_{kk} - k^{-1}\partial_k + m^2 k^{-2} + 1),$$

$$\Gamma_{41} = \frac{ik}{2}(\partial_{kk} + k^{-1}\partial_k - m^2 k^{-2} + 1), \qquad \Gamma_{51} = k\partial_k + \tfrac{1}{2}. \tag{4.8}$$

This action is irreducible and unitary equivalent to a single-valued representation of $SL(2,R)$, not $SO(2,1)$, from the negative discrete series $D^-_{|m|-\frac{1}{2}}$, as can be seen from (4.6) and (2.2). (Compare with Section 2.3.) Indeed, the eigenvalues of Γ_{45} in this model are $i(n+\tfrac{1}{2}), n = |m|, |m|+1, |m|+2, \ldots$. This model of D_l^- has been studied by a number of authors (e.g., [24, 96].)

Once the eigenfunctions $j_{m\mu}(k)$ of the second operators in Table 22 have been determined, the corresponding separable solutions $\Psi_{m\mu}$ of (1.1) can be computed from

$$\Psi_{m\mu}(x) = \exp\left[im(\theta - \pi/2)\right]\int_0^\infty \exp(ix_0 k)J_m(kr)j_{m\mu}(k)\,dk. \tag{4.9}$$

More generally one can study the EPD equation (4.5) for any real $m > 0$. The separable coordinate systems and model (4.8) are unchanged but the symmetry group becomes the universal covering group $\widetilde{SL}(2,R)$ of $SL(2,R)$, as in Section 2.3. The mapping from $L_2[0, \infty]$ to the solution space of (4.5) is

$$\Phi(x_0, r) = \exp(-im\pi/2)\int_0^\infty \exp(ix_0 k)J_m(kr)f(k)\,dk = U[f] \tag{4.10}$$

and the associated inner product is

$$(\Phi_1, \Phi_2) \equiv \langle f_1, f_2\rangle = i\int_0^\infty \Phi_1(x_0, r)\partial_0\overline{\Phi}_2(x_0, r)r\,dr$$

$$= -i\int_0^\infty \overline{\Phi}_2(x_0, r)\partial_0\Phi_1(x_0, r)r\,dr, \tag{4.11}$$

independent of x_0. Details concerning the spectral resolutions of the operators that determine the separated solutions can be found in [63].

We have characterized the solutions Φ_m of the EPD equation (4.5) as solutions of the wave equation (1.1) that are eigenfunctions of $L = -iM_{12}$: $L\Psi_m = m\Psi_m, \Psi_m = e^{im\varphi}\Phi_m(x_0, r)$. One can choose a basis $\{L_j\}$ for the complexification $so(3,2)^c \cong so(5, \mathcal{C})$ of the conformal symmetry algebra

ISBN-0-201-13503-5

such that $[L, L_j] = \alpha_j L_j$ where $\alpha_j = 0, \pm 1$. Indeed, the commutation relations

$$[L, P_1 \pm iP_2] = \pm (P_1 \pm iP_2), \qquad [L, M_{01} \pm iM_{02}] = \pm (M_{01} \pm iM_{02}),$$
$$[L, K_1 \pm iK_2] = \pm (K_1 \pm iK_2),$$

together with the fact that $[L, L'] = 0$ for $L' = D, P_0, K_0$ provides such a basis. It follows from these relations that $L_j \Psi_m$ is an eigenfunction of L with eigenvalue $m + \alpha_j = m, m \pm 1$; that is, $L_j(e^{im\varphi} \Phi_m) = \exp[i(m + \alpha_j)\varphi] \Phi_{m+\alpha_j}$. Factoring out the φ dependence, we see that each symmetry operator maps a solution of (4.5) for m to a solution for $m + \alpha_j$. Similarly, the operators (1.12) induce mappings from one EPD equation to another, as do certain of the group symmetry operators.

We see that this series of recurrence formulas relating distinct EPD equations to one another is a direct consequence of the conformal symmetry of the wave equation, from which the EPD equation arises by partial separation of variables. Weinstein [131, 132] has made use of two of these recurrence relations in his study of boundary value problems for the EPD equation. A complete group-theoretic discussion appears in [93], where it is also shown that quadratic transformation formulas for the $_2F_1$ [36] are related to the conformal symmetry of the wave equation.

We have mentioned all the semisplit systems for the wave equation with the exception of some curious nonorthogonal systems which correspond to diagonalization of the operator $\frac{1}{2} M_{12} + \frac{1}{4} K_0 - \frac{1}{4} P_0$ and are discussed in [60, 62], as well as some highly singular solutions, discussed in [62], that arise because diagonalization of a given first-order operator does not uniquely determine the corresponding coordinate. Orthogonal nonsplit coordinates are treated in [61].

4.5 The Wave Equation $(\partial_{tt} - \Delta_3)\Psi(x) = 0$

In many respects the real wave equation in four-dimensional space-time

$$(\partial_{00} - \partial_{11} - \partial_{22} - \partial_{33})\Psi(x) = 0 \qquad (5.1)$$

is the most important equation in this book. In addition to the well-known physical importance of (5.1), [12, 107], it is a fact that virtually every equation examined in the earlier chapters is either a special case of (5.1) or is obtained from (5.1) by a partial separation of variables. Moreover, whereas the three-space wave equation and its complexification are associated with the generating functions for Gegenbauer functions and polynomials, (5.1) is associated with generating functions for the general Gaussian hypergeometric function and Jacobi polynomials.

Although (5.1) is presently undergoing intensive study from a group-theoretic viewpoint, the results at this writing are still fragmentary. We shall

ISBN-0-201-13503-5

limit ourselves here to the indication of some general features of the separation of variables problem for (5.1) and a brief discussion of relevant published papers.

The 15-dimensional symmetry algebra $so(4,2)$ of (5.1) was computed in [14] and can be obtained in obvious analogy to that of (1.1). The symmetry group, locally isomorphic to $SO(4,2)$, is called the conformal group. It contains the homogeneous Lorentz group $SO(3,1)$, the Poincaré group $E(3,1)$, and the compact orthogonal group $SO(4,R)$ as proper subgroups. There is also an inversion symmetry analogous to I, (1.12). By utilizing the Fourier transform one can construct a Hilbert space $\mathcal{H}_+$ of positive energy solutions on which there is defined a unitary irreducible representation of the conformal group. This is carried out in analogy with (1.20) and details are presented in [44, 66, 118].

One expects the R-separable solutions of (5.1) to be characterized as simultaneous eigenfunctions of triplets of independent commuting operators that are at most second order in the enveloping algebra of $so(4,2)$. We will discuss a few of the special cases in which the details have been worked out.

By restricting the symmetry algebra of (1.1) to the compact subalgebra $so(3)$ we were led to the Laplace operator on the sphere S_2 and obtained two separable systems. Similarly, by restricting $so(4,2)$ to the compact subalgebra $so(4)$, we obtain the Laplace operator on the unit sphere S_3 in four-dimensional space. This operator is studied in [65], where it is shown that the eigenvalue equation separates in exactly six coordinate systems associated with six commuting pairs of second-order symmetry operators in the enveloping algebra of $so(4)$. The relationship between $so(4)$ and the Schrödinger equation for the Kepler problem in three space variables is also discussed.

Diagonalization of the symmetry operator $P_0 = \partial_0$ reduces (5.1) to the Helmholtz equation, which separates in eleven coordinate systems. Diagonalization of $P_3 = \partial_3$ reduces (5.1) to the Klein–Gordon equation

$$\left(\partial_{00} - \partial_{11} - \partial_{22} + \omega^2\right)\Phi = 0. \tag{5.2}$$

In [61], 53 Minkowski-orthogonal separable systems for (5.2) were classified. Diagonalization of the dilatation symmetry $\sum_{\alpha=0}^{3} x_\alpha \partial_\alpha$ reduces (5.1) to the eigenvalue equation for the Laplace operator on a hyperboloid in four-dimensional space. The reduced equation admits the homogeneous Lorentz group $SO(3,1)$ as its symmetry group and separates in 34 coordinate systems, each corresponding to a pair of second-order symmetric operators in the enveloping algebra of $so(3,1)$ [104, 64]. Diagonalization of $P_0 + P_1 = \partial_0 + \partial_1$ reduces (5.1) to the free-particle Schrödinger equation

$$(i\beta \partial_t + \partial_{22} + \partial_{33})\Phi = 0, \tag{5.3}$$

which separates in 17 corrdinate systems. Similarly, diagonalization of the symmetry $M_{23} = x_2 \partial_3 - x_3 \partial_2$ leads to a reduced EPD-like equation. Bateman has used the complexification of the reduced equation obtained by diagonalizing both M_{23} and $M_{01} = x_0 \partial_1 + x_1 \partial_0$ to derive generating functions for Jacobi polynomials [13, p. 392], and Koornwinder [68, 68a], has used it in connection with his study of the addition theorem for Jacobi polynomials. Henrici employed the same equation to derive generating functions for products of Gegenbauer polynomials [48].

Although the systems above were obtained in complete analogy with our treatment of (1.1), there are some novel types of nonsplit coordinates that appear for (5.1). For example, diagonalization of $P_2^2 + P_3^2$ reduces (5.1) to the two equations

$$(\partial_{00} - \partial_{11} + \omega^2)\Phi = 0, \qquad (\partial_{22} + \partial_{33} + \omega^2)\Theta = 0, \qquad (5.4)$$

where $\Psi = \Phi\Theta$. The possible separable systems for the reduced equations can be read off from Tables 1 and 2.

The explicit connection between the functions $_2F_1$ and the wave equation will be discussed in the following chapter.

Exercises

1. Compute the symmetry algebra of the wave equation (1.1).
2. Let $y_0 = \cos\sigma, y_1 = \sin\sigma\cos\alpha, y_2 = \sin\sigma\sin\alpha$ where (ψ, σ, α) are the R-separable coordinates (2.6) for the wave equation (1.1). Show that substitution of $\Psi = [\cos\sigma - \cos\psi]^{1/2} \exp[-i\psi(l + \frac{1}{2})]\Phi(y_1, y_2, y_3)$ into the wave equation leads to the reduced equation $(\Gamma_{12}^2 + \Gamma_{13}^2 + \Gamma_{23}^2)\Phi = -l(l+1)\Phi$, the eigenvalue equation for the Laplace operator on the sphere $y_0^2 + y_1^2 + y_2^2 = 1$. Here $\Gamma_{12}, \Gamma_{13}, \Gamma_{23}$ are the usual angular momentum operators on the sphere.
3. Show that the space of second-order symmetry operators in the enveloping algebra of $so(2, 1)$, modulo the Casimir operator, is decomposed into nine orbit types under the adjoint action of $SO(2,1)$. (Hint: This problem is equivalent to classifying the equivalence classes of 3×3 real symmetric matrices Q under the conjugacy transformations $Q \to A'QA, A \in SO(2, 1)$. For more details see [139].)
4. Show that the EPD equation (4.5) separates in the variables

$$x = \frac{1}{2}\left[(t + r)^{1/2} + (t - r)^{1/2}\right], \qquad y = \frac{1}{2}\left[(t + r)^{1/2} - (t - r)^{1/2}\right],$$

$$t \pm r > 0,$$

corresponding to the operators $\Gamma_{23}^2, \{\Gamma_{51}, \Gamma_{41} + \Gamma_{45}\}$. The separated solutions are products of Bessel functions [63].
5. As shown in the text, a function $\Phi(x_0, r)$ is a solution of the EPD equation

$$(\partial_{00} - \partial_{rr} - r^{-1}\partial_r + m^2 r^{-2})\Phi = 0$$

if and only if $\Psi_m = e^{im\varphi}\Phi$ is a solution of the wave equation (1.1), where $x_1 = r\cos\varphi, x_2 = r\sin\varphi$. Thus the solutions of the wave equation that are eigenfunctions of $M_{12} = \partial_\varphi$ correspond to solutions of the EPD equation. Use the expressions $[iM_{12}, \pm iM_{01} + M_{02}] = \mp(\pm iM_{01} + M_{02})$ to derive differential recurrence relations mapping solutions of the EPD equation for $m = m_0$ to those for $m = m_0 \mp 1$, respectively. Similarly, the other Lie symmetries of the wave equation yield mappings between EPD equations (see [93]).

ISBN-0-201-13503-5

CHAPTER 5

The Hypergeometric Function and Its Generalizations

5.1 The Lauricella Functions F_D

The Gaussian hypergeometric function $_2F_1$ is intimately associated with the Laplace and wave equations in four-dimensional space and their complexifications. The $_2F_1$ arises from these equations via separation of variables, and the conformal symmetry groups account for many of the properties of this function. Rather than pursue this differential equations approach, we shall instead treat the hypergeometric function directly, together with some of its generalizations that have been studied in the past 150 years.

The Lauricella functions F_A–F_D, (B.21)–(B.24), were defined by Lauricella [71] and studied in detail by Appell and Kampe de Feriet [4]. As can easily be seen from the power series definitions, these functions are generalizations of $_2F_1$ to n complex variables: When $n=1$, each of F_A–F_D reduces to $_2F_1$. Although the Lauricella functions were not originally obtained by separation of variables, we shall see that each of these functions can be obtained by a partial separation of variables in a system of n second-order partial differential equations. The F_D are most interesting generalizations of the $_2F_1$ from a group-theoretic view, and we shall devote much of our attention to these functions, obtaining results for $_2F_1$ by setting $n=1$. As can be seen from (B.24),

$$F_D[a; b_1, \ldots, b_n; c; z_1, \ldots, z_n]$$

depends on $n+2$ complex parameters a, b_j, c and n complex variables $z_1, \ldots, z_n$. For $n=1$ we have

$$F_D[a; b; c; z] \equiv {}_2F_1\left(\begin{matrix} a, b \\ c \end{matrix} \middle| z\right). \tag{1.1}$$

ENCYCLOPEDIA OF MATHEMATICS and Its Applications, Gian-Carlo Rota (ed.).
Vol. 4: Willard Miller, Jr., Symmetry and Separation of Variables

We can use (B.24) to derive differential recurrence relations obeyed by the functions F_{D} and then construct a Lie algebra from the recurrence relations. The details are similar to computations presented earlier, so here we list only the results. We define a family of functions

$$\Psi_c^{a,b_1,\ldots,b_n}(s,u_1,\ldots,u_n,t,z_1,\ldots,z_n)=\Psi_c^{a,b_j}(s,u_j,t,z_j)$$

$$=\frac{\Gamma(c-a)\Gamma(a)}{\Gamma(c)}F_{\mathrm{D}}(a;b_1,\ldots,b_n;c,z_1,\ldots,z_n)s^a u_1^{b_1}\cdots u_n^{b_n}t^c \quad (1.2)$$

where $c\neq 0,-1,-2,\ldots$, and s,u_j,t are complex variables. Furthermore, we define operators:

$$E^{\alpha}=s\left(\sum_j z_j\,\partial_{z_j}+s\,\partial_s\right),\qquad E^{\alpha\beta_k\gamma}=su_k t\,\partial_{z_k},$$

$$E^{\beta_k}=u_k(z_k\,\partial_{z_k}+u_k\,\partial_{u_k}),\qquad E_{\gamma}=t^{-1}\left(\sum_j z_j\,\partial_{z_j}+t\,\partial_t-1\right),$$

$$E^{\alpha\gamma}=st\left(\sum_j(1-z_j)\,\partial_{z_j}-s\,\partial_s\right),\quad E^{\gamma}=t\left(\sum_j(1-z_j)\,\partial_{z_j}+t\,\partial_t-s\,\partial_s-\sum_j u_j\,\partial_{u_j}\right),$$

$$E_{\alpha}=s^{-1}\left(\sum_j z_j(1-z_j)\,\partial_{z_j}+t\,\partial_t-s\,\partial_s-\sum_j z_j u_j\,\partial_{u_j}\right),$$

$$E_{\beta_k}=u_k^{-1}\left(z_k(1-z_k)\,\partial_{z_k}+z_k\sum_{j\neq k}(1-z_j)\,\partial_{z_j}+t\,\partial_t-z_k s\,\partial_s-\sum_j u_j\,\partial_{u_j}\right),$$

$$E^{\beta_k\gamma}=u_k t\left((z_k-1)\,\partial_{z_k}+u_k\,\partial_{u_k}\right),$$

$$E_{\alpha\gamma}=s^{-1}t^{-1}\left(\sum_j z_j(1-z_j)\,\partial_{z_j}-\sum_j z_j u_j\,\partial_{u_j}+t\,\partial_t-1\right)$$

$$E_{\alpha\beta_k\gamma}=s^{-1}u_k^{-1}t^{-1}\left(\sum_j z_j(z_j-1)\,\partial_{z_j}-t\,\partial_t+z_k s\,\partial_s+\sum_j z_j u_j\,\partial_{u_j}-z_k+1\right),$$

$$E_{\beta_k\gamma}=u_k^{-1}t^{-1}\left(z_k(z_k-1)\,\partial_{z_k}+\sum_{j\neq k}(z_k-1)z_j\,\partial_{z_j}+z_k s\,\partial_s-t\,\partial_t+1\right),$$

$$E_{\beta_p}^{\beta_k}=u_k u_p^{-1}\left((z_k-z_p)\,\partial_{z_k}+u_k\,\partial_{u_k}\right),\quad J_{\alpha}=s\,\partial_s-\tfrac{1}{2}t\,\partial_t,$$

$$J_{\beta_k}=u_k\,\partial_{u_k}-\tfrac{1}{2}t\,\partial_t+\tfrac{1}{2}\sum_{j\neq k}u_j\,\partial_{u_j},$$

$$J_{\gamma}=t\,\partial_t-\tfrac{1}{2}\left(s\,\partial_s+\sum_j u_j\,\partial_{u_j}+1\right),\qquad k,p=1,2,\ldots,n. \quad (1.3)$$

ISBN-0-201-13503-5

Unless otherwise indicated, j is summed from 1 to n. The action of the foregoing operators on the basis (1.2) is

$$E^{\alpha}\Psi_c^{a,b_j}=(c-a-1)\Psi_c^{a+1,b_j}, \qquad E^{\alpha\beta_k\gamma}\Psi_c^{a,b_j}=b_k\Psi_{c+1}^{a+1,\hat{b}_k},$$

$$E^{\beta_k}\Psi_c^{a,b_j}=b_k\Psi_c^{a,\hat{b}_k}, \qquad E_{\gamma}\Psi_c^{a,b_j}=(c-a-1)\Psi_{c-1}^{a,b_j},$$

$$E^{\alpha\gamma}\Psi_c^{a,b_j}=\left(\sum_j b_j-c\right)\Psi_{c+1}^{a+1,b_j}, \qquad E^{\gamma}\Psi_c^{a,b_j}=\left(c-\sum_j b_j\right)\Psi_{c+1}^{a,b_j},$$

$$E_{\alpha}\Psi_c^{a,b_j}=(a-1)\Psi_c^{a-1,b_j}, \qquad E_{\beta_k}\Psi_c^{a,b_j}=\left(c-\sum_j b_j\right)\Psi_c^{a,\hat{b}_k},$$

$$E^{\beta_k\gamma}\Psi_c^{a,b_j}=b_k\Psi_{c+1}^{a,\hat{b}_k}, \qquad E_{\alpha\gamma}\Psi_c^{a,b_j}=(a-1)\Psi_{c-1}^{a-1,b_j}$$

$$E_{\alpha\beta_k}\Psi_c^{a,b_j}=(1-a)\Psi_{c-1}^{a-1,\tilde{b}_k}, \qquad E_{\beta_k\gamma}\Psi_c^{a,b_j}=(a-c+1)\Psi_{c-1}^{a,\tilde{b}_k},$$

$$E_{\beta_p}^{\beta_k}\Psi_c^{a,b_j}=b_k\Psi_c^{a,b_1\cdots b_k+1\cdots b_p-1\cdots b_n}, \qquad J_{\alpha}\Psi_c^{a,b_j}=(a-c/2)\Psi_c^{a,b_j},$$

$$J_{\beta_k}\Psi_c^{a,b_j}=\left(b_k-\tfrac{1}{2}c+\tfrac{1}{2}\sum_{j\neq k}b_j\right)\Psi_c^{a,b_j},$$

$$J_{\gamma}\Psi_c^{a,b_j}=\left[c-\tfrac{1}{2}\left(a+\sum_j b_j+1\right)\right]\Psi_c^{a,b_j}, \qquad k,p=1,\ldots,n. \qquad (1.4)$$

The symbols $\hat{b}_k$ and $\tilde{b}_k$ are defined by

$$\hat{b}_k=b_1,\ldots,b_{k-1},b_k+1,b_{k+1},\ldots,b_n,$$
$$\tilde{b}_k=b_1,\ldots,b_{k-1},b_k-1,b_{k+1},\ldots,b_n. \qquad (1.5)$$

The differential recurrence relations for the F_D are obtained by factoring the dependence on s,u_j, and t from both sides of the expressions (1.4). Moreover, for $n=1$ these relations reduce exactly to the recurrence formulas (B.5) for the $_2F_1$.

Relations (1.4) can be verified by routine computation. Furthermore, it is straightforward to show that the operators (1.3) form a basis for the Lie algebra $sl(n+3,\mathcal{C})$ of dimension $(n+3)^2-1$. Recall that $SL(n+3,\mathcal{C})$ is the group of all $(n+3)\times(n+3)$ complex matrices A such that $\det A=1$. The Lie algebra $sl(n+3,\mathcal{C})$ of $SL(n+3,\mathcal{C})$ consists of all $(n+3)\times(n+3)$ complex matrices $\mathcal{Q}$ such that $\operatorname{tr}\mathcal{Q}=0$ [85]. Denoting by $\mathcal{E}_{ij}$ the matrix with a one in row i, column j, and zeros everywhere else (see (6.4), Section 3.6), we see that the matrices $\mathcal{E}_{ij},i\neq j$, and $\mathcal{E}_{ii}-\mathcal{E}_{33},1\leqslant i,j\leqslant n+3$, form a

ISBN-0-201-13503-5

basis for $sl(n+3,\mathcal{C})$. The commutation relations can be obtained from the general formula

$$[\mathcal{E}_{ij}, \mathcal{E}_{kl}] = \delta_{jk}\mathcal{E}_{il} - \delta_{li}\mathcal{E}_{kj}. \tag{1.6}$$

We can check that the appropriate commutation relations are satisfied if we make the identifications

$$E^\alpha = \mathcal{E}_{12}, \qquad\qquad E_\alpha = \mathcal{E}_{21}, \qquad\qquad E^{\beta_k} = \mathcal{E}_{k+3,3},$$

$$E_{\beta_k} = \mathcal{E}_{3,k+3}, \qquad\qquad E^{\beta_k}_{\beta_p} = \mathcal{E}_{k+3,p+3}, \qquad\qquad E^\gamma = \mathcal{E}_{31},$$

$$E_\gamma = -\mathcal{E}_{13}, \qquad\qquad E^{\alpha\gamma} = \mathcal{E}_{32}, \qquad\qquad E_{\alpha\gamma} = \mathcal{E}_{23},$$

$$E^{\beta_k\gamma} = -\mathcal{E}_{k+3,1}, \qquad\qquad E_{\beta_k\gamma} = -\mathcal{E}_{1,k+3}, \qquad\qquad E^{\alpha\beta_k\gamma} = -\mathcal{E}_{k+3,2},$$

$$E_{\alpha\beta_k\gamma} = -\mathcal{E}_{2,k+3}, \qquad\qquad J_\alpha = \tfrac{1}{2}(\mathcal{E}_{11} - \mathcal{E}_{22}),$$

$$J_{\beta_k} = \tfrac{1}{2}(\mathcal{E}_{k+3,k+3} - \mathcal{E}_{33}), \qquad J_\gamma = \tfrac{1}{2}(\mathcal{E}_{33} - \mathcal{E}_{11}), \quad 1 \leqslant k,p \leqslant n, \; k \neq p. \tag{1.7}$$

Let

$$C_k = E^\alpha E^{\beta_k} - E^{\alpha\beta_k\gamma}E_\gamma, \qquad 1 \leqslant k \leqslant n. \tag{1.8}$$

It is straightforward to check that the solution f of the simultaneous equations

$$C_k f = 0, \qquad J_\alpha f = (a - c/2)f, \qquad J_{\beta_k} f = \left(b_k - \tfrac{1}{2}c + \tfrac{1}{2}\sum_{j \neq k} b_j\right)f,$$

$$J_\gamma f = \left[c - \tfrac{1}{2}\left(a + \sum_j b_j + 1\right)\right]f, \qquad k = 1,\ldots,n, \tag{1.9}$$

analytic in a neighborhood of $z_1 = \cdots = z_n = 0$ is

$$f = F_D(a; b_1,\ldots,b_n; c; z_1,\ldots,z_n)s^a u_1^{b_1} \cdots u_n^{b_n} t^c, \tag{1.10}$$

unique to within a multiplicative constant. In fact, the last $n+2$ equations imply

$$f = F(z_1,\ldots,z_n)s^a u_1^{b_1} \cdots u_n^{b_n} t^c$$

and the first n imply

$$\left\{\left(\sum_{j=1}^n z_j \partial_{z_j} + a\right)(z_k \partial_{z_k} + b_k) - \partial_{z_k}\left(\sum_{j=1}^n z_j \partial_{z_j} + c - 1\right)\right\}F = 0, \tag{1.11}$$

$$k = 1,\ldots,n,$$

ISBN-0-201-13503-5

which are the partial differential equations for F_D. The operators C_k do not commute with all the elements of $sl(n+3,\mathcal{C})$, but each element leaves the solution space of the system of equations invariant. It follows from these remarks that if $\Psi(s,u_j,t,z_j)$ is a solution of $C_k\Psi=0, k=1,\ldots,n$, which has a Laurent expansion

$$\Psi = \sum_{a,b_j,c} g_{ab_jc}(z_j)s^au_1^{b_1}\cdots u_n^{b_n}t^c, \qquad (1.12)$$

and if Ψ is analytic at $z_1=\cdots=z_n=0$, then

$$g_{ab_jc} = k(ab_jc)F_D(a;b_1,\ldots,b_n;c;z_1,\ldots,z_n) \qquad (1.13)$$

where k is a constant. Moreover, the elements of $sl(n+3,\mathcal{C})$ map a solution of $C_k\Psi=0, 1\leqslant k\leqslant n$, into another solution.

We see now that the F_D arise as separable solutions of the system of n second-order partial differential equations $C_k\Psi=0, 1\leqslant k\leqslant n$, in the coordinates s,u_j,t,z_j. To simplify this system we perform the R-transformation $\Phi=t^{-1}\Psi$ to remove the multiplier term from E_γ. Then we transform to new variables v,v_j,w,w_j such that

$$E^\alpha = \partial_v, \qquad E^{\beta_k} = \partial_{v_k}, \qquad E^{\alpha\beta_k\gamma} = \partial_{w_k}, \qquad E_\gamma = \partial_w.$$

Explicitly,

$$s=-1/v, \quad u_j=-1/v_j, \quad t=w, \quad z_j=ww_j/vv_j, \qquad 1\leqslant j\leqslant n, \quad (1.14)$$

and the equations $C_k\Psi=0$ become

$$\left(\partial_v\partial_{v_k} - \partial_w\partial_{w_k}\right)\Phi=0, \qquad 1\leqslant k\leqslant n. \qquad (1.15)$$

From (1.2) it follows that the equations (1.15) have solutions

$$\Phi_c^{a,b_j}=t^{-1}\Psi_c^{a,b_j}=\frac{\Gamma(c-a)\Gamma(a)}{\Gamma(c)}F_D\left(a;b_1,\ldots,b_n;c;\frac{ww_1}{vv_1},\ldots,\frac{ww_n}{vv_n}\right)$$

$$\times\left(-\frac{1}{v}\right)^a\left(-\frac{1}{v_1}\right)^{b_1}\cdots\left(-\frac{1}{v_n}\right)^{b_n}w^c. \qquad (1.16)$$

In the special case where $n=1$ we can set $v=(z+t)/2, v_1=(z-t)/2, w=$

$(ix+y)/2, w_1 = (ix-y)/2$, and transform (1.15) to the complex wave equation

$$(\partial_{tt} - \partial_{xx} - \partial_{yy} - \partial_{zz})\Phi(t,x,y,z) = 0$$

for which (1.16) yields the $_2F_1$ as separated solutions. It is straightforward to show that the symmetry algebra of this equation is $o(6,\cancel{C}) \cong sl(4,\cancel{C})$.

Returning now to the operators (1.3), we will determine the group action of $SL(n+3,\cancel{C})$ induced by these operators. Rather than determine the global group action, we note that each of the triplets

$$\{J^+, J^-, J^0\} \equiv \{E^\alpha, E_\alpha, J_\alpha\}, \{E^{\beta_k}, E_{\beta_k}, J_{\beta_k}\},$$

$$\{E^\gamma, E_\gamma, J_\gamma\}, \{E^{\alpha\beta_k\gamma}, E_{\alpha\beta_k\gamma}, J_\alpha + J_{\beta_k} + J_\gamma\},$$

$$\{E^{\alpha\gamma}, E_{\alpha\gamma}, J_\alpha + J_\gamma\}, \{E^{\beta_k\gamma}, E_{\beta_k\gamma}, J_{\beta_k} + J_\gamma\},$$

$$\{E^{\beta_l}_{\beta_p}, E^{\beta_p}_{\beta_l}, J_{\beta_l} - J_{\beta_p}\}, \qquad 1 \leqslant k \leqslant n, 1 \leqslant l < p \leqslant n, \qquad (1.17)$$

satisfies the commutation relations

$$[J^0, J^\pm] = \pm J^\pm, [J^+, J^-] = 2J^0$$

and forms a basis for a subalgebra of $sl(n+3,\cancel{C})$ isomorphic to $sl(2,\cancel{C})$. Furthermore, each triplet generates a local Lie subgroup of $SL(n+3,\cancel{C})$ isomorphic to $SL(2,\cancel{C})$ and the subgroups so obtained suffice to generate the full group action of $SL(n+3,\cancel{C})$.

We pass from the Lie algebra action generated by $\{J^+, J^-, J^0\}$ to the group action via the relation

$$\mathbf{T}(A) = \exp(-bd^{-1}J^+)\exp(-cdJ^-)\exp(\tau J^0), \qquad \exp(\tau/2) = d^{-1}, \qquad (1.18)$$

where

$$A = \begin{pmatrix} a & b \\ c & d \end{pmatrix} \in SL(2,\cancel{C})$$

(see (4.14), Section 2.4). We find that the triplet $\{E^\alpha, E_\alpha, J_\alpha\}$ generates the group action

$$\mathbf{T}_1(A)\Psi(s, u_j, t, z_j)$$

$$= \Psi\left[\frac{as+c}{d+bs}, \frac{u_j(as+c)}{as+c(1-z_j)}, \frac{ts}{as+c}, \frac{z_j s}{(d+bs)(as-cz_j+c)}\right] \qquad (1.19)$$

ISBN-0-201-13503-5

and the triplet $\{E^{\beta_k}, E_{\beta_k}, J_{\beta_k}\}$ generates

$$\mathbf{T}_{2,k}(A)\Psi(s, u_j, u_k, t, z_j, z_k)$$

$$= \Psi\left(\frac{s(au_k + c)}{au_k + c(1 - z_k)}, \frac{u_j}{u_k}(au_k + c), \frac{au_k + c}{d + bu_k}, \frac{u_k t}{au_k + c}, \right.$$

$$\left. \frac{au_k z_j + c(z_j - z_k)}{au_k + c(1 - z_k)}, \frac{z_k u_k}{(d + bu_k)(au_k - cz_k + c)} \right), \qquad (1.20)$$

$$k = 1, \dots, n.$$

In (1.20) j runs from 1 to n excluding k. The triplet $\{E^\gamma, E_\gamma, J_\gamma\}$ generates

$$\mathbf{T}_3(A)\Psi(s, u_j, t, z_j) = \left(a + \frac{c}{t} \right)^{-1} \Psi\left(s(d + bt), u_j(d + bt), \right.$$

$$\left. \frac{at + c}{d + bt}, \left[dz_j - bt(1 - z_j) \right]\left(a + \frac{c}{t} \right) \right), \qquad (1.21)$$

the triplet $\{E^{\alpha\beta_k\gamma}, E_{\alpha\beta_k\gamma}, J_\alpha + J_{\beta_k} + J_\gamma\}$ generates

$$\mathbf{T}_{4,k}(A)\Psi(s, u_j, u_k, t, z_j, z_k) = \left(a + \frac{c(1 - z_k)}{u_k ts} \right)^{-1}$$

$$\Psi\left[as - \frac{cz_k}{u_k t}, u_j \left[\frac{asu_k t - cz_k}{asu_k t + cz_j - cz_k} \right], au_k - \frac{cz_k}{st}, t \left[\frac{asu_k t + c(1 - z_k)}{asu_k t - cz_k} \right], \right.$$

$$\left. z_j \left[\frac{asu_k t + c(1 - z_k)}{asu_k t + c(z_j - z_k)} \right], \left[z_k d - bsu_k t \right] \left[a + \frac{c(1 - z_k)}{su_k t} \right] \right], \qquad (1.22)$$

the triplet $\{E^{\alpha\gamma}, E_{\alpha\gamma}, J_\alpha + J_\gamma\}$ generates

$$\mathbf{T}_5(A)\Psi(s, u_j, t, z_j) = \left(a - \frac{c}{st} \right)^{-1} \Psi\left[\frac{s}{d - bst}, \frac{u_j st}{ast - cz_j}, \right.$$

$$\left. at - \frac{c}{s}, \frac{(dz_j - bst)(ast - c)}{(ast - cz_j)(d - bst)} \right], \qquad (1.23)$$

ISBN-0-201-13503-5

the triplet $\{E^{\beta_k\gamma}, E_{\beta_k\gamma}, J_{\beta_k} + J_\gamma\}$ generates

$$\mathbf{T}_{6,k}(A)\Psi(s, u_j, u_k, t, z_j, z_k)$$

$$= \left(a + \frac{c}{u_k t}\right)^{-1} \Psi\left(\frac{su_k t}{au_k t + cz_k}, u_j, \frac{u_k}{d + bu_k t}, at + \frac{c}{u_k}, \frac{z_j(au_k t + c)}{au_k t + cz_k},\right.$$

$$\left.\frac{(dz_k + bu_k t)(au_k t + c)}{(d + bu_k t)(au_k t + cz_k)}\right), \tag{1.24}$$

and the triplet $\{E_{\beta_p}^{\beta_k}, E_{\beta_k}^{\beta_p}, J_{\beta_k} - J_{\beta_p}\}$ generates

$$\mathbf{T}_{7,k,p}(A)\Psi(s, u_j, u_k, u_p, t, z_j, z_k, z_p)$$

$$= \Psi\left(s, u_j, \frac{u_k u_p}{du_p + bu_k}, \frac{u_p u_k}{au_k + cu_p}, t, z_j, \frac{dz_k u_p + bz_p u_k}{du_p + bu_k},\right.$$

$$\left.\frac{az_p u_k + cz_k u_p}{au_k + cu_p}\right), \qquad 1 \leqslant k < p \leqslant n. \tag{1.25}$$

Each of the operators $T_l(A)$ maps a solution Ψ of the system $C_k\Psi = 0$, $1 \leqslant k \leqslant n$, into another solution.

To compute the matrix elements of the group operators $\mathbf{T}_l(A)$ with respect to the basis $\{\Psi_c^{a,b_j}\}$ it is useful to construct a simpler model of relations (1.4). Such a model is provided by the functions

$$f_c^{a,b_j}(s, u_j, t) = s^a u_1^{b_1} \cdots u_n^{b_n} t^c$$

of $n + 2$ complex variables and the operators

$$E^\alpha = s(t\partial_t - s\partial_s - 1), \qquad E^{\alpha\beta_k\gamma} = su_k^2 t\partial_{u_k}, \qquad E^{\beta_k} = u_k^2 \partial_{u_k},$$

$$E_\gamma = t^{-1}(t\partial_t - s\partial_s - 1), \qquad E_\alpha = s^{-1}(s\partial_s - 1), \tag{1.26}$$

$$E_{\beta_k} = u_k^{-1}\left(t\partial_t - \sum_j u_j \partial_{u_j}\right), \qquad 1 \leqslant k \leqslant n,$$

which generate $sl(n + 3, \mathcal{C})$. For some examples of matrix elements computed in this way and associated generating functions for the F_D and the $_2F_1$ see [90] and [82, Chapter 5].

ISBN-0-201-13503-5

5.2 Transformation Formulas and Generating Functions for the F_D

We now show that the transformation formulas for the F_D are consequences of the $SL(n+3,\mathcal{C})$ symmetry. Let

$$I = \begin{pmatrix} 0 & 1 \\ -1 & 0 \end{pmatrix} \in SL(2,\mathcal{C}). \tag{2.1}$$

Expressions (1.2) and (1.19) imply

$$T_1(I)\Psi_c^{a,b_j} = (-1)^{a+c} \frac{\Gamma(c-a)\Gamma(a)}{\Gamma(c)} F_D\left(a; b_j; c; -\frac{z_j}{1-z_j}\right)$$

$$\times (1-z_1)^{-b_1} \cdots (1-z_n)^{-b_n} s^{c-a} u_1^{b_1} \cdots u_n^{b_n} t^c \tag{2.2}$$

However, $T_1(I)\Psi_c^{a,b_j}$ is a simultaneous eigenfunction of $J_\alpha, J_\beta, J_\gamma$ analytic at $z_1 = \cdots = z_n = 0$. Thus,

$$T_1(I)\Psi_c^{a,b_j} = k F_D\left(c-a; b_j; c; z_j\right) s^{c-a} u_1^{b_1} \cdots u_n^{b_n} t^c. \tag{2.3}$$

Setting $z_1 = \cdots = z_n = 0$ in (2.2), (2.3), we can evaluate the constant k and obtain the transformation formula

$$(1-z_1)^{-b_1} \cdots (1-z_n)^{-b_n} F_D\left(a; b_j; c; \frac{z_j}{z_j-1}\right) = F_D\left(c-a; b_j; c; z_j\right) \tag{2.4}$$

(see [4, Chapter VII]). Similarly, $T_{2,k}(I)\Psi_c^{a,b_j}$ yields the formulas

$$(1-z_k)^{-a} F_D\left(a; b_j, b_k; c; \frac{z_k-z_j}{z_k-1}, \frac{z_k}{z_k-1}\right)$$

$$= F_D\left(a; b_j, c - \sum_l b_l; c; z_j, z_k\right), \qquad k = 1, \ldots, n. \tag{2.5}$$

The remaining transformation formulas for the F_D can be obtained by composition from (2.4) and (2.5). The transformation formulas for $_2F_1$ follow from (2.4) for $n=1$.

Computing $T_3(I)\Psi_c^{a,b_j}$, we find that

$$F_D\left(a; b_j; a + \sum_l b_l - c + 1; 1 - z_j\right) \tag{2.6}$$

is a solution of equations (1.11), analytic at $z_1 = \cdots = z_n = 1$. Computing

$\mathbf{T}_5(I)\Psi_c^{a,b_j}$, we see that

$$z_1^{-b_1} \cdots z_n^{-b_n} F_D\left(\sum_l b_l - c + 1; b_j; \sum_l b_l - a + 1; z_j^{-1}\right) \qquad (2.7)$$

is another solution of (1.11). Similarly, $\mathbf{T}_{6,k}(I)\Psi_c^{a,b_j}$ yields the solution

$$z_k^{-a} F_D\left(a; b_j, a - c + 1; a - b_k + 1; \frac{z_j}{z_k}, \frac{1}{z_k}\right). \qquad (2.8)$$

For A close to the identity in $SL(2, \mathcal{C})$ the expressions $\mathbf{T}_j(A)\Psi_c^{a,b_j}$ can be expanded by use of matrix elements computed from the recurrence relations (1.4). However, for A far from the identity (e.g., $A = I$), these expansions are no longer valid. For example,

$$\exp(\lambda E^{\alpha\gamma})\Psi(s, u_j, t, z_j) = \Psi\left(\frac{s}{1 + \lambda st}, u_j, t, \frac{z_j + \lambda st}{1 + \lambda st}\right). \qquad (2.9)$$

For $|\lambda|$ small we find

$$\exp(\lambda E^{\alpha\gamma})\Psi_c^{a,b_j} = \sum_{h=0}^{\infty} \binom{\Sigma_l b_l - c}{h} \Psi_{c+h}^{a+h,b_j} \lambda^h,$$

that is,

$$(1 + \lambda)^{-a} F_D\left(a; b_j; c; \frac{z_j + \lambda}{1 + \lambda}\right)$$

$$= \sum_{h=0}^{\infty} \binom{\Sigma_l b_l - c}{h} \frac{(a)_h}{(c)_h} F_D(a + h; b_j; c + h; z_j)\lambda^h, \qquad (2.10)$$

$$|\lambda| < 1.$$

If $\lambda = 1$ and $|\tau| < 1$ where $\tau = s^{-1}t^{-1}$, then $\exp(E^{\alpha\gamma})\Psi_c^{a,b_j}$ is not analytic at $z_1 = \cdots = z_n = \tau = 0$. However, we can apply $\exp(E^{\alpha\gamma})$ to the solution (2.6) and use (1.12), (1.13) to obtain

$$(1 + \tau)^{-a} F_D\left(a; b_j; a + \sum_l b_l - c + 1; \frac{\tau(1 - z_j)}{1 + \tau}\right)$$

$$= \sum_{h=0}^{\infty} B_h F_D(-h; b_j; c - a - h; z_j)\tau^h. \qquad (2.11)$$

ISBN-0-201-13503-5

To evaluate the constants B_h we set $z_1 = \cdots = z_n = 0$:

$$(1+\tau)^{-a} F_D\left(a; b_j; a + \sum_l b_l - c + 1; \tau/(1+\tau)\right)$$

$$= (1+\tau)^{-a}\,_2F_1\left(a, \sum_l b_l; a + \sum_l b_l - c + 1; \tau/(1+\tau)\right) = \sum_{h=0}^{\infty} B_h \tau^h.$$

Thus,

$$B_h = \binom{-a}{h}\,_2F_1\left[\begin{array}{c|c} -h, \sum b_l \\ a + \sum b_l - c + 1 \end{array}\, 1\right] = \binom{-a}{h} \frac{(a-c+1)_h}{\left(a + \sum b_l - c + 1\right)_h} \quad (2.12)$$

from [82, p. 211], and Vandermonde's theorem [120, p. 28].

Expanding $\mathbf{T}_1(A)\Psi_\gamma^{\alpha,\beta_j}$ as a power series in $\tau = s^{-1}$, we obtain

$$a^{\alpha-\gamma}b^{-\alpha}\left(1 + \frac{c\tau}{a}\right)^{\alpha+\Sigma\beta_l-\gamma}\left(1 + \frac{d\tau}{b}\right)^{-\alpha}\left(1 + \frac{c\tau(1-z_1)}{a}\right)^{-\beta_1}$$

$$\cdots \left(1 + \frac{c\tau(1-z_n)}{a}\right)^{-\beta_n} F_D\left(\alpha; \beta_j; \gamma; \frac{z_j\tau}{(b+d\tau)[a+c\tau(1-z_j)]}\right)$$

$$= \sum_{h=0}^{\infty} B_h F_D(-h; \beta_j; \gamma; z_j)\tau^h. \quad (2.13)$$

Setting $z_1 = \cdots = z_n = 0$ and using identity (5.124) [82, p. 206], we find

$$B_h = \left(\frac{a}{b}\right)^{\alpha} a^{-\gamma-h} c^h \binom{-\gamma}{h}\,_2F_1\left(\begin{array}{c|c} -h, \alpha \\ \gamma \end{array} -\frac{1}{bc}\right), \quad ad - bc = 1. \quad (2.14)$$

If $a = b = d = 1$ and $c = 0$, the identity becomes

$$(1+\tau)^{-\alpha} F_D\left(\alpha; \beta_j; \gamma; \frac{z_j\tau}{1+\tau}\right) = \sum_{h=0}^{\infty} \binom{-\alpha}{h} F_D(-h; \beta_j; \gamma; z_j)\tau^h, \quad (2.15)$$

$$|\tau| < 1,$$

ISBN-0-201-13503-5

and, if $a = c = 1, b = -w^{-1}$ it reduces to

$$(1+\tau)^{\alpha + \Sigma \beta_l - \gamma} [1 + (1-w)\tau]^{-\alpha} \prod_{l=1}^{n} [1 + (1-z_l)\tau]^{-\beta_l}$$

$$\times F_D \left(\alpha; \beta_j; \gamma; \frac{-z_j \tau w}{[1 + (1-w)\tau][1 + (1-z_j)\tau]} \right)$$

$$= \sum_{h=0}^{\infty} \binom{-\gamma}{h} {}_2F_1 \left(\begin{matrix} -h, \alpha \\ \gamma \end{matrix} \middle| w \right) F_D(-h; \beta_j; \gamma; z_j) \tau^k, \qquad (2.16)$$

$$|\tau| < \min(1, |1 - z_j|^{-1}, |1 - w|^{-1}).$$

More generally, we can derive generating functions for the F_D through the characterization of a solution Ψ of $C_k \Psi = 0, 1 \leqslant k \leqslant n$, by the requirement that Ψ is a simultaneous eigenfunction of $n+2$ commuting (or almost commuting) operators constructed from the enveloping algebra of $sl(n+3, \mathbb{C})$. Such a characterization of $\Psi_c^{a,b}$ is given by (1.9).

As an example we compute the solution Ψ of the simultaneous equations

$$E^\alpha \Psi = \Psi, \qquad J_{\beta_k} \Psi = \left(\beta_k + \tfrac{1}{2} \sum_{j \neq k} \beta_j - \tfrac{1}{2}\gamma \right) \Psi,$$

$$\left(J_\gamma + \tfrac{1}{2} J_\alpha \right) \Psi = \left(\tfrac{3}{4}\gamma - \tfrac{1}{2} \sum_l \beta_l - \tfrac{1}{2} \right) \Psi, \qquad C_k \Psi = 0, \qquad k = 1, \ldots, n, \qquad (2.17)$$

which is analytic at $z_1 = \cdots = z_n = 0$. The first $n+2$ equations have the general solution

$$\Psi = f \left(\frac{z_j}{s} \right) \exp(-s^{-1}) u_1^{\beta_1} \cdots u_n^{\beta_n} t^\gamma$$

where f is arbitrary. Substitution of this expression into $C_k \Psi = 0, 1 \leqslant k \leqslant n$, yields

$$f(x_j) = \Phi(\beta_j; \gamma; x_j)$$

$$= \sum_{m_j = 0}^{\infty} \frac{(\beta_1)_{m_1} \cdots (\beta_n)_{m_n}}{(\gamma)_{m_1 + \cdots + m_n}} \frac{x_1^{m_1} \cdots x_n^{m_n}}{m_1! \cdots m_n!}$$

$$= \lim_{\alpha \to \infty} F_D \left(\alpha; \beta_j; \gamma; \frac{z_j}{\alpha} \right). \qquad (2.18)$$

ISBN-0-201-13503-5

Expanding $\mathbf{T}_1(A)\Psi$ as a power series in $\tau = s^{-1}$, we obtain

$$\exp\left[-\left(\frac{d\tau + b}{a + c\tau}\right)\right](a + c\tau)^{\Sigma\beta_l - \gamma}\prod_{l=1}^{n}\left[a + c\tau(1 - z_l)\right]^{-\beta_l}$$

$$\times \Phi\left(\beta_j; \gamma; \frac{z_j\tau}{(a + c\tau)\left[a + c\tau(1 - z_j)\right]}\right)$$

$$= \sum_{h=0}^{\infty} B_h F_D(-h; \beta_j; \gamma; z_j)\tau^h, \qquad ad - bc = 1. \tag{2.19}$$

Setting $z_1 = \cdots = z_n = 0$ and using the generating function (4.11), Section 2.4, for Laguerre polynomials, we find

$$B_h = a^{-\gamma}e^{-b/a}(c/a)^h L_h^{(\gamma-1)}((ac)^{-1}), \tag{2.20}$$

where $L_n^{(\alpha)}(z)$ is a generalized Laguerre polynomial. If $b = c = 0, a = d = 1$, the identity simplifies to

$$\exp(-\tau)\Phi(\beta_j; \gamma; z_j\tau) = \sum_{h=0}^{\infty} F_D(-h; \beta_j; \gamma; z_j)(-\tau)^h/h!. \tag{2.21}$$

If $a = c = d^{-1} = w^{-1/2}, b = 0$, we find

$$\exp\left[-\frac{w\tau}{(1+\tau)}\right](1+\tau)^{\Sigma\beta_l - \gamma}[1 + \tau(1 - z_1)]^{-\beta_1}$$

$$\cdots[1 + \tau(1 - z_n)]^{-\beta_n}\Phi\left(\begin{array}{c}\beta_j \\ \gamma\end{array}\middle|\frac{z_j w\tau}{(1+\tau)\left[1 + \tau(1 - z_j)\right]}\right)$$

$$= \sum_{h=0}^{\infty} L_h^{(\gamma-1)}(w)F_D\left(\begin{array}{c}-h; \beta_j \\ \gamma\end{array}\middle|z_j\right)\tau^h, \qquad |\tau| < \min(1, |z_j - 1|^{-1}). \tag{2.22}$$

If $b = -c = 1, a = d = 0$, then $\mathbf{T}_1(A)\Psi$ becomes

$$e^s(1 - z_1)^{-\beta_1}\cdots(1 - z_n)^{-\beta_n}s^\gamma\Phi\left(\begin{array}{c}\beta_j \\ \gamma\end{array}\middle|\frac{z_j s}{1 - z_j}\right)u_1^{\beta_1}\cdots u_n^{\beta_n}t^\gamma.$$

Expanding this function in powers of s, we obtain

$$e^s(1 - z_1)^{-\beta_1}\cdots(1 - z_n)^{-\beta_n}\Phi\left(\begin{array}{c}\beta_j \\ \gamma\end{array}\middle|\frac{z_j s}{1 - z_j}\right) = \sum_{h=0}^{\infty} F_D\left(\begin{array}{c}\gamma + h; \beta_j \\ \gamma\end{array}\middle|z_j\right)\frac{s^h}{h!}.$$

$$\tag{2.23}$$

ISBN-0-201-13503-5

Note that Φ, (2.18), is a confluent form of F_D. For $n = 1$ we have

$$\Phi\left(\begin{matrix}\beta\\\gamma\end{matrix}\middle|z\right) = {}_1F_1\left(\begin{matrix}\beta\\\gamma\end{matrix}\middle|z\right).$$

Note also that we have demonstrated that the system of equations (1.15) admits a partial R-separation of variables in terms of the coordinates $z_j / s, s, u_j, t$ and the partially separated solutions are characterized by operator equations of the form (2.17). An exhaustive classification of generating functions for the F_D awaits the classification of partially separable coordinate systems for the equations (1.15).

We can use the differential recurrence relations obeyed by the functions Φ and other confluent forms of the F_D to obtain a Lie algebraic theory of these functions. The corresponding Lie algebras can also be derived as contractions of the symmetry algebra of the F_D [90].

The Lauricella functions F_A, F_B, F_C are other n-variable generalizations of the ${}_2F_1$ which can be treated by similar Lie algebraic methods (see [91, 92]). However, not all of the recurrence relations obeyed by the ${}_2F_1$ can be extended to these functions, which therefore seem somewhat less interesting than the F_D. Similarly, the generalized hypergeometric functions ${}_pF_q$ can be treated by Lie algebraic methods [88].

Exercises

1. Show that the $2(p+q) + 1$ operators

$$E^{\alpha_l} = t_l\left(z\partial_z + t_l\partial_{t_l}\right), \qquad E_{\beta_k} = u_k^{-1}(z\,\partial_z + u_k\,\partial_{u_k} - 1),$$

$$E^{\alpha_1\cdots\beta_q} = t_1\cdots t_p u_1\cdots u_q\partial_z, \qquad T_l = t_l\partial_{t_l}, \qquad U_k = u_k\,\partial_{u_k},$$

$$l = 1,\ldots,p, \qquad k = 1,\ldots,q,$$

form a basis for a Lie algebra $\mathcal{G}_{p,q}$ by working out the commutation relations.

2. Making use of the differential recurrence formulas (B.20) for the generalized hypergeometric functions ${}_pF_q$, determine the action of $\mathcal{G}_{p,q}$ on the basis functions

$$\Psi_{b_j}^{a_i}(t_i, u_j, z) = {}_pF_q\left(\begin{matrix}a_i\\b_j\end{matrix}\middle|z\right)t_1^{a_1}\cdots t_p^{a_p}u_1^{b_1}\cdots u_q^{b_q}.$$

3. Show that the differential equation (B.19) for the ${}_pF_q$ is equivalent to $L_{p,q}\Psi_{b_j}^{a_i} = 0$ where

$$L_{p,q} = E^{\alpha_1}\cdots E^{\alpha_p} - E^{\alpha_1\cdots\beta_q}E_{\beta_1}\cdots E_{\beta_q}.$$

ISBN-0-201-13503-5

4. Show that the $\Psi_{b_j}^{a_i}$ can be characterized as the solutions $\Psi(t_i, u_j, z)$ of the equations

$$L_{p,q}\Psi = 0, \qquad T_l\Psi = a_l\Psi, \qquad U_k\Psi = b_k\Psi, \qquad 1 \leqslant l \leqslant p, \quad 1 \leqslant k \leqslant q,$$

which are analytic at $t=0$. (This proves that the $_pF_q$ arise from the partial differential equation $L_{p,q}\Psi = 0$ by a separation of variables.)

5. Prove that $\mathcal{G}_{p,q}$ is a symmetry algebra of the equation $L_{p,q}\Psi = 0$.

6. Determine the special function identities for the $_pF_q$ which are associated with the expressions

$$\exp(cE^{\alpha_1})\Psi_{b_j}^{a_i}, \qquad \exp(cE_{\beta_1})\Psi_{b_j}^{a_i} \quad \text{and} \quad \exp(cE^{\alpha_1 \cdots \beta_q})\Psi_{b_j}^{a_i}.$$

7. Use Weisner's principle and the expression $\exp(1 E^{\alpha_1})\Psi_{b_j}^{\sigma, a_i}$ to derive the identity

$$(1-\tau)^{-\sigma} {}_pF_q\left(\begin{matrix} \sigma, a_i \\ b_j \end{matrix} \middle| \frac{-z\tau}{1-\tau}\right) = \sum_{n=0}^{\infty} \frac{\Gamma(\sigma+n)}{\Gamma(\sigma)n!} {}_pF_q\left(\begin{matrix} -n, a_i \\ b_j \end{matrix} \middle| z\right)\tau^n, \qquad |\tau| < 1.$$

8. Show that a solution Ψ of $L_{p,q}\Psi = 0$ such that

$$E^{\alpha_1}\Psi = \Psi, \qquad T_l\Psi = a_l\Psi, \qquad 1 \leqslant l \leqslant p-1;$$
$$U_k\Psi = b_k\Psi, \qquad 1 \leqslant k \leqslant q;$$

takes the form

$$\Psi = f(z/t_p)\exp(-t_p^{-1})t_1^{a_1} \cdots t_{p-1}^{a_{p-1}}u_1^{b_1} \cdots u_q^{b_q};$$

that is, Ψ is R-separable in the coordinates $t_1, \ldots, t_{p-1}, u_1, \ldots, u_q, z/t_p$. Show that if Ψ is analytic at $z=0$, then to within a constant multiple,

$$f(x) = {}_{p-1}F_q\left(\begin{matrix} a_i \\ b_j \end{matrix} \middle| x\right).$$

Apply Weisner's principle to derive the identity

$$e^\tau {}_{p-1}F_q\left(\begin{matrix} a_i \\ b_j \end{matrix} \middle| -z\tau\right) = \sum_{n=0}^{\infty} {}_pF_q\left(\begin{matrix} a_i, -n \\ b_j \end{matrix} \middle| z\right)\frac{\tau^n}{n!}.$$

(For further identities for the $_pF_q$ derived by group-theoretic methods, see [88].)

9. Use the differential recurrence relations (B.5) for the $_2F_1$ to verify relations (1.4) in the special case where $n=1$.

10. Work out the identities (2.21), (2.22), and (2.23) when $n=1$, in which case they become generating functions for the $_2F_1$.

ISBN-0-201-13503-5

APPENDIX A

Lie Groups and Algebras

We list here some of the basic facts concerning Lie groups and algebras that are needed in this book. Complete proofs and further details can be found in [85]. Since almost all Lie groups that arise in mathematical physics are groups of matrices, we shall confine our attention to local linear Lie groups.

Let W be an open, connected set containing $\mathbf{e}=(0,\ldots,0)$ in the space R^n of all real n-tuples $\mathbf{g}=(g_1,\ldots,g_n)$.

DEFINITION. An n-dimensional (*real*) *local linear Lie group* G is a set of $m\times m$ nonsingular complex matrices $A(\mathbf{g})=A(g_1,\ldots,g_n)$ defined for each $\mathbf{g}\in W$ such that

1. $A(\mathbf{e})=E_m$ (the identity matrix).
2. The matrix elements of $A(\mathbf{g})$ are analytic functions of the parameters $g_1,\ldots,g_n$ and the map $\mathbf{g}\rightarrow A(\mathbf{g})$ is one to one.
3. The n matrices $\partial A(\mathbf{g})/\partial g_j, j=1,\ldots,n$, are linearly independent for each $\mathbf{g}\in W$.
4. There exists a neighborhood W' of $\mathbf{e}$ in R^n, $W'\subset W$, with the property that for every pair of n-tuples $\mathbf{g},\mathbf{h}$ in W' there is an n-tuple $\mathbf{k}$ in W satisfying $A(\mathbf{g})A(\mathbf{h})=A(\mathbf{k})$ where the operation on the left is matrix multiplication.

A local Lie group can be considered as a neighborhood of the identity in a global Lie group. (For the theory of global Lie groups see [46,47].) If in the foregoing definition W and W' are neighborhoods of $\mathbf{e}$ in $\mathcal{C}^n$, then G is a *complex* local linear Lie group.

The parameters $\mathbf{g}=(g_1,\ldots,g_n)$ define *local coordinates* on G and it can be shown that group multiplication can be expressed in terms of local coordinates by $\mathbf{k}=\varphi(\mathbf{g},\mathbf{h})$ where φ is an analytic vector-valued function of its $2n$

ENCYCLOPEDIA OF MATHEMATICS and Its Applications, Gian-Carlo Rota (ed.).
Vol. 4: Willard Miller, Jr., Symmetry and Separation of Variables

ISBN-0-201-13503-5

arguments for $\mathbf{g},\mathbf{h}$ sufficiently close to $\mathbf{e}$ and $\boldsymbol{\varphi}(\mathbf{e},\mathbf{g})=\boldsymbol{\varphi}(\mathbf{g},\mathbf{e})=\mathbf{g}$. Any local coordinate transformation $\mathbf{g}'=\mathbf{f}(\mathbf{g})$ leads to a new Lie group which we identify with G.

Let $\mathbf{g}(t)$ be an analytic curve in R^n such that $\mathbf{g}(0)=\mathbf{e}$. (Here t is a real parameter and $\mathbf{g}(t)$ is defined and analytic in t for $|t|<1$.) The *Lie algebra* $\mathcal{G}$ of G is the set of all $m\times m$ matrices $\mathcal{C}=(d/dt)A(\mathbf{g}(t))|_{t=0}$ where $\mathbf{g}$ ranges over all analytic curves through $\mathbf{e}$. It follows easily that every $\mathcal{C}\in\mathcal{G}$ is a linear combination of the n linearly independent matrices

$$\mathcal{C}_j=\frac{\partial A(\mathbf{g})}{\partial g_j}\bigg|_{\mathbf{g}=\mathbf{e}}.$$

Indeed, $\mathcal{C}=\sum_{j=1}^n\alpha_j\mathcal{C}_j$ where $\alpha_j=(dg_j/dt)(t)|_{t=0}$. This shows that $\mathcal{G}$ is an n-dimensional real vector space under addition and scalar multiplication of matrices. The matrices $\mathcal{C}_j$ form a basis for $\mathcal{G}$.

Furthermore, the matrix *commutator* $[\mathcal{C},\mathcal{B}]=\mathcal{C}\mathcal{B}-\mathcal{B}\mathcal{C}$ belongs to $\mathcal{G}$ for any $\mathcal{C},\mathcal{B}\in\mathcal{G}$. In particular, $[\mathcal{C}_l,\mathcal{C}_s]=\sum_{j=1}^n c_j^{ls}\mathcal{C}_j,1\leqslant l,s\leqslant n$, where $c_j^{ls}=c_{j,ls}-c_{j,ls}$ and

$$c_{j,ls}=\frac{\partial^2}{\partial g_l\,\partial h_s}\varphi_j(\mathbf{g},\mathbf{h})\bigg|_{\mathbf{g}=\mathbf{h}=\mathbf{e}}\quad,\qquad \boldsymbol{\varphi}=(\varphi_1,\dots,\varphi_n).$$

The *matrix exponential* $\exp(\mathcal{C})$ of an $m\times m$ matrix $\mathcal{C}$ is the $m\times m$ matrix

$$\exp(\mathcal{C})=\sum_{p=0}^{\infty}(p!)^{-1}\mathcal{C}^p.\tag{A.1}$$

This series is convergent and analytic in the matrix elements of $\mathcal{C}$. Here $\exp(\mathcal{C})\exp(-\mathcal{C})=E_m$ and $\exp(\mathcal{C})\exp(\mathcal{B})=\exp(\mathcal{C}+\mathcal{B})$ for $m\times m$ matrices $\mathcal{C},\mathcal{B}$ with $\mathcal{C}\mathcal{B}=\mathcal{B}\mathcal{C}$.

Denote the matrix elements of $\mathcal{C}$ by $\mathcal{C}_{ij},1\leqslant i,j\leqslant m$, and define the *norm* of $\mathcal{C}$ by $\|\mathcal{C}\|=\max_{i,j}|\mathcal{C}_{ij}|$. There exist positive numbers ε,δ such that (1) $\exp(\mathcal{C})\in G$ for each $\mathcal{C}\in\mathcal{G}$ with $\|\mathcal{C}\|<\varepsilon$ and (2) each $A\in G$ with $\|A-E_m\|<\delta$ can be expressed as $A=\exp(\mathcal{C})$ for a unique $\mathcal{C}\in\mathcal{G}$ with $\|\mathcal{C}\|<\varepsilon$. This is a one-to-one analytic mapping of a neighborhood of the zero matrix in $\mathcal{G}$ onto a neighborhood of E_m in G. Writing $A=\exp(\mathcal{C})$ where $\mathcal{C}=\sum_{j=1}^n\alpha_j\mathcal{C}_j$, we can use the *canonical coordinates* $\alpha_1,\dots,\alpha_n$ to parametrize G.

Let U be an open connected set in $\mathcal{C}^p$. Any $\mathbf{z}\in U$ can be designated by its coordinates $\mathbf{z}=(z_1,\dots,z_p),z_j\in\mathcal{C}$. Let $\mathbf{Q}$ be a mapping that associates to each pair $(\mathbf{z},A),\mathbf{z}\in U,A\in G$, an element $\mathbf{Q}(\mathbf{z},A)$ in $\mathcal{C}^p$. We write $\mathbf{Q}(z,A)=\mathbf{z}^A\in\mathcal{C}^p$.

ISBN-0-201-13503-5

DEFINITION. The n-dimensional local linear Lie group G acts on the manifold U as a *local Lie transformation group* if $\mathbf{Q}$ satisfies the properties

1. $\mathbf{z}^A$ is analytic in the $p+n$ coordinates of $\mathbf{z}$ and A;
2. $\mathbf{z}^{E_m}=\mathbf{z}$ all $\mathbf{z}\in U$;
3. if $\mathbf{z}^A\in U$, then $(\mathbf{z}^A)^B=\mathbf{z}^{(AB)}$ for all $A,B\in G$ such that $AB\in G$.

Suppose G is a local Lie transformation group on U and let $\mathcal{F}$ be the space of all functions $f(\mathbf{z})$ analytic in a neighborhood of a fixed $\mathbf{z}^0\in U$. (Here the neighborhood is allowed to vary with the function.) A *local multiplier* ν for this transformation group is a scalar-valued function $\nu(\mathbf{z},A)$ analytic in the $p+n$ coordinates $\mathbf{z}\in U,A\in G$ such that (1) $\nu(\mathbf{z},E_m)=1$ and (2) $\nu(\mathbf{z},AB)=\nu(\mathbf{z},A)\nu(\mathbf{z}^A,B)$ for $A,B,AB\in G$. Note that $\nu(\mathbf{z},A)\equiv 1$ is a (trivial) local multiplier. (For the general theory of local multipliers see [82, Chapter 8].)

A *local multiplier representation* $\mathbf{T}$ corresponding to $G,U,\mathcal{F},\nu$ is a mapping $\mathbf{T}(A)$ of $\mathcal{F}$ onto $\mathcal{F}$ defined for $A\in G$ and $f\in\mathcal{F}$ by

$$[\mathbf{T}(A)f](\mathbf{z})=\nu(\mathbf{z},A)f(\mathbf{z}^A).\tag{A.2}$$

Since ν is a local multiplier, it follows that

1. $\mathbf{T}(E_m)f=f$ for all $f\in\mathcal{F}$;
2. $\mathbf{T}(AB)f=\mathbf{T}(A)[\mathbf{T}(B)f]$ for all $A,B\in G$ sufficiently close to E_m.

Let $A(\mathbf{g}(t))$ be a one-parameter curve in G with Lie algebra element

$$\mathcal{Q}=\frac{d}{dt}A(\mathbf{g}(t))\bigg|_{t=0}=\sum_{j=1}^{n}\alpha_j\mathcal{C}_j\tag{A.3}$$

as defined earlier. Let $\mathbf{T}$ be a multiplier representation of G and $f\in\mathcal{F}$.

DEFINITION. The *generalized Lie derivative* $D_\mathcal{Q}f$ of f is the analytic function

$$D_\mathcal{Q}f(\mathbf{z})=\frac{d}{dt}\big[\mathbf{T}(A\mathbf{g}(t))f\big](\mathbf{z})\bigg|_{t=0}.\tag{A.4}$$

Direct computation yields

$$D_\mathcal{Q}=\sum_{i=1}^{p}\sum_{j=1}^{n}P_{ij}(\mathbf{z})\alpha_j\partial_{z_i}+\sum_{j=1}^{n}\alpha_jP_j(\mathbf{z}).\tag{A.5}$$

Here $D_\mathcal{Q}$ depends only on $\mathcal{Q}\in\mathcal{G}$, not on the possible curves $\mathbf{g}(t)$ that lead to $\mathcal{Q}$. The $P_{ij}(\mathbf{z})$ can be computed uniquely from $\mathbf{Q}(\mathbf{z},A)$, while the $P_j(\mathbf{z})$ follow from $\nu(\mathbf{z},A)$. In particular, if $\nu\equiv1$, then $P_j\equiv0$ and $D_\mathcal{Q}$ is an *ordinary* Lie derivative.

ISBN-0-201-13503-5

The following theorems, which are used frequently in this book, are essentially due to Sophus Lie [82, 85].

THEOREM $A.1$. *The generalized Lie derivatives of a local multiplier representation form a Lie algebra under the operations of addition of derivatives and Lie bracket*

$$[D_{\mathcal{A}}, D_{\mathcal{B}}] = D_{\mathcal{A}}D_{\mathcal{B}} - D_{\mathcal{B}}D_{\mathcal{A}}. \qquad (A.6)$$

This algebra is a homomorphic image of $\mathcal{G}$:

$$D_{(a\mathcal{A} + b\mathcal{B})} = aD_{\mathcal{A}} + bD_{\mathcal{B}}, \qquad D_{[\mathcal{A}, \mathcal{B}]} = [D_{\mathcal{A}}, D_{\mathcal{B}}], \qquad (A.7)$$

$$\mathcal{A}, \mathcal{B} \in \mathcal{G}, a, b \in R.$$

(The equalities in (A.6), (A.7) are meant in the sense that both sides of the equation yield the same result when applied to a fixed $f \in \mathcal{F}$.)

THEOREM $A.2$

$$[\mathbf{T}(\exp(t\mathcal{A}))f](\mathbf{z}) = \sum_{j=0}^{\infty} (j!)^{-1}t^{j}D_{\mathcal{A}}^{j}f(\mathbf{z})$$

$$= (\exp D_{\mathcal{A}})f(\mathbf{z}), \qquad \mathcal{A} \in \mathcal{G}. \qquad (A.8)$$

(*This result is valid for all* $t \in R$ *with* $|t|$ *sufficiently small.*)

THEOREM $A.3$. *Let*

$$D_{j} = \sum_{i=1}^{p} P_{ij}(\mathbf{z})\partial_{z_{i}} + P_{j}(\mathbf{z}), \qquad j = 1, \ldots, n, \qquad (A.9)$$

be n linearly independent differential operators defined and analytic in an open set $U \subseteq \mathcal{C}^{p}$. If there exist real constants c_{jk}^{l} such that

$$[D_{j}, D_{k}] = \sum_{l=1}^{n} c_{jk}^{l}D_{l}, \qquad 1 \leqslant j, k \leqslant n, \qquad (A.10)$$

then the D_{j} form a basis for a Lie algebra that is the algebra of generalized Lie derivatives for a local multiplier representation $\mathbf{T}$ of a local Lie group G. There is a basis $\{\mathcal{C}_{j}\}$ for the Lie algebra $\mathcal{G}$ of G such that

$$[\mathcal{C}_{j}, \mathcal{C}_{k}] = \sum_{l=1}^{n} c_{jk}^{l}\mathcal{C}_{l}.$$

ISBN-0-201-13503-5

The action of G is obtained by integration of the equations

$$\frac{dz_i(t)}{dt} = \sum_{j=1}^{n} P_{ij}(\mathbf{z}(t))\alpha_j, \qquad \frac{d}{dt}\ln\nu(\mathbf{z}^0,\exp t\mathcal{Q}) = \sum_{j=1}^{n} \alpha_j P_j(\mathbf{z}(t)) \quad (A.11)$$

where

$$\mathbf{z}(0)=\mathbf{z}^0, \qquad \nu(\mathbf{z}^0, E_m)=1, \qquad \mathbf{z}(t)=\mathbf{z}^{0(\exp t\mathcal{Q})}, \qquad 1\leqslant i\leqslant p, \quad (A.12)$$

and $\mathcal{Q}$ is given by (A.3).

ISBN-0-201-13503-5

APPENDIX B.

Basic Properties of Special Functions

As a convenient reference we collect here some fundamental definitions and relations for those special functions which appear most frequently in this book. With the exception of the gamma and elliptic functions, all these functions arise as solutions of differential equations obtained by separating variables in the partial differential equations of mathematical physics. The notation used here is the same as that adopted in the Bateman project [36, 37], and the reader can find many additional properties of these functions in those references.

1. The Gamma Function

Defining integral:

$$\Gamma(z) = \int_0^\infty e^{-t} t^{z-1} dt, \qquad \operatorname{Re} z > 0.$$

By analytic continuation $\Gamma(z)$ can be extended to a function analytic in the whole complex plane, with the exception of simple poles at $z = -n, n = 0, 1, 2, \ldots$.

Functional equations:

$$\Gamma(z+1) = z\Gamma(z), \qquad \Gamma(z)\Gamma(1-z) = \pi/\sin \pi z.$$

ENCYCLOPEDIA OF MATHEMATICS and Its Applications, Gian-Carlo Rota (ed.).
Vol. 4: Willard Miller, Jr., Symmetry and Separation of Variables

ISBN-0-201-13503-5

Special values:

$$\Gamma(n+1)=n!, \qquad n=0,1,2,\ldots; \quad \Gamma(\tfrac{1}{2})=\sqrt{\pi} \; .$$

The binomial coefficients are defined by

$$\binom{\mu}{n}=\mu(\mu-1)\cdots(\mu-n+1)/n!=\Gamma(\mu+1)/\Gamma(\mu-n+1)n!. \quad \text{(B.1)}$$

2. The Hypergeometric Function

The hypergeometric series, convergent for $|z|<1$, is given by

$$_2F_1\left(\genfrac{}{}{0pt}{}{a,b}{c}\bigg|z\right)=\sum_{n=0}^{\infty}\frac{(a)_n(b)_n}{(c)_n n!}z^n \qquad \text{(B.2)}$$

where

$$(a)_0=1, \qquad (a)_n=a(a+1)\cdots(a+n-1), \qquad n=1,2,\ldots, \qquad \text{(B.3)}$$

is Pochhammer's symbol. By analytic continuation the $_2F_1$ can be extended to define a function analytic and single valued in the complex z plane cut along the positive real axis from $+1$ to $+\infty$.

Integral representation:

$$_2F_1\left(\genfrac{}{}{0pt}{}{a,b}{c}\bigg|z\right)=\frac{\Gamma(c)}{\Gamma(b)\Gamma(c-b)}\int_0^1 t^{b-1}(1-t)^{c-b-1}(1+tz)^{-a}\,dt,$$

$$\operatorname{Re}c>\operatorname{Re}b>0,|\arg(1-z)|<\pi.$$

For fixed $z, {}_2F_1\left(\genfrac{}{}{0pt}{}{a,b}{c}\big|z\right)/\Gamma(c)$ is an entire function of the parameters a,b,c. If a or b is a negative integer and c is not a negative integer, then the hypergeometric series becomes a polynomial in z. Differential equation:

$$z(1-z)\frac{d^2u}{dz^2}+[c-(a+b+1)z]\frac{du}{dz}-abu=0. \qquad \text{(B.4)}$$

This equation has the solution $u={}_2F_1\left(\genfrac{}{}{0pt}{}{a,b}{c}\big|z\right)$. For c not an integer the equation admits the linearly independent solution $u=$

ISBN-0-201-13503-5

$z^{1-c}{}_2F_1\left({a-c+1, b-c+1 \atop 2-c}\Big|z\right)$. Differential recurrence formulas:

$$\frac{d}{dz}\,{}_2F_1\left({a,b \atop c}\Big|z\right)=\frac{ab}{c}\,{}_2F_1\left({a+1,b+1 \atop c+1}\Big|z\right),$$

$$\left[z\frac{d}{dz}+a\right]{}_2F_1=a\,{}_2F_1\left({a+1,b \atop c}\Big|z\right),$$

$$\left[z\frac{d}{dz}+c-1\right]{}_2F_1=(c-1)\,{}_2F_1\left({a,b \atop c-1}\Big|z\right),$$

$$\left[z(1-z)\frac{d}{dz}-bz+c-a\right]{}_2F_1=(c-a)\,{}_2F_1\left({a-1,b \atop c}\Big|z\right),$$

$$\left[(1-z)\frac{d}{dz}-(a+b-c)\right]{}_2F_1=(c-a)(c-b)c^{-1}{}_2F_1\left({a,b \atop c+1}\Big|z\right),$$

$$\left[z(1-z)\frac{d}{dz}-bz+c-1\right]{}_2F_1=(c-1)\,{}_2F_1\left({a-1,b \atop c-1}\Big|z\right),$$

$$\left[(1-z)\frac{d}{dz}-a\right]{}_2F_1=a(b-c)c^{-1}{}_2F_1\left({a+1,b \atop c+1}\Big|z\right),$$

$$\left[z(1-z)\frac{d}{dz}-(b+a-1)z+c-1\right]{}_2F_1=(c-1)\,{}_2F_1\left({a-1,b-1 \atop c-1}\Big|z\right). \quad \text{(B.5)}$$

Symmetry relation:

$$_2F_1\left({a,b \atop c}\Big|z\right)={}_2F_1\left({b,a \atop c}\Big|z\right).$$

Transformation formulas:

$$_2F_1\left({a,b \atop c}\Big|z\right)=(1-z)^{-a}{}_2F_1\left({a,c-b \atop c}\Big|\frac{z}{z-1}\right)$$

$$=(1-z)^{c-a-b}{}_2F_1\left({c-a,c-b \atop c}\Big|z\right).$$

Special cases.   (i)   Legendre polynomials:

$$P_n(x)={}_2F_1\left({n+1,\,-n \atop 1}\Big|\frac{1-x}{2}\right), \qquad n=0,1,2,\dots;$$

(ii)   Gegenbauer polynomials:

$$C_n^\nu(x)=\frac{\Gamma(2\nu+n)}{\Gamma(2\nu)n!}\,{}_2F_1\left({2\nu+n,\,-n \atop \nu+\frac{1}{2}}\Big|\frac{1-x}{2}\right), \qquad n=0,1,2,\dots;$$

(iii)   Jacobi polynomials:

$$P_n^{(\alpha,\beta)}(x)=\binom{n+\alpha}{n}{}_2F_1\left(\begin{array}{c}n+\alpha+\beta+1,-n\\\alpha+1\end{array}\bigg|\frac{1-x}{2}\right),\qquad n=0,1,2,\ldots;$$

(iv)   Legendre functions:

$$P_\nu^\mu(z)=\left(\frac{z+1}{z-1}\right)^{\mu/2}\frac{{}_2F_1\left(\begin{array}{c}\nu+1,-\nu\\1-\mu\end{array}\bigg|\frac{1-z}{2}\right)}{\Gamma(1-\mu)},$$

$$Q_\nu^\mu(z)=e^{i\mu\pi}2^{-\nu-1}\pi^{1/2}\Gamma(\nu+\mu+1)z^{-\nu-\mu-1}(z^2-1)^{\mu/2}$$

$$\times\frac{{}_2F_1\left(\begin{array}{c}\nu/2+\mu/2+1,\nu/2+\mu/2+1/2\\\nu+3/2\end{array}\bigg|z^{-2}\right)}{\Gamma(\nu+3/2)}.\qquad(B.6)$$

3. The Confluent Hypergeometric Function.

The function ${}_1F_1\left(\begin{array}{c}a\\c\end{array}\big|z\right)$ is defined by the series

$${}_1F_1\left(\begin{array}{c}a\\c\end{array}\bigg|z\right)=\sum_{n=0}^\infty\frac{(a)_n}{(c)_n}\frac{z^n}{n!}$$

convergent for all z.
    Integral representation:

$${}_1F_1\left(\begin{array}{c}a\\c\end{array}\bigg|z\right)=\frac{\Gamma(c)}{\Gamma(a)\Gamma(c-a)}\int_0^1 e^{zt}t^{a-1}(1-t)^{c-a-1}\,dt,\qquad \operatorname{Re}c>\operatorname{Re}a>0.$$

For fixed z, ${}_1F_1\left(\begin{array}{c}a\\c\end{array}\big|z\right)/\Gamma(c)$ is an entire function of a and c. Differential equation:

$$z\frac{d^2u}{dz^2}+(c-z)\frac{du}{dz}-au=0.\qquad(B.7)$$

This equation has the solution $u={}_1F_1\left(\begin{array}{c}a\\c\end{array}\big|z\right)$ and for c not an integer the equation admits the independent solution $u=z^{1-c}{}_1F_1\left(\begin{array}{c}a-c+1\\2-c\end{array}\big|z\right).$

ISBN-0-201-13503-5

Differential recurrence formulas:

$$\frac{d}{dz}\,{}_1F_1\!\left(\genfrac{}{}{0pt}{}{a}{c}\bigg|z\right)=\frac{a}{c}\,{}_1F_1\!\left(\genfrac{}{}{0pt}{}{a+1}{c+1}\bigg|z\right),$$

$$\left[z\frac{d}{dz}-z+c-1\right]{}_1F_1=(c-1)\,{}_1F_1\!\left(\genfrac{}{}{0pt}{}{a-1}{c-1}\bigg|z\right),$$

$$\left[z\frac{d}{dz}+a\right]{}_1F_1=a\,{}_1F_1\!\left(\genfrac{}{}{0pt}{}{a+1}{c}\bigg|z\right),$$

$$\left[z\frac{d}{dz}-z+c-a\right]{}_1F_1=(c-a)\,{}_1F_1\!\left(\genfrac{}{}{0pt}{}{a-1}{c}\bigg|z\right),$$

$$\left[\frac{d}{dz}-1\right]{}_1F_1=\frac{a-c}{c}\,{}_1F_1\!\left(\genfrac{}{}{0pt}{}{a}{c+1}\bigg|z\right),$$

$$\left[z\frac{d}{dz}+c-1\right]{}_1F_1=(c-1)\,{}_1F_1\!\left(\genfrac{}{}{0pt}{}{a}{c-1}\bigg|z\right). \qquad\text{(B.8)}$$

Transformation formula:

$$\,{}_1F_1\!\left(\genfrac{}{}{0pt}{}{a}{c}\bigg|z\right)=e^z\,{}_1F_1\!\left(\genfrac{}{}{0pt}{}{c-a}{c}\bigg|-z\right).$$

Special cases.   (i)   Laguerre polynomials:

$$L_n^{(\alpha)}(x)=\frac{\Gamma(\alpha+n+1)}{\Gamma(\alpha+1)n!}\,{}_1F_1\!\left(\genfrac{}{}{0pt}{}{-n}{\alpha+1}\bigg|x\right),\qquad n=0,1,2,\dots;$$

(ii)   Bessel functions:

$$J_\nu(x)=\frac{e^{-ix}(x/2)^\nu}{\Gamma(\nu+1)}\,{}_1F_1\!\left(\genfrac{}{}{0pt}{}{\nu+1/2}{2\nu+1}\bigg|2ix\right);$$

(iii)   Parabolic cylinder functions:

$$D_\nu(x)=2^{1/2}\exp\!\left(\frac{-x^2}{4}\right)\left[\frac{\Gamma(1/2)}{\Gamma(1/2-\nu/2)}\,{}_1F_1\!\left(\genfrac{}{}{0pt}{}{-\nu/2}{1/2}\bigg|\frac{x^2}{2}\right)\right.$$

$$\left.+x2^{-1/2}\frac{\Gamma(-1/2)}{\Gamma(-\nu/2)}\,{}_1F_1\!\left(\genfrac{}{}{0pt}{}{1/2-\nu/2}{3/2}\bigg|\frac{x^2}{2}\right)\right]. \qquad\text{(B.9)}$$

ISBN-0-201-13503-5

4. Parabolic Cylinder Functions

The function $u = D_\nu(x)$, (B.9iii), is a solution of the equation

$$\frac{d^2u}{dz^2} + \left(\nu + \frac{1}{2} - \frac{z^2}{4} \right) u = 0. \tag{B.10}$$

A linearly independent solution is $u = D_{-\nu-1}(iz)$ and, for ν not an integer, $D_\nu(-z)$. If $\nu = n = 0, 1, 2, \dots$, then

$$D_n(z) = 2^{-n/2} \exp(-z^2/4) H_n(2^{-1/2}z) \tag{B.11}$$

where

$$H_n(z) = (-1)^n \exp(z^2) \frac{d^n}{dz^n} \exp(-z^2) \tag{B.12}$$

is the Hermite polynomial of order n.
Differential recurrence relations:

$$\left[\frac{d}{dz} + \frac{z}{2} \right] D_\nu(z) = \nu D_{\nu-1}(z), \qquad \left[-\frac{d}{dz} + \frac{z}{2} \right] D_\nu(z) = D_{\nu+1}(z). \tag{B.13}$$

5. Bessel Functions

The Bessel function $J_\nu(z)$ is given by (B.9ii) or by

$$J_\nu(z) = \frac{(z/2)^\nu}{\Gamma(\nu+1)} {}_0F_1\left(\nu + 1 \Big| \frac{-z^2}{4} \right), \qquad |\arg z| < \pi, \tag{B.14}$$

where

$$ {}_0F_1(c|x) = \sum_{n=0}^{\infty} \frac{x^n}{(c)_n n!}, \tag{B.15}$$

convergent for all x. Here, $z^{-\nu}J_\nu(z)$ is an entire function of z.
Differential equation:

$$\frac{d^2u}{dz^2} + \frac{1}{z}\frac{du}{dz} + \left(1 - \frac{\nu^2}{z^2} \right) u = 0. \tag{B.16}$$

This equation has solutions $u_1 = J_\nu(z)$ and $u_2 = J_{-\nu}(z)$, linearly indepen-dent for ν not an integer n. However, $J_{-n}(z) = (-1)^n J_n(z)$ and for $\nu = n$, $J_n(z)$ is the only solution of (B.16) which is bounded near $z = 0$. Differen-

ISBN-0-201-13503-5

tial recurrence formulas:

$$\left[-\frac{d}{dz}+\frac{\nu}{z}\right]J_\nu(z)=J_{\nu+1}(z), \qquad \left[\frac{d}{dz}+\frac{\nu}{z}\right]J_\nu(z)=J_{\nu-1}(z). \quad \text{(B.17)}$$

The functions in these classes (Sections 2–5) are either hypergeometric functions $_2F_1$ or various special and limiting cases of the $_2F_1$. However, the functions in Sections 6 and 7 are generalizations of the $_2F_1$, the first to differential equations of higher order and the second to functions of several variables.

6. Generalized Hypergeometric Functions

The functions $_pF_q$ are defined by the series

$$_pF_q\left(\begin{matrix}a_1,a_2,\ldots,a_p\\b_1,b_2,\ldots,b_q\end{matrix}\middle|z\right)=\,_pF_q(a_i;b_j;z)$$

$$=\sum_{n=0}^{\infty}\frac{(a_1)_n\cdots(a_p)_n}{(b_1)_n\cdots(b_q)_n}\frac{z^n}{n!}. \qquad \text{(B.18)}$$

Unless the parameters a_i, b_j are chosen such that the series terminates or becomes undefined, it can be shown that the series converges for all z if $p \leqslant q$, converges for $|z| < 1$ if $p = q+1$, and diverges for all $z \neq 0$ if $p > q+1$.

Differential equation:

$$\left(z\frac{d}{dz}+a_1\right)\cdots\left(z\frac{d}{dz}+a_p\right)u-\frac{d}{dz}\left(z\frac{d}{dz}+b_1-1\right)\cdots\left(z\frac{d}{dz}+b_q-1\right)u=0.$$

$$\text{(B.19)}$$

This equation has the solution $u=\,_pF_q(a_i;b_j;z)$ and, except for special choices of the parameters a_i, b_j, this is the only solution of (B.19) which is bounded in a neighborhood of $z=0$.

Differential recurrence formulas:

$$\left(z\frac{d}{dz}+a_1\right)_pF_q(a_i;b_j;z)=a_{1\,p}F_q\left(\begin{matrix}a_1+1,a_2,\ldots,a_p\\b_j\end{matrix}\middle|z\right),$$

$$\left(z\frac{d}{dz}+b_1-1\right)_pF_q(a_i;b_j;z)=(b_1-1)_pF_q\left(\begin{matrix}a_i\\b_1-1,b_2,\ldots,b_q\end{matrix}\middle|z\right),$$

$$\frac{d}{dz}\,_pF_q(a_i;b_j;z)=\frac{a_1\cdots a_p}{b_1\cdots b_q}\,_pF_q(a_i+1;b_j+1;z). \quad \text{(B.20)}$$

ISBN-0-201-13503-5

Symmetry relation: The $_pF_q(a_i; b_j; z)$ is a symmetric function of $a_1,\ldots,a_p$ and of $b_1,\ldots,b_q$.

7. The Lauricella Functions

The Lauricella functions are generalizations of the $_2F_1$ to n variables $z_1,\ldots,z_n$. They fall into four classes:

$$F_A[a; b_1,\ldots,b_n; c_1,\ldots,c_n; z_1,\ldots,z_n]$$

$$= \sum_{m_1=0}^{\infty} \cdots \sum_{m_n=0}^{\infty} \frac{(a)_{m_1+\cdots+m_n}(b_1)_{m_1}\cdots(b_n)_{m_n}}{(c_1)_{m_1}\cdots(c_n)_{m_n}} \frac{z_1^{m_1}\cdots z_n^{m_n}}{m_1!\cdots m_n!}, \quad (B.21)$$

$$|z_1| + \cdots + |z_n| < 1,$$

$$F_B[a_1,\ldots,a_n; b_1,\ldots,b_n; c; z_1,\ldots,z_n]$$

$$= \sum_{m_1=0}^{\infty} \cdots \sum_{m_n=0}^{\infty} \frac{(a_1)_{m_1}\cdots(a_n)_{m_n}(b_1)_{m_1}\cdots(b_n)_{m_n}}{(c)_{m_1+\cdots+m_n}} \frac{z_1^{m_1}\cdots z_n^{m_n}}{m_1!\cdots m_n!}, \quad (B.22)$$

$$|z_1| < 1,\ldots,|z_n| < 1,$$

$$F_C[a; b; c_1,\ldots,c_n; z_1,\ldots,z_n]$$

$$= \sum_{m_1=0}^{\infty} \cdots \sum_{m_n=0}^{\infty} \frac{(a)_{m_1+\cdots+m_n}(b)_{m_1+\cdots+m_n}}{(c_1)_{m_1}\cdots(c_n)_{m_n}} \frac{z_1^{m_1}\cdots z_n^{m_n}}{m_1!\cdots m_n!}, \quad (B.23)$$

$$|z_1|^{1/2} + \cdots + |z_n|^{1/2} < 1,$$

and

$$F_D[a; b_1,\ldots,b_n; c; z_1,\ldots,z_n]$$

$$= \sum_{m_1=0}^{\infty} \cdots \sum_{m_n=0}^{\infty} \frac{(a)_{m_1+\cdots+m_n}(b_1)_{m_1}\cdots(b_n)_{m_n}}{(c)_{m_1+\cdots+m_n}} \frac{z_1^{m_1}\cdots z_n^{m_n}}{m_1!\cdots m_n!}, \quad (B.24)$$

$$|z_1| < 1,\ldots,|z_n| < 1.$$

The functions in the following sections cannot be obtained as special cases or limits of functions of hypergeometric type.

8. Mathieu Functions

Mathieu's differential equation is

$$\frac{d^2u}{dx^2} + (a - 2q\cos 2x)u = 0 \quad (B.25)$$

ISBN-0-201-13503-5

where usually the variable x is real and q is a given real nonzero parameter. If we impose the periodic boundary condition $u(x) = u(x + 2\pi)$ on the solutions of (B.25), we can cast this equation into the form of a regular Sturm–Liouville eigenvalue problem for the eigenvalues a. It follows from the general theory of such problems that there exist countably infinitely many such eigenvalues, all real, each of multiplicity one, and bounded below but increasing to $+\infty$. Moreover, due to the symmetry properties of (B.25) this equation has four types of periodic solutions (called Mathieu functions of the first kind, or just Mathieu functions):

(i) $\quad \mathrm{ce}_{2n}(x,q) = \displaystyle\sum_{m=0}^{\infty} A_{2m}^{(2n)} \cos 2mx,$

(ii) $\quad \mathrm{ce}_{2n+1}(x,q) = \displaystyle\sum_{m=0}^{\infty} A_{2m+1}^{(2n+1)} \cos(2m+1)x,$

(iii) $\quad \mathrm{se}_{2n+1}(x,q) = \displaystyle\sum_{m=0}^{\infty} B_{2m+1}^{(2n+1)} \sin(2m+1)x,$

(iv) $\quad \mathrm{se}_{2n+2}(x,q) = \displaystyle\sum_{m=0}^{\infty} B_{2m+2}^{(2n+2)} \sin(2m+2)x, \qquad n = 0,1,2,\ldots . \quad$ (B.26)

The coefficients A, B depend on q and recurrence relations for these coefficients can be easily obtained by substituting expressions (B.26) into (B.25) [7]. The eigenvalues a of $\mathrm{ce}_{2n}, \mathrm{ce}_{2n+1}, \mathrm{se}_{2n+1}, \mathrm{se}_{2n+2}$ are denoted $a_{2n}, a_{2n+1}, b_{2n+1}, b_{2n+2}$, respectively. The eigenvalues are just those values of a such that the functions (B.26) whose coefficients are determined by the recurrence relations belong to $L_2[-\pi, \pi]$; that is, such that the functions are square integrable. The coefficients can always be chosen to be real and the Mathieu functions are normalized so that

$$\int_{-\pi}^{\pi} [u(x)]^2 \, dx = \pi. \qquad (B.27)$$

Furthermore, the normalization is such that

$$\lim_{q \to 0} \mathrm{ce}_0(x,q) = 2^{-1/2}, \qquad \lim_{q \to 0} \mathrm{ce}_n(x,q) = \cos nx, \qquad n \neq 0,$$

$$\lim_{q \to 0} \mathrm{se}_n(x,q) = \sin nx. \qquad\qquad\qquad\qquad\qquad (B.28)$$

APPENDIX C

Elliptic Functions

We list here the basic properties of elliptic functions that are needed in this book. For further details see [7, 37, 136a].

Elliptic functions depend on a complex variable z and a real parameter k (the *modulus*) which in this book will always satisfy $0 \leqslant k \leqslant 1$. The *complementary modulus* is $k' = (1 - k^2)^{1/2}, 1 \geqslant k' \geqslant 0$. The *elliptic functions* $\mathrm{sn}(z, k), \mathrm{cn}(z, k), \mathrm{dn}(z, k)$, or briefly $\mathrm{sn}\, z, \mathrm{cn}\, z, \mathrm{dn}\, z$, are defined by

$$z = \int_0^{\mathrm{sn}\, z} \left[(1 - t^2)(1 - k^2 t^2) \right]^{-1/2} dt = \int_{\mathrm{cn}\, z}^1 \left[(1 - t^2)(k'^2 + k^2 t^2) \right]^{-1/2} dt$$

$$= \int_{\mathrm{dn}\, z}^1 \left[(1 - t^2)(t^2 - k'^2) \right]^{-1/2} dt. \tag{C.1}$$

The values of the integrals depend on the integration contours and this is reflected in the periodicity properties of elliptic functions.

As $k \to 0$ we have

$$\mathrm{sn}(z, k) \to \sin z, \qquad \mathrm{cn}(z, k) \to \cos z, \qquad \mathrm{dn}(z, k) \to 1,$$

and as $k \to 1$

$$\mathrm{sn}(z, k) \to \tanh z, \qquad \mathrm{cn}(z, k) \to \mathrm{sech}\, z, \qquad \mathrm{dn}(z, k) \to \mathrm{sech}\, z.$$

Periodicity:

$$\mathrm{sn}(z + 2K) = -\mathrm{sn}\, z, \qquad \mathrm{sn}(z + 2iK') = \mathrm{sn}\, z,$$

$$\mathrm{cn}(z + 2K) = -\mathrm{cn}\, z, \qquad \mathrm{cn}(z + 2iK') = -\mathrm{cn}\, z, \tag{C.2}$$

$$\mathrm{dn}(z + 2K) = \mathrm{dn}\, z, \qquad \mathrm{dn}(z + 2iK') = -\mathrm{dn}\, z.$$

ENCYCLOPEDIA OF MATHEMATICS and Its Applications, Gian-Carlo Rota (ed.).
Vol. 4: Willard Miller, Jr., Symmetry and Separation of Variables

ISBN-0-201-13503-5

Here K, K' are defined by

$$K(k) = \int_0^{\pi/2} (1 - k^2 \sin^2 \theta)^{-1/2} d\theta, \qquad K' = K(k'). \qquad (C.3)$$

Special relations:

$$\operatorname{sn}(-z) = -\operatorname{sn}(z), \quad \operatorname{cn}(-z) = \operatorname{cn} z, \quad \operatorname{dn}(-z) = \operatorname{dn} z,$$
$$\operatorname{sn}^2 z + \operatorname{cn}^2 z = 1, \qquad k^2 \operatorname{sn}^2 z + \operatorname{dn}^2 z = 1. \qquad (C.4)$$

Special values:

$$\begin{aligned}
\operatorname{sn} 0 &= 0, & \operatorname{sn} K &= 1, & \operatorname{sn}(K + iK') &= 1/k, \\
\operatorname{cn} 0 &= 1, & \operatorname{cn} K &= 0, & \operatorname{cn}(K + iK') &= -ik'/k, & (C.5) \\
\operatorname{dn} 0 &= 1, & \operatorname{dn} K &= k', & \operatorname{dn}(K + iK') &= 0.
\end{aligned}$$

The elliptic functions all have simple poles at $z = iK'$. As z increases from 0 to K, $\operatorname{sn} z$ increases from 0 to 1, $\operatorname{cn} z$ decreases from 1 to 0, and $\operatorname{dn} z$ decreases from 1 to k'. As z varies from K to $K + iK'$, $\operatorname{sn} z$ increases from 1 to k^{-1}, $\operatorname{cn} z$ is pure imaginary and varies from 0 to $-ik'/k$, and $\operatorname{dn} z$ decreases from k' to 0. As z varies from $K + iK'$ to iK', $\operatorname{sn} z$ increases from $1/k$ to $+\infty$, $\operatorname{cn} z$ is pure imaginary and varies from $-ik'/k$ to $-i\infty$, and $\operatorname{dn} z$ is pure imaginary and varies from 0 to $-i\infty$.

Derivatives:

$$\frac{d}{dz} \operatorname{sn} z = \operatorname{cn} z \operatorname{dn} z, \qquad \frac{d}{dz} \operatorname{cn} z = -\operatorname{sn} z \operatorname{dn} z, \qquad \frac{d}{dz} \operatorname{dn} z = -k^2 \operatorname{sn} z \operatorname{cn} z.$$
$$(C.6)$$

REFERENCES

1. L. Ahlfors, *Complex Analysis*. McGraw-Hill, New York, 1953.
2. N. Akheizer, and I. Glazman, *Theory of Linear Operators in Hilbert Space*, Vol. II (transl. from the Russian). Ungar, New York, 1963.
3. R. Anderson, S. Kumei, and C. Wulfman, "Invariants of the equations of wave mechanics, I and II," *Rev. Mexicana Fis.* **21** (1972), 1–33; 35–57.
4. P. Appell, and J. Kampe de Feriet, *Functions Hypergéométriques et Hypersphériques*. Gauthiers-Villars, Paris, 1926.
5. L. Armstrong, Jr., "Group properties of radial wavefunctions," *J. Phys. Colloq.* **C4**, *Suppl.* **31** (1970), 17–23.
6. L. Armstrong, Jr., "$O(2,1)$ and the harmonic oscillator radial function," *J. Math. Phys.* **12** (1971), 953–957.
7. F. Arscott, *Periodic Differential Equations*. Macmillan (Pergamon), New York, 1964.

ISBN-0-201-13503-5

8. F. Arscott, "The Whittaker–Hill equation and the wave equation in paraboloidal coordinates," *Proc. Roy. Soc. Edinburgh* **A67** (1967), 265–276.

9. V. Bargmann, "Zur Theorie des Wasserstoffatoms," *Z. Physik* **99** (1936), 576–582.

10. V. Bargmann, "Irreducible unitary representations of the Lorentz group," *Ann. of Math.* **48** (1947), 568–640.

11. V. Bargmann, "On a Hilbert space of analytic functions and an associated integral transform, I," *Comm. Pure Appl. Math.* **14** (1961), 187–214.

12. H. Bateman, *Electrical and Optical Wave-Motion* (reprint of 1914 ed.). Dover, New York, 1955.

13. H. Bateman, *Partial Differential Equations of Mathematical Physics* (1st ed., 1932). Cambridge Univ. Press, London and New York, 1969.

14. H. Bateman, "The transformation of the electrodynamical equations," *J. London Math. Soc.* **8** (1909), 223–264.

15. G. Blumen and J. Cole, "The general similarity solution of the heat equation," *J. Math. and Mech.* **18** (1969), 1025–1042.

16. G. Blumen and J. Cole, *Similarity Methods for Differential Equations* (*Applied Mathematical Sciences*, Vol. 13). Springer, New York, 1974.

17. M. Bôcher, *Die Reihenentwickelungen der Potentialtheorie*. Leipzig, 1894.

18. C. Boyer, "The maximal kinematical invariance group for an arbitrary potential," *Helv. Phys. Acta* **47** (1974), 589–605.

19. C. Boyer, "Lie theory and separation of variables for the equation $iU_t + \Delta_2 U - (\alpha/x_1^2 + \beta/x_2^2)U = 0$," *SIAM J. Math. Anal.* **7** (1976), 230–263.

20. C. Boyer, E. Kalnins, and W. Miller, Jr., "Lie theory and separation of variables, 6: The equation $iU_t + \Delta_2 U = 0$," *J. Math. Phys.* **16** (1975), 499–511.

21. C. Boyer, E. Kalnins, and W. Miller, Jr., "Lie theory and separation of variables, 7: The harmonic oscillator in elliptic coordinates and Ince polynomials," *J. Math. Phys.* **16** (1975), 512–517.

22. C. Boyer, E. Kalnins, and W. Miller, Jr., "Symmetry and separation of variables for the Helmholtz and Laplace equations," *Nagoya Math. J.* **60** (1976), 35–80.

23. C. Boyer and W. Miller, Jr., "A classification of second-order raising operators for Hamiltonians in two variables," *J. Math. Phys.* **15** (1974), 1484–1489.

24. C. Boyer and B. Wolf, "Finite $SL(2, R)$ representation matrices of the D_k^+ series for all subgroup reductions," *Rev. Mexicana Fis.* **25**, (1976), 31–45.

25. L. Bragg, "The radial heat polynomials and related functions," *Trans. Amer. Math. Soc.* **119** (1965), 270–290.

26. H. Buchholz, *The Confluent Hypergeometric Function*. Springer, New York, 1969.

27. E. Cartan, "Sur la détermination d'un système orthogonal complet dans un espace de Riemann symétrique clos," *Rend. Circ. Math. Palermo* **53** (1929), 217–252.

28. T. Cherry, "Expansions in terms of parabolic cylinder functions," *Proc. Edinburgh Math. Soc.* (2), **8** (1949), 50–65.

29. F. Cholewinski and D. Haimo, "The dual Poisson–Laguerre transform," *Trans. Amer. Math. Soc.* **144** (1969), 271–300.

30. C. Coulson and A. Joseph, "A constant of the motion for the two-center Kepler problem," *Internat. J. Quant. Chem.* **1** (1967), 337–347.

31. A. Davydov, *Quantum Mechanics* (transl. from the Russian). Pergamon, Oxford, England, 1965.

32. P. Dirac, "Discussion of the infinite distribution of electrons in the theory of the positron," *Proc. Cambridge Phil. Soc.* **30** (1934), 150–163.

33. N. Dunford and J. Schwartz, *Linear Operators*, Parts I and II. Wiley (Interscience), New York, 1958, 1963.

34. L. Eisenhart, *Continuous Groups of Transformations* (reprint). Dover, New York, 1961.

35. A. Erdélyi, "Generating functions of certain continuous orthogonal systems," *Proc. Roy. Soc. Edinburgh* **A61**, (1941), 61–70.

ISBN-0-201-13503-5

36. A. Erdélyi *et al.*, *Higher Transcendental Functions*, Vol. I. McGraw-Hill, New York, 1953.

37. A. Erdélyi *et al.*, *Higher Transcendental Functions*, Vol. II. McGraw-Hill, New York, 1953.

38. F. Estabrook and B. Harrison, "Geometric approach to invariance groups and solution of partial differential systems," *J. Math. Phys.* **12** (1971), 653–666. F. Estabrook and H. Wahlquist, "Prolongation structures of nonlinear evolution equations," *J. Math. Phys.* **16** (1975), 1–7.

39. B. Friedman and J. Russek, "Addition theorems for spherical waves," *Quart. Appl. Math.* **12** (1954), 13–23.

40. I. Gel'fand, R. Minlos, and Z. Shapiro, *Representations of the Rotation and Lorentz Groups and Their Applications* (transl. from the Russian). Macmillan, New York, 1963.

41. I. Gel'fand and M. Naimark, "Unitary representations of the classical groups" (in Russian), *Trudy Mat. Inst. Steklov* **36** (1950).

42. I. Gel'fand and N. Vilenkin, *Generalized Functions*, Vol. 4: *Application of Harmonic Analysis* (transl. from the Russian). Academic Press, New York, 1964.

43. R. Gilmore, *Lie Groups, Lie Algebras and Some of their Applications*. Wiley, New York, 1974.

44. L. Gross, "Norm invariance of mass-zero equations under the conformal group," *J. Math. Phys.* **5** (1964), 687–695.

45. M. Hamermesh, *Group Theory and its Applications to Physical Problems*. Addison-Wesley, Reading, Mass., 1962.

46. M. Hausner and J. Schwartz, *Lie Groups and Lie Algebras*. Gordon & Breach, New York, 1968.

47. S. Helgason, *Differential Geometry and Symmetric Spaces*. Academic Press, New York, 1962.

48. P. Henrici, "Addition theorems for general Legendre and Gegenbauer functions," *J. Rat. Mech. Anal.* **4** (1955), 983–1018.

49. T. Hida, *Brownian Motion* (in Japanese). Iwanami Book Co., Tokyo, 1975.

50. H. Hochstadt, "Addition theorems for solutions of the wave equation in parabolic coordinates," *Pacific J. Math.* **7** (1957), 1365–1380.

51. E. Ince, *Ordinary Differential Equations* (reprint). Dover, New York, 1956.

52. L. Infeld and T. Hull, "The factorization method," *Revs. Mod. Phys.* **23** (1951), 21–68.

53. T. Inoui, "Unified theory of recurrence formulas," *Progr. Theoret. Phys.* **3** (1948), 169–187, 244–261.

54. E. Kalnins, "Mixed-basis matrix elements for the subgroup reductions of $SO(2,1)$," *J. Math. Phys.* **14** (1973), 654–657.

55. E. Kalnins, "On the separation of variables for the Laplace equation in two- and three-dimensional Minkowski space," *SIAM J. Math. Anal.* **6** (1975), 340–374.

56. E. Kalnins and W. Miller, Jr., "Symmetry and separation of variables for the heat equation," *Proc. Conf. on Symmetry, Similarity and Group-Theoretic Methods in Mechanics*, pp. 246–261. Univ. of Calgary, Calgary, Canada, 1974.

57. E. Kalnins and W. Miller, Jr., "Lie theory and separation of variables, 3: The equation $f_{tt} - f_{ss} = \gamma^2 f$," *J. Math. Phys.* **15** (1974), 1025–1032: "Erratum," *J. Math. Phys.* **16** (1975), 1531.

58. E. Kalnins and W. Miller, Jr., "Lie theory and separation of variables, 4: The groups $SO(2,1)$ and $SO(3)$," *J. Math. Phys.* **15** (1974), 1263–1274.

59. E. Kalnins and W. Miller, Jr., "Lie theory and separation of variables, 5: The equations $iU_t + U_{xx} = 0$ and $iU_t + U_{xx} - c/x^2 U = 0$," *J. Math. Phys.* **15** (1974), 1728–1737.

60. E. Kalnins and W. Miller, Jr., "Lie theory and separation of variables, 8: Semisubgroup coordinates for $\Psi_{tt} - \Delta_2 \Psi = 0$," *J. Math. Phys.* **16** (1975), 2507–2516.

61. E. Kalnins and W. Miller, Jr., "Lie theory and separation of variables, 9: Orthogonal R-separable coordinate systems for the wave equation $\Psi_{tt} - \Delta_2 \Psi = 0$," *J. Math. Phys.* **17** (1976), 331–355.

ISBN-0-201-13503-5

62. E. Kalnins and W. Miller, Jr., "Lie theory and separation of variables, 10: Nonorthogonal R-separable solutions of the wave equation $\Psi_{tt} = \Delta_2\Psi$," *J. Math. Phys.* **17** (1976), 356–368.

63. E. Kalnins and W. Miller, Jr., "Lie theory and separation of variables, 11: The EPD equation," *J. Math. Phys.* **17** (1976), 369–377.

64. E. Kalnins and W. Miller, Jr., "Lie theory and the wave equation in space-time, 1: The Lorentz group," *J. Math. Phys.* **18**, (1977), 1–16.

65. E. Kalnins, W. Miller, Jr., and P. Winternitz, "The group $O(4)$, separation of variables and the hydrogen atom," *SIAM J. Appl. Math.* **30** (1976), 630–664.

66. H. Kastrup, "Conformal group and its connection with an indefinite metric in Hilbert space," *Phys. Rev.* **140** (1965), B183–186.

67. T. Kato, *Perturbation Theory for Linear Operators.* Springer, New York, 1966.

68. T. Koornwinder, "The addition formula for Jacobi polynomials and spherical harmonics," *SIAM J. Appl. Math.* **25** (1973), 236–246.

68a. T. Koornwinder, "Jacobi polynomials, II: An analytic proof of the product formula," *SIAM J. Math. Anal.* **5** (1974), 125–137.

69. J. Korevaar, *Mathematical Methods*, Vol. 1. Academic Press, New York, 1968.

70. L. Landau and E. Lifshitz, *Quantum Mechanics, Non-Relativistic Theory* (transl. from the Russian). Addison-Wesley, Reading, Mass., 1958.

71. G. Lauricella, "Sulle funzioni ipergemetriche a pui variabili," *Rend. Circ. Mat. Palermo* **7** (1893), 111–113.

72. B. Levitan and I. Sarsjan, *Introduction to Spectral Theory of Selfadjoint Ordinary Differential Operators* (transl. from the Russian; *Translations of Mathematical Monographs*, Vol. 39). Amer. Math. Soc., Providence, R. I., 1975.

73. J.-M. Levy-Leblond, "Galilei group and Galilean invariance," in *Group Theory and Its Applications* (E. Loebl, Ed.), Vol. II. Academic Press, New York, 1971.

74. N. Macfadyen and P. Winternitz, "Crossing symmetric expansions of physical scattering amplitudes; the $O(2,1)$ group and Lamé functions," *J. Math. Phys.* **12** (1971), 281–293.

75. G. Mackey, *Induced Representations of Groups and Quantum Mechanics.* W. A. Benjamin, New York, 1968.

76. A. Makarov, J. Smorodinsky, K. Valiev, and P. Winternitz, "A systematic search for nonrelativistic systems with dynamical symmetries, Part I: The integrals of motion," *Nuovo Cimento* **52**A (1967), 1061–1084.

77. K. Maurin, *General Eigenfunction Expansions and Unitary Representations of Topological Groups.* PWN-Polish Scientific Publishers, Warsaw, 1968.

78. E. McBride, *Obtaining Generating Functions.* Springer, Berlin, 1971.

79. J. Meixner and F. Schäfke, *Mathieushe Funktionen und Sphäroidfunktionen.* Springer, Berlin, 1965.

80. W. Miller, Jr., *On Lie Algebras and Some Special Functions of Mathematical Physics* (Amer. Math. Soc. Memoir No. 50). Amer. Math. Soc., Providence, R.I., 1964.

81. W. Miller, Jr., "Confluent hypergeometric functions and representations of a four-parameter Lie group," *Comm. Pure Appl. Math.* **19** (1966), 251–259.

82. W. Miller, Jr., *Lie Theory and Special Functions.* Academic Press, New York, 1968.

83. W. Miller, Jr., "Special functions and the complex Euclidean group in 3-space, I," *J. Math. Phys.* **9** (1968), 1163–1175.

84. W. Miller, Jr., "Special functions and the complex Euclidean group in 3-space, III," *J. Math. Phys.* **9** (1968), 1434–1444.

85. W. Miller, Jr., *Symmetry Groups and Their Applications.* Academic Press, New York, 1972.

86. W. Miller, Jr., "Clebsch–Gordan coefficients and special function identities, I: The harmonic oscillator group," *J. Math. Phys.* **13** (1972), 648–655.

87. W. Miller, Jr., "Clebsch–Gordan coefficients and special function identities, II: The rotation and Lorentz groups," *J. Math. Phys.* **13** (1972), 827–833.

ISBN-0-201-13503-5

88. W. Miller, Jr., "Lie theory and generalized hypergeometric functions," *SIAM J. Math. Anal.* **3** (1972), 31–44.

89. W. Miller, Jr., "Lie theory and Meijer's *G* function," *SIAM J. Math. Anal.* **5** (1974), 309–318.

90. W. Miller, Jr., "Lie theory and the Lauricella functions F_D," *J. Math. Phys.* **13** (1972), 1393–1399.

91. W. Miller, Jr., "Lie theory and generalizations of the hypergeometric functions," *SIAM J. Appl. Math.* **25** (1973), 226–235.

92. W. Miller, Jr., "Lie algebras and generalizations of the hypergeometric function," in *Harmonic Analysis on Homogeneous Spaces* (*Proc. Symp. Pure Math.* **26**), pp. 355–356, Amer. Math. Soc., Providence, R.I., 1973.

93. W. Miller, Jr., "Symmetries of differential equations: The hypergeometric and Euler–Darboux equations," *SIAM J. Math. Anal.* **4** (1973), 314–328.

94. W. Miller, Jr., "Lie theory and separation of variables, 1: Parabolic cylinder coordinates," *SIAM J. Math. Anal.* **5** (1974), 626–643.

95. W. Miller, Jr., "Lie theory and separation of variables, 2: Parabolic coordinates," *SIAM J. Math. Anal.* **5** (1974), 822–836.

96. W. Montgomery and L. O'Raifeartaigh, "Noncompact Lie-algebraic approach to the unitary representations of $\widetilde{SU}(1,1)$," *J. Math. Phys.* **15** (1974), 380–382.

97. P. Moon and D. Spencer, *Field Theory Handbook*. Springer, Berlin, 1961.

98. P. Morse and H. Feshbach, *Methods of Theoretical Physics*, Part I. McGraw-Hill, New York, 1953.

99. M. Moshinsky, T. Seligman, and K. Wolf, "Canonical transformations and the radial oscillator and Coulomb problems," *J. Math. Phys.* **13** (1972), 901–907.

100. M. Naimark, *Linear Differential Operators*, Part II. Ungar, New York, 1968.

101. A. Naylor and G. Sell, *Linear Operator Theory*. Holt, New York, 1971.

102. U. Niederer, "The maximal kinematical invariance group of the harmonic oscillator," *Helv. Phys. Acta.* **46** (1973), 191–200.

103. U. Niederer, Universität Zurich preprint, December 1973.

104. P. Olevski, "The separation of variables in the equation $\Delta_3 u + \lambda u = 0$ for spaces of constant curvature in two and three dimensions," *Mat. Sb.* **27** (69) (1950), 379–426.

105. L. Ovsjannikov, *Group Properties of Differential Equations* (in Russian). Acad. Sci. USSR, Novasibirsk, 1962.

106. J. Patera and P. Winternitz, "A new basis for the representations of the rotation group: Lamé and Heun polynomials," *J. Math. Phys.* **14** (1973), 1130–1139.

107. I. Petrovsky, *Lectures on Partial Differential Equations* (transl. from the Russian). Wiley (Interscience), New York, 1954.

108. A. Pham Ngoc Dinh, "Opérateurs diagonaux associés a l'équation de Mathieu et applications," *C. R. Acad. Sci. Paris* **A279** (1974), 557–560.

109. E. Prugovecki, *Quantum Mechanics in Hilbert Space*. Academic Press, New York, 1971.

110. E. Rainville, "The contiguous function relations for $_pF_q$ with applications," *Bull. Amer. Math. Soc.* **51** (1945), 714–723.

111. M. Reed and B. Simon, *Methods of Modern Mathematical Physics*, Vol. I: *Functional Analysis*. Academic Press, New York, 1972.

112. F. Riesz and B. Sz-Nagy, *Functional Analysis* (transl.). Ungar, New York, 1955.

113. P. Rosenbloom and D. Widder, "Expansions in terms of heat polynomials and associated functions," *Trans. Amer. Math. Soc.* **92** (1959), 220–266.

114. S. Rosencrans, "Perturbation algebra of an elliptic operator," *J. Math. Anal. Appl.* **56**, (1976), 317–329.

115. P. Sally, *Analytic Continuation of the Irreducible Unitary Representations of the Universal Covering Group of $SL(2,R)$* (Amer. Math. Soc. Mem. No. 69). Amer. Math. Soc., Providence, R.I., 1967.

116. F. Schafke, *Einfuhrung in die Theorie der Speziellen Funktion der Mathematischen Physik*. Springer, Berlin, 1963.

ISBN-0-201-13503-5

117. E. Schrödinger, "On solving eigenvalue problems by factorization," *Proc. Roy. Irish Acad.* **A46** (1940), 9–16.

118. S. Schweber, *Relativistic Quantum Field Theory*. Harper, New York, 1961.

119. R. Shapiro, "Special functions related to representations of the group $SU(n)$, of class I with respect to $SU(n-1)$ $(n \geqslant 3)$," *Izv. Vyss. Učebn. Zaved. Matematika* **4** (71) (1968), 97–107 (in Russian).

120. L. Slater, *Generalized Hypergeometric Functions*. Cambridge Univ. Press, London and New York, 1966.

121. Y. Smorodinsky and I. Tugov, "On complete sets of observables," *Soviet Physics JETP* **1966**, 434–436.

122. I. Stakgold, *Boundary Value Problems of Mathematical Physics*, Vol. 1. Macmillan, New York, 1967.

123. G. Szegö, *Orthogonal Polynomials* (*Amer. Math. Soc. Colloq.* **23**). Amer. Math. Soc., Providence, R.I., 1959.

124. J. Talman, *Special Functions: A Group Theoretic Approach*. W. A. Benjamin, New York, 1968.

125. E. Titchmarsh, *Eigenfunction Expansions*, Part I (2nd ed.), Oxford Univ. Press, London and New York, 1962.

126. C. Truesdell, *An Essay Toward a Unified Theory of Special Functions* (Ann. of Math. Studies No. 18). Princeton Univ. Press, Princeton, N. J., 1948.

127. K. Urwin and F. Arscott, "Theory of the Whittaker–Hill equation," *Proc. Roy. Soc. Edinburgh* **A69** (1970), 28–44.

128. N. Vilenkin, *Special Functions and the Theory of Group Representations* (transl. from the Russian; Amer. Math. Soc. Transl., Vol 22). Amer. Math. Soc., Providence, R. I., 1968

129. B. Viswanathan, "Generating functions for ultraspherical functions," *Canad. J. Math.* **20** (1968), 120–134.

130. G. Warner, *Harmonic Analysis on Semi-Simple Lie Groups*, Vols. I and II. Springer, New York, 1972.

130a. G. N. Watson, *A Treatise on the Theory of Bessel Functions*. Cambridge Univ. Press, London and New York, 1966.

131. A. Weinstein, "The generalized radiation problem and the Euler–Poisson–Darboux equation," *Summa Brasil. Math.* **3** (1955), 125–146.

132. A. Weinstein, "On a Cauchy problem with subharmonic initial values," *Ann. Mat. Pura Appl.* (4), **43** (1957), 325–340.

133. L. Weisner, "Group-theoretic origin of certain generating functions," *Pacific J. Math.* **5** (1955), 1033–1039.

134. L. Weisner, "Generating functions for Bessel functions," *Canad. J. Math.* **11** (1959), 148–155.

135. L. Weisner, "Generating functions for Hermite functions," *Canad. J. Math.* **11** (1959), 141–147.

136. E. Whittaker, "On Hamilton's principal function in quantum mechanics," *Proc. Roy. Soc. Edinburgh* **A61** (1941), 1–19.

136a. E. T. Whittaker and G. N. Watson, *A Course of Modern Analysis* (4th ed.). Cambridge Univ. Press, London and New York, 1958. p. 366

137. E. Wigner, *Group Theory and Its Application to the Quantum Mechanics of Atomic Spectra*. Academic Press, New York, 1959.

138. P. Winternitz and I. Fris, "Invariant expansions of relativistic amplitudes and subgroups of the proper Lorentz group," *Soviet Physics JNP* **1** (1965), 636–643.

139. P. Winternitz, I. Lukač, and Y. Smorodinsky, "Quantum numbers in the little groups of the Poincaré group," *Soviet Physics JNP* **7** (1968), 139–145.

140. P. Winternitz, Y. Smorodinsky, M. Uhlir, and I. Fris, "Symmetry groups in classical and quantum mechanics," *Soviet Physics JNP* **4** (1967), 444–450.

141. K. Yosida, *Lectures on Differential and Integral Equations*. Wiley (Interscience), New York, 1960.

ISBN-0-201-13503-5

Subject Index

Subject Index